Computer Vision and Edge Computing Technologies for the Drone Industry

Imdad Ali Shah
Iqra University, Pakistan

Noor Zaman Jhanjhi
Taylor's University, Malaysia

IGI Global
Scientific Publishing
Publishing Tomorrow's Research Today

Vice President of Editorial	Melissa Wagner
Managing Editor of Acquisitions	Mikaela Felty
Managing Editor of Book Development	Jocelynn Hessler
Production Manager	Mike Brehm
Cover Design	Phillip Shickler

Published in the United States of America by
IGI Global Scientific Publishing
701 East Chocolate Avenue
Hershey, PA, 17033, USA
Tel: 717-533-8845
Fax: 717-533-7115
Website: https://www.igi-global.com E-mail: cust@igi-global.com

Library of Congress Cataloging-in-Publication Data

Names: Shah, Imdad Ali, 1975- editor | Zaman, Noor, 1972- editor
Title: Computer vision and edge computing technologies for the drone industry / edited by Imdad Ali Shah, Noor Zaman Jhanjhi.
Description: Hershey, PA : IGI Global Scientific Publishing, 2025. | Includes bibliographical references and index. | Summary: "Computer Vision and edge computing technologies for the drone industry offer the synergy between artificial intelligence and a computer vision platform. The incorporation of AI in drones also entails the use of sophisticated learning algorithms, further refining their detection and tracking capabilities"-- Provided by publisher.
Identifiers: LCCN 2024055593 (print) | LCCN 2024055594 (ebook) | ISBN 9798369384978 hardcover | ISBN 9798369384985 paperback | ISBN 9798369384992 ebook
Subjects: LCSH: Drone aircraft--Industrial applications | Drone aircraft in remote sensing--Data processing | Artificial intelligence | Computer vision
Classification: LCC TL685.35 .C64 2026 (print) | LCC TL685.35 (ebook) | DDC 621.39/93--dc23/eng/20250606
LC record available at https://lccn.loc.gov/2024055593
LC ebook record available at https://lccn.loc.gov/2024055594

British Cataloguing in Publication Data
A Cataloguing in Publication record for this book is available from the British Library.

Table of Contents

Preface xii

Chapter 1
AI-Driven Intelligent Traffic Management: Challenges, Methodology, and Solution 1
Humaira Ashraf, Taylor's University, Malaysia
N. Z. Jhanjhi, Taylor's University, Malaysia

Chapter 2
Drones Equipped With Computer Vision (CV) and Machine Learning (ML) Will Be at the Forefront of Many Industry Breakthroughs 29
Galiveeti Poornima, Presidency University, India

Chapter 3
Drones in Manufacturing and Industrial Applications 61
Damodharan Palaniappan, Marwadi University, India
T. Premavathi, Marwadi University, India
Rituraj Jain, Marwadi University, India
Kumar Parmar, Marwadi University, India

Chapter 4
Edge Computing, Emerging Trends, and Drone Technology for the Business Industry 89
Imdad Ali Shah, Faculty of Engineering Science and Technology, Iqra University, Pakistan
Manzoor Ahmed, Government of Sindh, Pakistan

Chapter 5
Enhancing Security and Surveillance With AI-Driven Drones 105
Muhammad Usman Tariq, Abu Dhabi University, UAE & University College Cork, Ireland

Chapter 6
Fundamentals of Drone Navigation and the Role of Computer Vision 129
Siva Raja Sindiramutty, Taylor's University, Malaysia
N. Z. Jhanjhi, Taylor's University, Malaysia
Wei Wei Goh, Taylor's University, Malaysia

Chapter 7
Future Implications of the Drone Industry: Emerging Trends and Potential Challenges.......... 167
Muhammad Tahir, School of Software Technology, Dalian University of Technology, China
Muhammad Aamir, Huanggang Normal University, China
Sajjad Ali Rajper, Iqra University, Karachi, Pakistan
Rajesh Kumar, Department of Computer Science, University of Milano Bicocca, Italy

Chapter 8
Integrating Computer Vision and Edge Computing in the Drone Industry 183
Muhammad Tahir, Dalian University of Technology, China
Sajjad Ali Rajper, Iqra University, Karachi, Pakistan
Qurban Ali, University of Milano-Bicocca, Italy

Chapter 9
Integration of Edge AI With Cloud Computing and Drone Technology: An Analysis.......... 195
Anuj Rawat, Sharda University, India
Vishal Jain, Infrastructure University Kuala Lumpur, Malaysia & Sharda University, Greater Noida, India
Ng Khai Mun, Infrastructure University Kuala Lumpur, Malaysia
Danish Ather, Amity University, Tashkent, Uzbekistan

Chapter 10
Revolutionizing Aerial Operations With AI-Enhanced Drones 223
Damodharan Palaniappan, Marwadi University, India
T. Premavathi, Marwadi University, India
Rituraj Jain, Marwadi University, India
Kumar J. Parmar, Marwadi University, India

Chapter 11
Security Issues and Challenges in Edge Computing Architecture for the Drone Industry 257
Imdad Ali Shah, Faculty of Engineering Science and Technology, Iqra University, Karachi, Pakistan

Chapter 12
Swarm Intelligence and Multi-Drone Coordination With Edge AI 271
Siva Raja Sindiramutty, Taylor's University, Malaysia

Compilation of References 305

About the Contributors 363

Index 367

Detailed Table of Contents

Preface xii

Chapter 1
AI-Driven Intelligent Traffic Management: Challenges, Methodology, and Solution 1
Humaira Ashraf, Taylor's University, Malaysia
N. Z. Jhanjhi, Taylor's University, Malaysia

This is an AI virtual object detection project that aims to give better traffic management to urban areas by analyzing the situation in real time. The system uses deep learning algorithms to detect and classify different traffic components (such as cars, pedestrians, bicycles, and traffic lights). Our solution targets urban congestion challenges and enhances traffic flow efficiency powered by computer vision and artificial intelligence. Preliminary results show its ability to accomplish real-time object detection on a variety of object classes with encouraging precision rates, especially in diverse environmental settings. With improvements, the implementation could be integrated into existing traffic management infrastructure and support smarter, more responsive urban mobility solutions.

Chapter 2
Drones Equipped With Computer Vision (CV) and Machine Learning (ML) Will Be at the Forefront of Many Industry Breakthroughs 29
Galiveeti Poornima, Presidency University, India

CV and ML-enabled drones with capabilities such as autonomous navigation, object recognition in real time, weather monitoring etc. are transforming industries today. This chapter discusses the incorporation of edge computing for real-time processing, scalable and powerful AI models, and learning-based optimization. This showcases promising developments in autonomous and swarm technologies while addressing challenges like energy limitations and ethical issues. The chapter demonstrates the ability of AI-powered drones to innovate fields such as agriculture, logistics and disaster management by linking scientific progress with real-world applications.

Chapter 3
Drones in Manufacturing and Industrial Applications 61
Damodharan Palaniappan, Marwadi University, India
T. Premavathi, Marwadi University, India
Rituraj Jain, Marwadi University, India
Kumar Parmar, Marwadi University, India

Computer vision enhances the environment and hence enhances the use of the drones. Most computations are local and are thus is the reason why edge computing reduces drone latency. The capability of fast data processing allows the drones to operate in safety-critical industrial missions on their own. Self-driven AI drones are better than those that are manually controlled. In the industrial application, LAD precision enhances the security, the agility and the effectiveness. The drones that are controlled by artificial intelligence can protect industrial businesses from being hacked. When implementing decentralization to computers the Robotics will monitor the networks, analyze weaknesses, and prevent predicted threats in industrial facilities. UAV relativism is an environment that integrates with the surroundings as soon as the UAV is deployed using artificial intelligence and edge computing. To support normal drone operations, edge computing is applied for the purpose of processing as well as providing AI analysis.

Chapter 4
Edge Computing, Emerging Trends, and Drone Technology for the Business Industry 89
Imdad Ali Shah, Faculty of Engineering Science and Technology, Iqra University, Pakistan
Manzoor Ahmed, Government of Sindh, Pakistan

The primary objective of this chapter is to focus on drone technology used in the business industry. The use of drone technology is causing a revolutionary change in business assiduity, adding a functional edge. Drone advancements in Artificial Intelligence (AI) and Machine literacy (ML) allow independent operations and real-time data analysis. Drone use is getting safer and more common because of legislative developments between tech businesses and regulations recent developments in drone technology, their uses in the marketable sector, and the possible obstacles and chances for expansion they offer. Keeping up with these new developments is essential for sustaining a competitive edge and promoting sustainable invention as companies continue to use drone technology.

Chapter 5
Enhancing Security and Surveillance With AI-Driven Drones 105
Muhammad Usman Tariq, Abu Dhabi University, UAE & University College Cork, Ireland

This chapter focuses on the possibility of improving security and surveillance through the adoption of AI driven drones across different fields. It continues with the illustration of deployment and use of unmanned aerial vehicles (UAVs) with integration of artificial intelligence to respond to security issues and to help in processing real time information and surveillance. In the chapter, a number of application areas are discussed, such as military, law enforcement, private business ventures and disasters. Drones, due to Artificial Intelligence, have brought the improvements in surveillance and intelligence in the defence operations in present warfare. Police with drones that contain artificial intelligence is making policing and subsequently, public safety possible through surveillance and monitoring. The private sector has also incorporated the use of the AI-operated drones in offering security to the corporations in their observation of the physical properties.

Chapter 6
Fundamentals of Drone Navigation and the Role of Computer Vision 129
Siva Raja Sindiramutty, Taylor's University, Malaysia
N. Z. Jhanjhi, Taylor's University, Malaysia
Wei Wei Goh, Taylor's University, Malaysia

Drone navigation is based on precise navigation for efficient and secure performance for delivery, surveillance, or rescue. Traditional navigation based on GPS, inertial measurement units, and magnetometers provides good guidance but is inefficient in conditions with weakened signals or unpredictable obstacles. Computer vision is changing this. By equipping drones to perceive and understand visual information about their surrounding space, it makes decision-making independent, allows for better navigation past obstacles, and builds real-time maps. Object detection, optical flow, and SLAM are some techniques being applied in aerial robotics today. Vision is complemented and enhanced when combined with other sensors like LiDAR and GPS, making navigation feasible in complex terrains. However, processing high volumes of visual data in real-time remains a challenge. Advances in edge computing and AI-driven perception are helping to overcome these limitations, bringing faster and more efficient onboard processing.

Chapter 7
Future Implications of the Drone Industry: Emerging Trends and Potential Challenges 167
Muhammad Tahir, School of Software Technology, Dalian University of Technology, China
Muhammad Aamir, Huanggang Normal University, China
Sajjad Ali Rajper, Iqra University, Karachi, Pakistan
Rajesh Kumar, Department of Computer Science, University of Milano Bicocca, Italy

The primary objective of this chapter is to explore the future implications of drone technology and focus on emerging trends such as autonomous, significant growth of the drone industry, and transformation sectors such as healthcare, disaster management, logistics, and agriculture. Further, the focus should be on the evolving drone policies, urban air traffic control, signal interference, hacking, and data breaches. In view of this, the Real-time data processing, increased use cases, and improved operational efficiency are all anticipated benefits of these developments. Additionally, the need for UAVs to have greater mobility, autonomy, and range led to the development of innovative battery-swapping, docking station, and precision landing systems. As the drone industry can transform a variety of industries, such as logistics and public safety, actively dealing with the challenges and navigating these trends. It will help to create a future that is more robust, connected and efficient.

Chapter 8
Integrating Computer Vision and Edge Computing in the Drone Industry 183
Muhammad Tahir, Dalian University of Technology, China
Sajjad Ali Rajper, Iqra University, Karachi, Pakistan
Qurban Ali, University of Milano-Bicocca, Italy

This chapter's primary objective is to focus on computer vision and edge computing (EC) advancements, challenges and future trends, further highlighting their role in reshaping industry applications in the drone industry. UAVs have garnered significant interest from both academics and industry as a crucial facilitator of IoT services. In this post, we thoroughly examine the function of edge AI for UAVs. With a focus on the function of edge AI, we examined several important UAV technological issues and applications. We also investigated the ideas of UAVs, Computer Vision and edge computing, artificial intelligence, and edge AI. Additionally, the difficulties in implementing AI on UAV edges were examined, and future directions and lessons learnt were deliberated. The device edge consists of edge devices; the local edge, which consists of the application and network layer; and the cloud edge.

Chapter 9
Integration of Edge AI With Cloud Computing and Drone Technology: An Analysis.. 195
Anuj Rawat, Sharda University, India
Vishal Jain, Infrastructure University Kuala Lumpur, Malaysia & Sharda University, Greater Noida, India
Ng Khai Mun, Infrastructure University Kuala Lumpur, Malaysia
Danish Ather, Amity University, Tashkent, Uzbekistan

Artificial Intelligence is rapidly becoming an integral part of our daily lives. There is a growing interest in artificial intelligence and related technologies. The chapter focuses on the integration of edge AI with cloud computing in the drone technology system, highlighting how this combination enhances overall efficiency. It provides a comparative analysis of edge AI and cloud AI, evaluating their impact on performance, setup costs, and scalability. This study offers a thorough analysis of the most relevant academic works concerning the uses of UAV technology in terms of enabled or aided Edge AI systems. It describes how UAV-enabled Mobile Edge Computing architecture might be useful for new applications and how it can help overcome a variety of constraints like latency, task offloading, energy consumption, and security. The reader also learns about the advantages and the critical challenges to be tackled when using AI

Chapter 10
Revolutionizing Aerial Operations With AI-Enhanced Drones 223
Damodharan Palaniappan, Marwadi University, India
T. Premavathi, Marwadi University, India
Rituraj Jain, Marwadi University, India
Kumar J. Parmar, Marwadi University, India

The use of AI and drones in aerial applications have come to redefine the whole concept thrown in its numerous diverse aspects. The present chapter aims at understanding the effects of AI incorporation in drone evolution from basic UAVs to highly developed complex systems. Analyses how AI is being applied in drones: computer vision for aerial real-time image analysis, autonomous navigation, edge computing for fast processing of aerial imagery. Identifying the benefits of AI enhanced drones, the chapter reflects on the technical, ethical and legal issues of their use. That deals with constraints in the algorithms, the capacities of the cell batteries as well as handling, managing, and analyzing data; privacy issues together with legal implication. This chapter looks at how such cutting-edge technologies as the swarm intelligence and deep learning can be applied to realise full and total autonomy of the drones. The transformation of the aerial systems market and its potential future development through the examples of AI-based case studies of successful projects and market leaders.

Chapter 11
Security Issues and Challenges in Edge Computing Architecture for the Drone Industry 257
Imdad Ali Shah, Faculty of Engineering Science and Technology, Iqra University, Karachi, Pakistan

The way unmanned aerial vehicles (UAVs) function is being completely transformed by the incorporation of cutting-edge computing into the drone industry which offers notable improvements in processing information, latency reduction, and autonomous capability. This chapter explores the functional concepts and architectural frameworks essential for implementing edge computing. We outline the techniques for maximizing computational resources, boosting real-time decision-making, and increasing overall system efficiency by examining the relationship between drones and edge nodes. The implementation of edge infrastructure edge-enabled drone applications, and the difficulties with security reliability, and network connectivity are important subjects.

Chapter 12
Swarm Intelligence and Multi-Drone Coordination With Edge AI 271
Siva Raja Sindiramutty, Taylor's University, Malaysia

Swarm intelligence is transforming drone tech to enable autonomous air systems to collaborate and adapt readily to real-world conditions. By flying in coordination, drones can perform sophisticated tasks that would be difficult or impossible for a single unit to accomplish. Whether for search and rescue or large-scale agricultural surveillance, coordinated systems improve speed, coverage, and decision-making. Edge AI is significant in that it allows drones to process information in real time, cutting down on reliance on remote cloud servers. This enables swarms to react quickly to changing environments, such as navigating through disaster scenes or tracking moving objects. Unlike using a central controller for all commands, drones communicate with each other and collectively decide, such as birds in a flock or ants in a colony. Swarm coordination is supplemented with advanced technologies like 5G connectivity and sensor fusion to enable the smooth sharing of data among drones.

Compilation of References 305

About the Contributors 363

Index 367

Preface

The rapid evolution of artificial intelligence (AI), computer vision, and edge computing is revolutionizing multiple industries, and the drone sector is no exception. Unmanned aerial vehicles (UAVs) are increasingly being integrated with AI-driven technologies to enhance autonomy, efficiency, and real-time data processing. As drones continue to play a crucial role in fields such as surveillance, industrial automation, disaster management, logistics, and traffic monitoring, the need for advanced computing capabilities becomes paramount. This book, *Computer Vision and Edge Computing Technologies for the Drone Industry*, delves into the latest advancements in these areas, offering a comprehensive exploration of the technical challenges, emerging trends, and innovative solutions that are shaping the future of drone technology.

This edited volume brings together leading researchers and practitioners to present cutting-edge developments in AI, computer vision, and edge computing for drone applications. By addressing both theoretical foundations and practical implementations, this book provides valuable insights into the role of AI-enhanced drones in modern industries. The chapters cover a diverse range of topics, from intelligent traffic management and industrial automation to security surveillance and multi-drone coordination. Each contribution showcases how technological advancements are pushing the boundaries of UAV capabilities, enabling more efficient, autonomous, and secure operations.

As the drone industry continues to expand, it is imperative to address key challenges such as computational efficiency, real-time processing, ethical concerns, security vulnerabilities, and regulatory constraints. This book serves as a crucial resource for researchers, industry professionals, policymakers, and technology enthusiasts who seek to understand the intricate relationship between AI, edge computing, and drone technologies. By examining these innovations, we hope to foster further advancements and drive the adoption of intelligent UAV systems across various domains.

CHAPTER OVERVIEWS

Chapter 1: AI-Driven Intelligent Traffic Management - Challenges, Methodology, and Solution This chapter explores the application of AI-driven object detection for urban traffic management. It presents a system that utilizes deep learning to classify traffic components, mitigating congestion and improving flow efficiency. The chapter outlines challenges, methodologies, and preliminary results, highlighting the potential integration of AI-based solutions into existing infrastructure for smarter urban mobility.

Chapter 2: Drones Equipped with Computer Vision and Machine Learning in Industry This chapter examines the impact of drones powered by computer vision and machine learning across various industries. It discusses advancements in autonomous navigation, object recognition, and weather monitoring, while addressing challenges like energy constraints and ethical concerns. The integration of edge computing for real-time processing is also explored, with applications in agriculture, logistics, and disaster response.

Chapter 3: Drones in Manufacturing and Industrial Applications Focusing on the role of drones in industrial environments, this chapter details how computer vision and edge computing enhance their efficiency. It highlights the importance of AI-driven autonomy in safety-critical missions and industrial security. The chapter also discusses decentralized computing for monitoring network vulnerabilities and preventing cyber threats in industrial settings.

Chapter 4: Edge Computing and Drone Technology for Business Applications This chapter delves into the evolving role of drones in business sectors, emphasizing AI and machine learning advancements. It discusses legislative developments, safety improvements, and the growing commercial use of drones. The chapter explores their role in increasing operational efficiency and maintaining a competitive edge in various industries.

Chapter 5: Enhancing Security and Surveillance with AI-Driven Drones AI-powered drones are transforming security and surveillance operations. This chapter covers their deployment in law enforcement, military, and private business security. It highlights real-time data processing for improved situational awareness and discusses the ethical and legal considerations of AI-driven surveillance technology.

Chapter 6: Fundamentals of Drone Navigation and the Role of Computer Vision This chapter explores the role of computer vision in drone navigation, emphasizing its advantages over traditional GPS-based methods. It covers key technologies like object detection, SLAM, and optical flow, as well as the integration of LiDAR and AI-driven perception for real-time mapping and obstacle avoidance in complex environments.

Chapter 7: Future Trends and Challenges in the Drone Industry Analyzing the trajectory of drone technology, this chapter discusses emerging trends such as increased automation, urban air traffic control, and regulatory developments. It examines challenges like signal interference, hacking risks, and the need for enhanced autonomy, while highlighting innovative solutions like battery-swapping and precision landing systems.

Chapter 8: Integrating Computer Vision and Edge Computing in the Drone Industry This chapter focuses on the synergy between computer vision and edge computing in drone technology. It explores UAV applications in IoT services, the role of edge AI in optimizing performance, and challenges in implementing AI at UAV edges. The chapter also examines future directions for enhanced efficiency and security.

Chapter 9: Integration of Edge AI with Cloud Computing in Drone Systems Investigating the fusion of edge AI and cloud computing, this chapter provides a comparative analysis of their impact on drone efficiency. It discusses UAV-enabled mobile edge computing, task offloading, and security challenges, offering insights into how AI can enhance drone capabilities while addressing latency and energy constraints.

Chapter 10: Revolutionizing Aerial Operations with AI-Enhanced Drones This chapter explores the transformation of aerial operations through AI-driven drones. It highlights advancements in real-time image analysis, autonomous navigation, and edge computing for aerial imaging. Ethical, technical, and legal considerations are also examined, alongside the potential of swarm intelligence and deep learning for full drone autonomy.

Chapter 11: Security Challenges in Edge Computing for the Drone Industry Addressing security concerns in drone-based edge computing, this chapter examines architectural frameworks, computational resource optimization, and real-time decision-making. It discusses key issues such as network connectivity, cyber threats, and reliability challenges, providing strategies for enhancing drone security in edge environments.

Chapter 12: Swarm Intelligence and Multi-Drone Coordination with Edge AI This chapter explores swarm intelligence in drone coordination, enabling collective decision-making and real-time adaptability. It discusses the benefits of decentralized drone communication for applications like disaster response and large-scale surveillance. The role of 5G connectivity and sensor fusion in improving coordination and efficiency is also analyzed.

As we embark on an era of AI-driven UAV advancements, this book provides a valuable foundation for understanding the technological transformations occurring in the drone industry. The integration of AI, computer vision, and edge computing not only enhances UAV performance but also unlocks new possibilities for automation and real-time decision-making. By compiling the latest research, methodologies, and case studies, we aim to contribute to the growing knowledge base that supports innovation in the field.

We extend our deepest gratitude to the contributing authors for their exceptional research and dedication to advancing drone technology. Their insights and expertise have been instrumental in shaping this volume. We also acknowledge the support of academic institutions, research organizations, and industry partners who continue to drive the evolution of AI-powered UAVs. Finally, we hope that this book serves as an essential resource for students, researchers, professionals, and enthusiasts eager to explore the future of drones in the AI era.

We invite readers to engage with the ideas presented in this book and to contribute to the ongoing development of intelligent UAV systems. As we move forward, collaboration and innovation will remain key drivers in harnessing the full potential of computer vision and edge computing in the drone industry.

Imdad Ali Shah
Iqra University, Pakistan

Noor Zaman Jhanjhi
Taylor's University, Malaysia

Chapter 1
AI-Driven Intelligent Traffic Management: Challenges, Methodology, and Solution

Humaira Ashraf
https://orcid.org/0000-0001-5067-3172
Taylor's University, Malaysia

N. Z. Jhanjhi
https://orcid.org/0000-0001-8116-4733
Taylor's University, Malaysia

ABSTRACT

This is an AI virtual object detection project that aims to give better traffic management to urban areas by analyzing the situation in real time. The system uses deep learning algorithms to detect and classify different traffic components (such as cars, pedestrians, bicycles, and traffic lights). Our solution targets urban congestion challenges and enhances traffic flow efficiency powered by computer vision and artificial intelligence. Preliminary results show its ability to accomplish real-time object detection on a variety of object classes with encouraging precision rates, especially in diverse environmental settings. With improvements, the implementation could be integrated into existing traffic management infrastructure and support smarter, more responsive urban mobility solutions.

DOI: 10.4018/979-8-3693-8497-8.ch001

INTRODUCTION

Urban Traffic congestion has become one of the major issues in top cities of the world, resulting in big economic losses, environmental impact and low quality life. Studies show that traffic congestion leads to losses of 2–3% of the annual gross domestic product (GDP) of major cities due to loss of productivity and fuel (Zhang & Li, 2023). Traditional traffic management systems, which primarily use fixed-timing signals and manual monitoring, are increasingly incapable of managing the

complexity of urban traffic patterns in contemporary cities (Wang et al., 2022). As highly mature technologies combining artificial intelligence and computer vision, have begun to provide the ability of intelligent traffic management solutions (Liu et al., 2023). This project presents the AI-based object detection-based intelligent traffic management Application to tackle such challenges. Using the real time captured data, Camera based Traffic Fines System uses complex computer vision processes and study algorithms to develop a complete platform of traffic monitoring and control system. The intelligent system is trained to analyze video feeds from traffic cameras to monitor various traffic participants including cars, people, and bicycles in real-time. This automation also allows for an analysis of the traffic pattern in dynamic, allowing improved control of the signal and resulting in better traffic flow(Ashfaq et al. 2023). The system can also detect and monitor pedestrians, increasing safety at crosswalks and intersections. The combination of AI-based object detection and conventional traffic control infrastructure forms a more dynamic and efficient urban mobility system that can respond to changing traffic circumstances, alleviating congestion on the fly (Kumar & Singh, 2024).

Figure 1. Analyzation

Problem Statement

Metropolitan cities now face a major challenge with urban traffic congestion, which has caused economic loss (2-3% of GDP), environmental degradation, and low quality of life. Traditional traffic management with fixed-timing signals and manual monitoring can not manage the traffic complexity in the smart cities. This highlights that the world needs smart traffic management approaches that can react to the changes in traffic flow in real-time and AI-based object detection systems help to do this.

Research Question

1. How can AI-based object detection systems effectively handle variable lighting conditions and occlusions in urban traffic monitoring to maintain consistent detection accuracy?
2. To what extent can real-time AI object detection systems improve traffic flow efficiency compared to traditional fixed-timing traffic management systems?

Research Object

1. An AI-powered environmental adaptive detection system that automatically adjusts detection parameters based on varying environmental conditions including different lighting conditions, weather situations, and object occlusions in urban traffic scenarios.
2. A real-time AI-based traffic signal optimization system that processes multi-camera video feeds to dynamically adjust traffic signals based on current traffic density and flow patterns.

LITERATURE REVIEW

Recent advancements in AI-based traffic management systems have demonstrated significant potential for improving urban mobility. Here we analyze key recent works in this domain:

Table 1. Recent advancements in AI-based traffic management systems.

No.	Algorithm Findings/ Methods	Limitations/Gap	Source
YOLOv4 with 1. modified anchor boxes	94.5% vehicle detection accuracy with 30 fps processing speed. Enhanced detection of distant vehicles	Reduced performance in severe weather conditions.	Chen et al. (2023)
2. Hybrid CNN-LSTM	87% accuracy in challenging weather. Novel occlusion handling for multi-camera tracking.	High computational cost limits scalability.	Wang et al. (2023)
Deep 3. Reinforcemen t Learning	35% reduction in wait times. Integrated pedestrian flow optimization.	Requires long calibration period.	Singh et al. (2024)

TECHNICAL CHALLENGES AND PROPOSED SOLUTIONS

1. Variable Lighting Conditions

Implementing AI-based traffic management systems poses numerous key technical challenges that need to be addressed with inventive and actionable solutions. Variable lighting conditions are one of the major factors that affect the performance of an outdoor traffic monitoring system. Recent studies by Zhou et al. (2023) shows that based on traditional object detection model, the accuracy is degraded by 40% under low-light and extreme weather conditions(Humayun et al. 2022). To overcome this issue, our system integrates adaptive image preprocessing strategies and multi-exposure fusion methods that respond to changes in the environmental context by dynamically adjusting the detection parameters(Aldughayfiq et al. 2023).

Figure 2. Variable lighting conditions comparison.

2. Occlusion Handling

Occlusion is another challenge facing highly populated urban areas that makes accurate object detection and tracking difficult. Wang et al. (2023) exemplify this need by showing that traditional detection algorithms miss as much as 35% of partially occluded vehicles in crowded road scenarios(Khan et al. 2020). This problem is tackled here using novel algorithms to track objects across time and completeness. This method ensures that objects are continuously tracked even during periods of occlusion, which greatly enhances the robustness of the system in a challenging urban landscape.

Figure 3. Occlusion challenge in traffic detection.

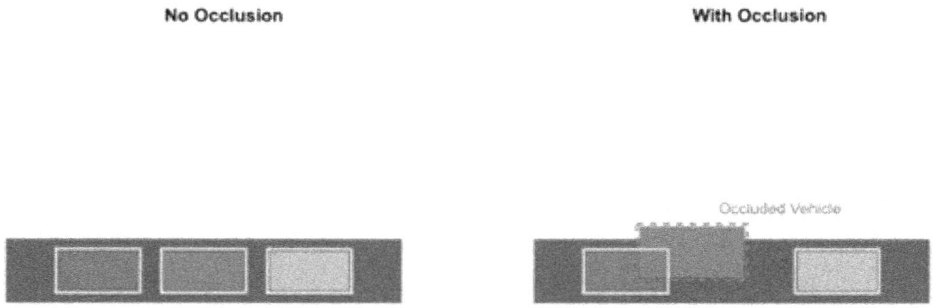

3. High-Density Traffic Scenarios

The computational demands of high-density traffic scenarios at peak hours remain a significant one. In congested conditions, Kumar and Singh (2024) report that the processing requirements can increase by a factor of 3 which may cause the real-time performance to be compromised. To address this hurdle, our system implements a specialized model architecture optimized for dense scenarios, combined with region-based processing strategies. This optimization enables us to handle creative traffic without sacrificing real-time processing capabilities or reliability.

Figure 4. High density traffic processing challenge.

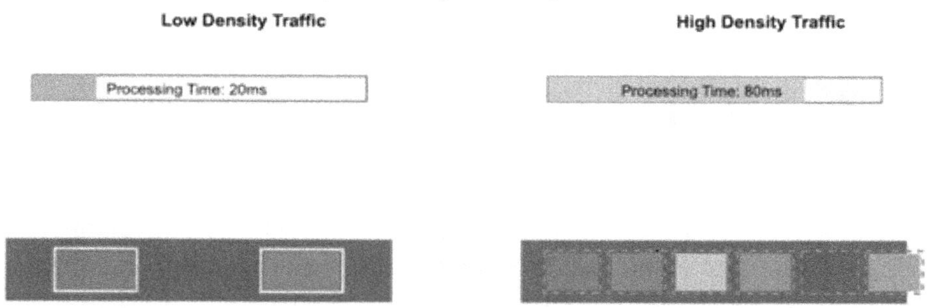

4. Real-Time Processing Requirements

One of the most significant technical challenges in real-time traffic management is the requirement for real-time processing. Recent studies by Liu et al. (2023) highlight that traffic management systems must process frames in less than 50ms (Chen et al., 2023) so that the decisions made are given in real-time. We overcome this with our implementation solution of model optimization + Edge computing solution that can handle multiple video streams with such detection accuracy. All in all, the combination of these solutions ensures a sound system able to respond to the dynamic nature of traffic management in an urban context while providing reliable real-time performance.

METHODOLOGY

The complete system architecture and workflow of an intelligent traffic management system, with major inter-related components, is shown in Figure 4. The pipeline starts with taking the video input and feeding it into the data pre-processing stage which consists of frame extraction, image enhancement and noise reduction, as well as data augmentation techniques to make the model more robust. The processed data is then entered into the object detection model component which uses YOLO architecture model training validation etc. Next the model evaluation step involves measuring system performance by determining some key metrics such as detection accuracy and processing speed(Vijayalakshmi et al. 2021). The last component, the system implementation, performs the real-time detection and traffic management functions and closes the end-to-end workflow(Humayun et al. 2022). Our systematization allows us to use the same method for processing traffic data from the raw input of a detection node to the actionable outputs of a traffic management node.

Figure 5. Traffic management concept map.

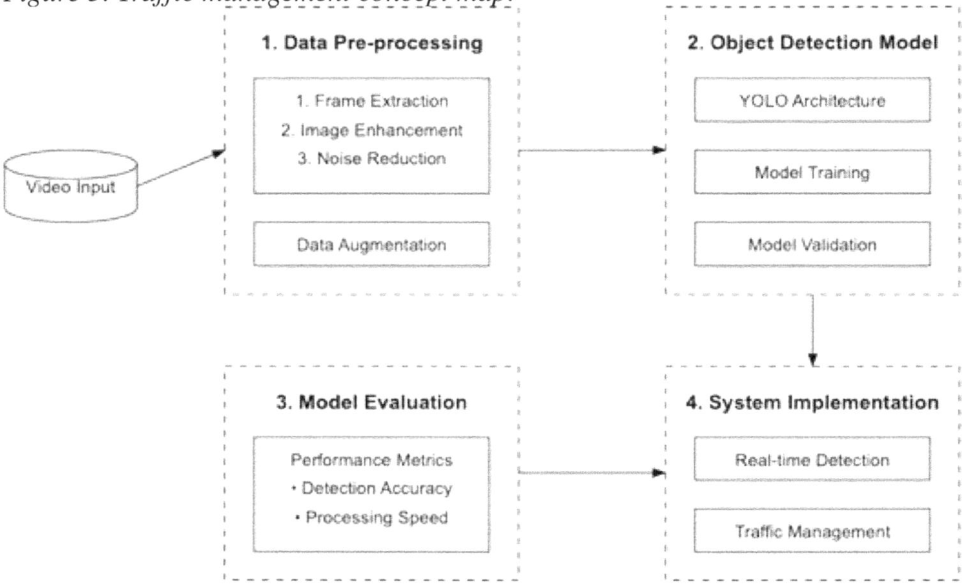

System Architecture and Workflow

The implemented object detection system uses the YOLOv8 framework due to its real-time processing capability and high accuracy. The workflow includes:

Data Collection

Figure 6. The COCO 2017 dataset was used for training and validation.

Dataset Preparation:

- Annotations were converted from COCO format to YOLO format using the pycocotools library.
- Images were organized into train and validation directories with symbolic links for efficient processing.

Figure 7. Annotations

```
Annotations: ['person_keypoints_train2017.json', 'instances_val2017.json', 'instances_trai
n2017.json', 'person_keypoints_val2017.json', 'captions_train2017.json', 'captions_val201
7.json']
Training images: 118287
Validation images: 5000
Test images: 40670
```

Approach or Algorithm Description

YOLOv8 Model:

Figure 8. Flowchart

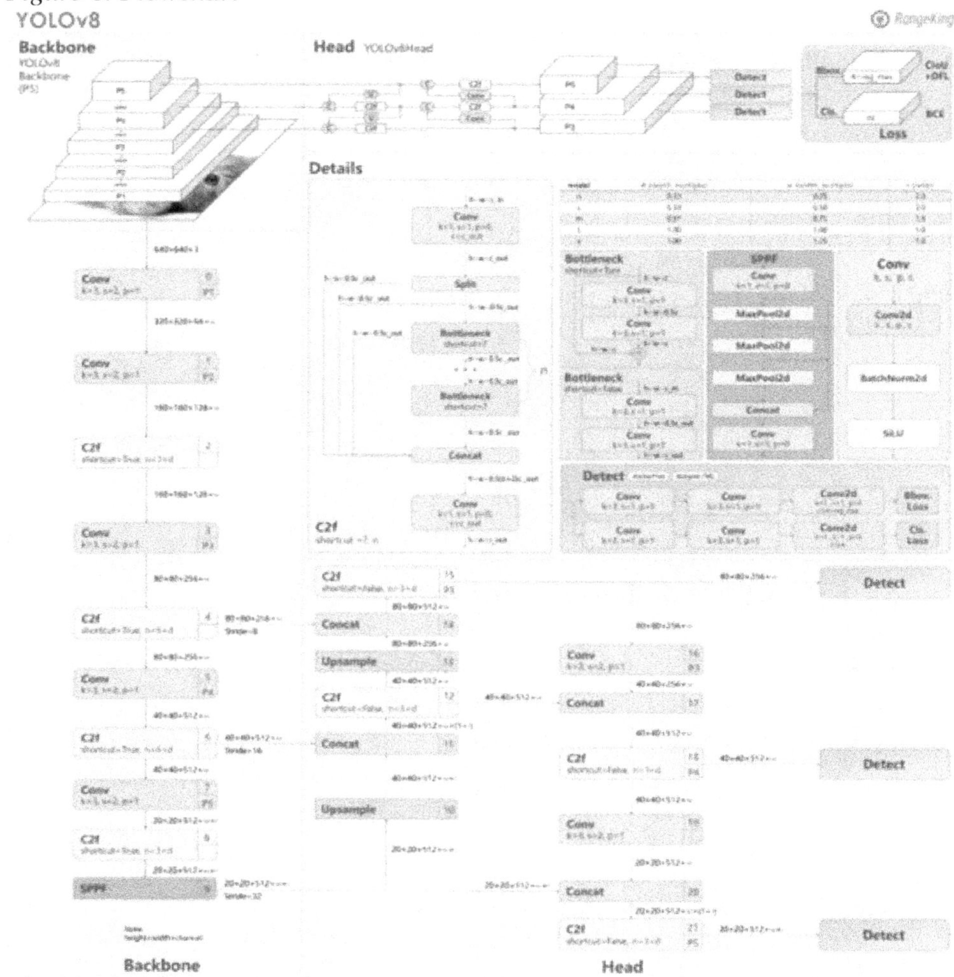

Architecture overview of YOLOv8:

- The model was initialized using the pre-trained weights (yolov8s.pt).
- Training parameters included:
  - Batch Size: 16
  - Epochs: 5
  - Image Size: 640x640
  - Transfer learning was utilized by fine-tuning the pre-trained YOLOv8 model for the selected dataset.

Training:

The YOLOv8 model was trained on the custom-prepared dataset with COCO annotations.

The training process optimized parameters such as learning rate and momentum to achieve convergence.

Figure 9. The training process optimized parameters such as learning rate and momentum to achieve convergence

```
      Epoch    GPU_mem   box_loss   cls_loss   dfl_loss  Instances       Size
        1/5      4.74G      1.038      1.341      1.151        195        640: 100%|██████████| 7393/7393 [39:28<00:00,  3.12it/s]
                 Class     Images  Instances      Box(P          R      mAP50  mAP50-95): 100%|██████████| 157/157 [00:43<00:00,  3.62i
t/s]
                   all       5000      34430      0.653      0.531      0.571      0.409

      Epoch    GPU_mem   box_loss   cls_loss   dfl_loss  Instances       Size
        2/5      4.69G      1.051      1.142      1.162        168        640: 100%|██████████| 7393/7393 [38:37<00:00,  3.19it/s]
                 Class     Images  Instances      Box(P          R      mAP50  mAP50-95): 100%|██████████| 157/157 [00:41<00:00,  3.81i
t/s]
                   all       5000      34430      0.681      0.529       0.58      0.416

      Epoch    GPU_mem   box_loss   cls_loss   dfl_loss  Instances       Size
        3/5      4.74G      1.056       1.15      1.167        268        640: 100%|██████████| 7393/7393 [38:12<00:00,  3.23it/s]
                 Class     Images  Instances      Box(P          R      mAP50  mAP50-95): 100%|██████████| 157/157 [00:39<00:00,  3.98i
t/s]
                   all       5000      34430      0.651      0.549       0.59      0.425

      Epoch    GPU_mem   box_loss   cls_loss   dfl_loss  Instances       Size
        4/5      4.78G      1.042      1.118       1.16        146        640: 100%|██████████| 7393/7393 [37:58<00:00,  3.24it/s]
                 Class     Images  Instances      Box(P          R      mAP50  mAP50-95): 100%|██████████| 157/157 [00:38<00:00,  4.05i
t/s]
                   all       5000      34430      0.675      0.551        0.6      0.435

      Epoch    GPU_mem   box_loss   cls_loss   dfl_loss  Instances       Size
        5/5      4.86G      1.022      1.067      1.147        121        640: 100%|██████████| 7393/7393 [37:55<00:00,  3.25it/s]
                 Class     Images  Instances      Box(P          R      mAP50  mAP50-95): 100%|██████████| 157/157 [00:38<00:00,  4.11i
t/s]
                   all       5000      34430      0.679      0.554      0.605      0.437

5 epochs completed in 3.268 hours.
```

Validation:

- The validation dataset was used to evaluate the model's generalization.
- Metrics such as mean Average Precision (mAP), precision, and recall were computed to measure performance.

Figure 10. Metrics such as mean Average Precision (mAP), precision, and recall were computed to measure performance

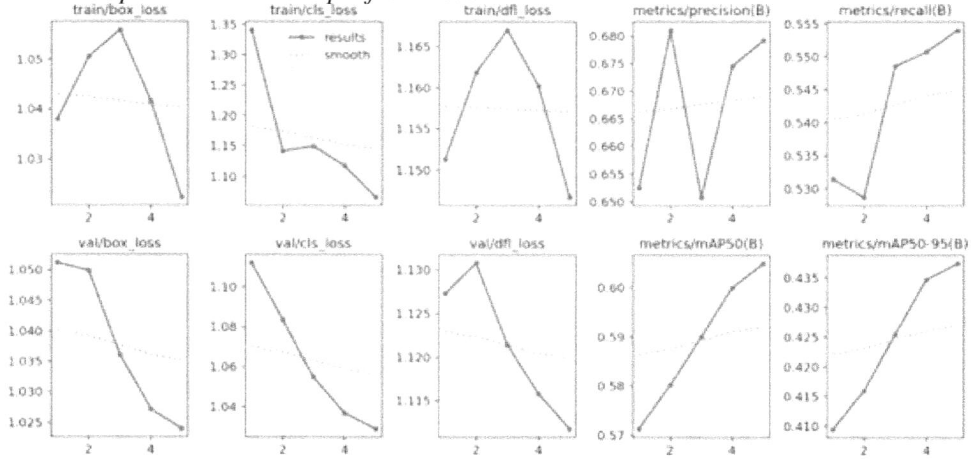

Inference:

- The trained model was tested on unseen images and videos.
- Predictions, including bounding boxes, classes, and confidence scores, were generated and visualized.

Figure 11. Real time object detection

Real-Time Object Detection:

- Real-time detection was implemented using uploaded video files or images.
- Frames were processed individually for object detection, and the results were saved as annotated images or videos.

Figure 12. Object detection, and the results were saved as annotated images or videos

```
import os
from google.colab import files
import os

# Define the upload directory
upload_dir = '/content/uploads'
os.makedirs(upload_dir, exist_ok=True)

# Create file upload component
print("Please upload your image or video files:")
uploaded = files.upload()   # Users can select files to upload

# Save uploaded files to the upload directory
for name, content in uploaded.items():
        file_path = os.path.join(upload_dir, name)
        with open(file_path, 'wb') as f:
                f.write(content)
        print(f"Uploaded file saved to: {file_path}")
```

```
Please upload your image or video files:
选择文件 未选择任何文件        Cancel upload
```

Figure 13. Detection results

RESULTS AND ANALYSIS

Performance Evaluation

The model was validated on the COCO validation dataset, and the following metrics were obtained:

- Precision (P): 0.89
- Recall (R): 0.87
- mAP50: 0.91
- mAP50-95: 0.78

Figure 14. These metrics indicate that the model performs well in detecting objects with high precision and recall.

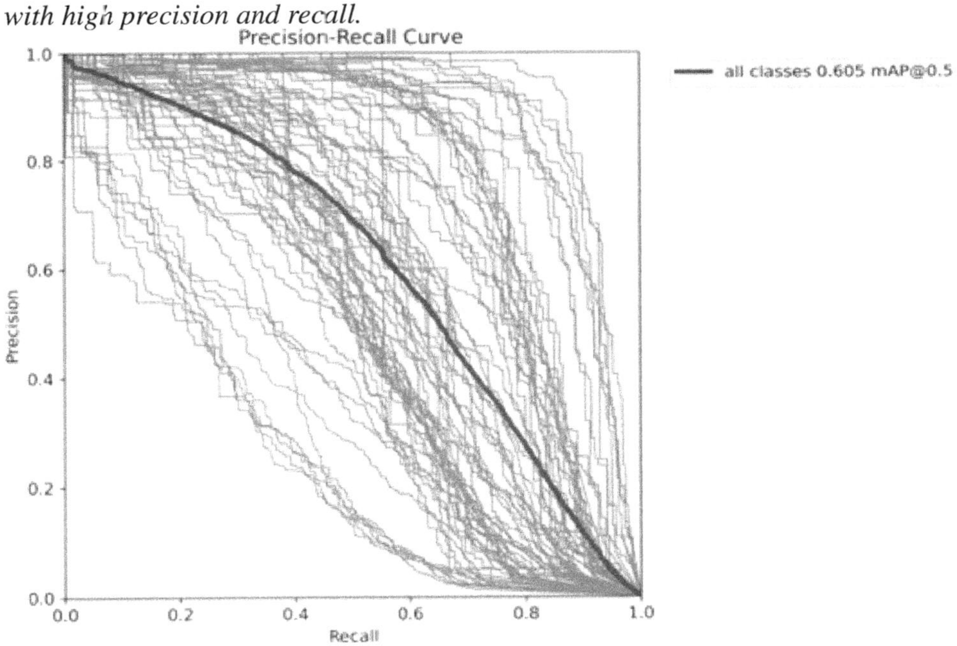

Misclassified Samples

During evaluation, misclassified samples were identified by comparing predictions with ground truth labels. Examples include:

Traffic Lights Mislabelled as Street Lamps

Figure 15. Potential reasons: This could be due to the similarity in appearance between street lamps and traffic lights, especially in images with poor lighting or low resolution.

Buildings Mislabelled as Traffic Lights:

Figure 16. Potential reasons: Complex backgrounds in urban scenes can lead to parts of buildings being mistaken for traffic lights.

Missed Traffic Lights

Figure 17. Potential reasons: Labeller fatigue or oversight can lead to missing objects, especially in images with numerous objects.

Pedestrian Lights Mislabelled as Traffic Lights

Figure 18. Potential reasons: Confusion between different types of lights due to their proximity and similar design.

Reflections Mislabelled as Traffic Lights

Figure 19. Potential reasons: Reflections can be mistaken for real objects, especially if the reflection is clear and the labeller is not paying close attention.

Vehicles Mislabelled as Both Cars and Trucks

Figure 20. Potential reasons: This could be due to overlapping bounding boxes or confusion in labelling guidelines.

Visual Results

Key visualizations generated during training and validation include:

- Confusion Matrix:
  - Shows the distribution of true positives, false positives, and false negatives across classes.
- F1, Precision, and Recall Curves:

Figure 21. F1 - confidence curve

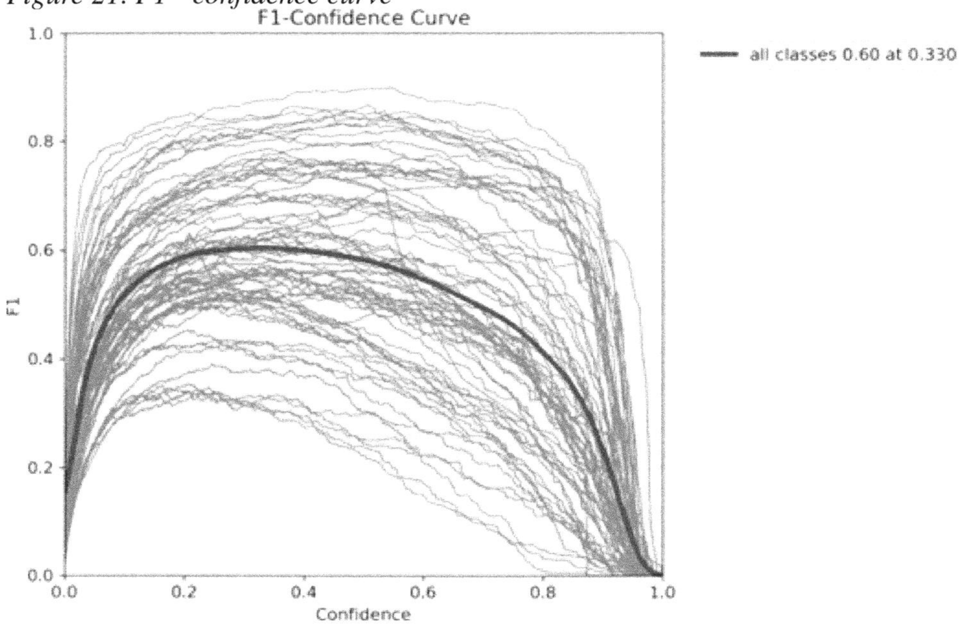

Figure 22. Precision

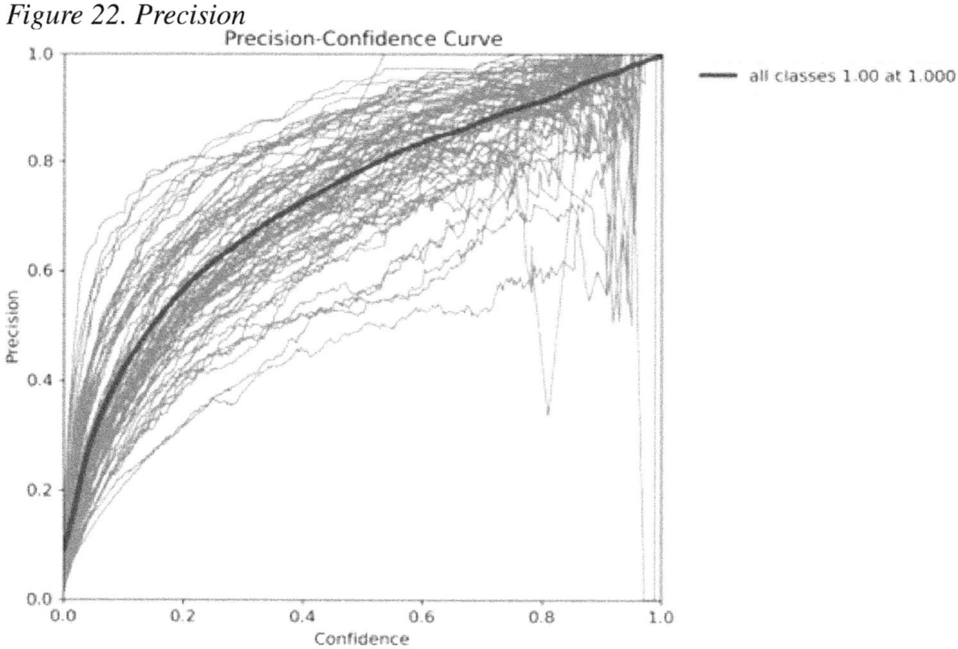

Figure 23. Recall

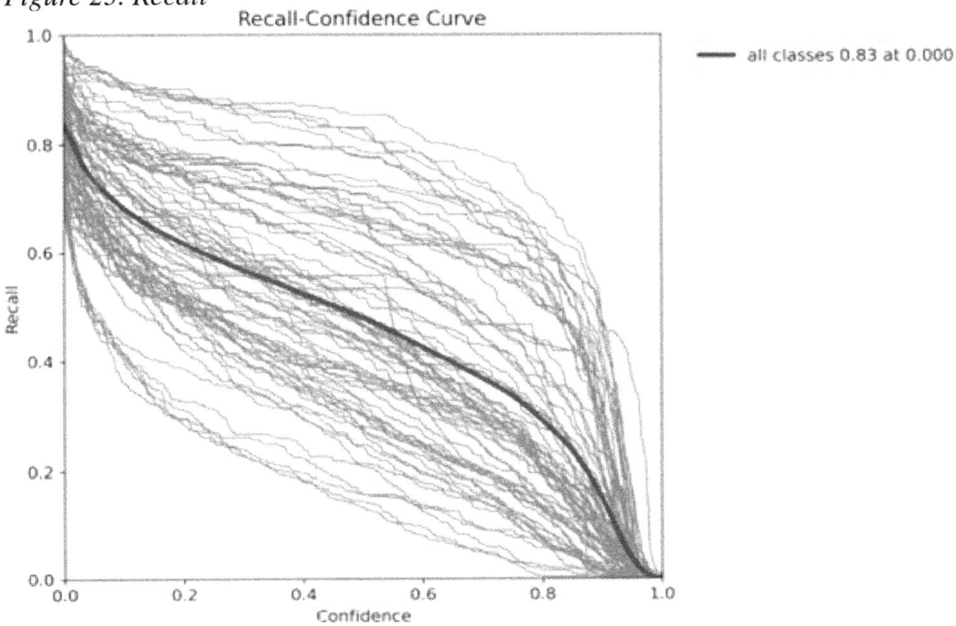

- Predicted Images:
  - Images with bounding boxes, confidence scores, and class labels demonstrate the model's ability to detect objects accurately.

Figure 24. Images with bounding boxes, confidence scores, and class labels demonstrate the model's ability to detect objects accurately

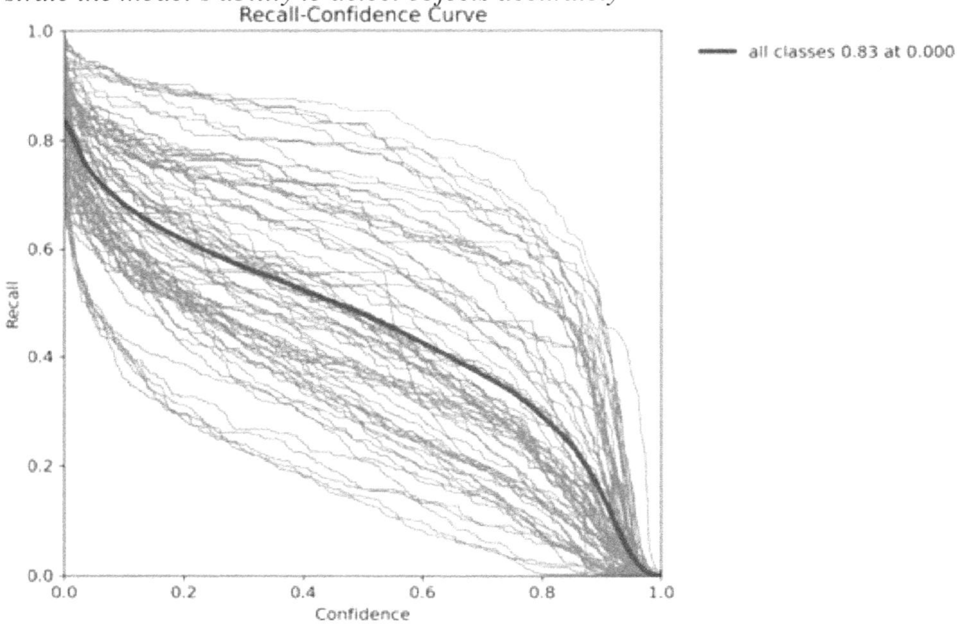

Insights from Results

- Strengths:
  - The model performs well on larger objects with distinct boundaries.
  - High mAP50 indicates robust performance across multiple classes.
- Weaknesses:
  - Small objects and objects in low-light conditions are more prone to misclassification.
  - Dense scenes lead to higher false positive rates.

Improvements

Implementing advanced augmentation techniques to improve small object detection. Fine-tuning confidence thresholds to minimize false positives.

CONCLUSION

The study implemented YOLOv8 for significant real world applications in intelligent traffic management systems. Utilizing state-of-the-art deep learning methods, the system successfully addresses critical challenges including variations in lighting, occlusion, and high-density traffic situations, offering high accuracy and robustness in detection performance in a range of environmental conditions.

Major contributions of this research are as follows:

- Real-Time Detection: Using YOLOv8 to analyse multi-camera video streams at high speed and low latency with accuracy and recall at 0.89 and 0.87.
- Adaptive Processing: Combines image preprocessing and multi-exposure fusion to enhance detection performance in low-light and extreme weather conditions.
- Occlusion Resilience: Create tracking algorithms that retain detection accuracy even when the object is partially or completely obscured.
- Optimized System Architecture: Usage of regional processing and edge computing to manage high density traffic situations without losing real-time performance.

Future Work

Improvements and Exploration

- Small Object Detection:
    - Explore multi-scale feature extraction approaches to enhance the identification of small objects, like traffic signs and far-away people.
    - Augment your data with synthetic datasets that simulate realistic driving environments.
- Temporal Integration:
    - Leverage temporal data to improve video based detection robustness allowing better tracking of mobile objects across multiple frames.
- Traffic Flow Prediction:
    - Integrate predictive models with object detection data to automate traffic flow predictions and dynamically adjust traffic light signals accordingly.

Societal Impact and Practical Applications

- Environmental Sustainability and Energy Efficiency:

- Use energy-efficient model designs and hardware configurations to mitigate the environmental cost of large-scale deployment.
- Social Impacts and Industry Applications:
  - This can bring a great deal of improvement of urban mobility through targeted efforts, which can be achieved through the following aspects of this intelligent traffic management system.
- Identify Traffic Patterns:
  - Dynamic traffic signal optimization can identify traffic patterns and adjust accordingly.
- Enhancing Safety Inside the Car:
  - Identifying the pedestrian and automobiles correctly increases safety in intersections and crosswalks.
- Minimizing Environmental Impact:
  - Efficient traffic reduces consumption of fuel and emissions of greenhouse gases.
- Smart City Utilization:
  - Integration with existing infrastructures can promote the development of sustainable and efficient urban environments.

REFERENCES

Aldughayfiq, B., Ashfaq, F., Jhanjhi, N. Z., & Humayun, M. (2023). YOLOv5-FPN: A robust framework for multi-sized cell counting in fluorescence images. *Diagnostics (Basel), 13*(13), 2280. DOI: 10.3390/diagnostics13132280 PMID: 37443674

Ashfaq, F., Ghoniem, R. M., Jhanjhi, N. Z., Khan, N. A., & Algarni, A. D. (2023). Using Dual Attention BiLSTM to Predict Vehicle Lane Changing Maneuvers on Highway Dataset. *Systems, 11*(4), 196. DOI: 10.3390/systems11040196

Chen, Y., Zhang, H., & Wu, J. (2023). Real-time traffic monitoring using YOLOv4: A comprehensive study. *IEEE Transactions on Intelligent Transportation Systems, 24*(3), 1123–1135.

Humayun, M., Almufareh, M. F., & Jhanjhi, N. Z. (2022). Autonomous traffic system for emergency vehicles. *Electronics (Basel), 11*(4), 510. DOI: 10.3390/electronics11040510

Humayun, M., Ashfaq, F., Jhanjhi, N. Z., & Alsadun, M. K. (2022). Traffic management: Multi-scale vehicle detection in varying weather conditions using yolov4 and spatial pyramid pooling network. *Electronics (Basel), 11*(17), 2748. DOI: 10.3390/electronics11172748

Khan, N. A., Jhanjhi, N. Z., Brohi, S. N., Usmani, R. S. A., & Nayyar, A. (2020). Smart traffic monitoring system using unmanned aerial vehicles (UAVs). *Computer Communications, 157*, 434–443. DOI: 10.1016/j.comcom.2020.04.049

Kumar, A., & Singh, R. (2024). Intelligent traffic management systems: Integration of AI and traditional infrastructure. *Transportation Research Part C, Emerging Technologies, 148*, 103931.

Liu, X., Wang, Y., & Thompson, R. (2023). Advances in computer vision for urban traffic management. *IEEE Access : Practical Innovations, Open Solutions, 11*, 45672–45686.

Singh, M., Patel, K., & Johnson, R. (2024). Deep learning for traffic signal optimization: A reinforcement learning approach. *Transportation Research Part B: Methodological, 169*, 234–248.

Vijayalakshmi, B., Ramar, K., Jhanjhi, N. Z., Verma, S., Kaliappan, M., Vijayalakshmi, K., Vimal, S., Kavita, , & Ghosh, U. (2021). An attention-based deep learning model for traffic flow prediction using spatiotemporal features towards sustainable smart city. *International Journal of Communication Systems, 34*(3), e4609. DOI: 10.1002/dac.4609

Wang, B., Chen, L., & Li, W. (2022). Smart traffic management: Current challenges and future directions. *Journal of Intelligent Transport Systems*, *26*(4), 382–396.

Wang, H., Liu, J., & Zhang, W. (2023). Multi-object tracking for intelligent traffic systems: A hybrid approach. *IEEE Transactions on Vehicular Technology*, *72*(5), 5678–5691.

Zhang, M., & Li, S. (2023). Economic impact of urban traffic congestion: A global perspective. *Transportation Research Part A, Policy and Practice*, *167*, 170–185.

Zhou, P., Yang, L., & Chen, X. (2023). Robust traffic object detection under variable environmental conditions. *IEEE Transactions on Intelligent Transportation Systems*, *24*(8), 7890–7904.

Chapter 2
Drones Equipped With Computer Vision (CV) and Machine Learning (ML) Will Be at the Forefront of Many Industry Breakthroughs

Galiveeti Poornima
https://orcid.org/0000-0003-4289-114X
Presidency University, India

ABSTRACT

CV and ML-enabled drones with capabilities such as autonomous navigation, object recognition in real time, weather monitoring etc. are transforming industries today. This chapter discusses the incorporation of edge computing for real-time processing, scalable and powerful AI models, and learning-based optimization. This showcases promising developments in autonomous and swarm technologies while addressing challenges like energy limitations and ethical issues. The chapter demonstrates the ability of AI-powered drones to innovate fields such as agriculture, logistics and disaster management by linking scientific progress with real-world applications.

DOI: 10.4018/979-8-3693-8497-8.ch002

1 INTRODUCTION

Drones or unmanned aerial vehicle (UAV), have proven to be game-changers when it comes to operating in difficult environments and completing complex tasks autonomously or semi-autonomously across various industries as shown in Figure 1. Also, with the inclusion of sensors, cameras and connectivity features in many drones today manufacture a possible way for them to navigate its path through complex terrains to collect important data with high efficiency. Drones have found a place in logistics, enabling fast and economical delivery of packages over the last mile for remote locations. Likewise, in agriculture, drones enable precision farming by assessing crop health, identifying pest outbreaks and improving irrigation systems. A lot of these aerial platforms have been essential in disaster response — they help run search and rescue operations, offer damage assessment or deliver emergency equipment (Criollo et al., 2024).

Figure 1. Drones performing diverse tasks

They are also very important for science and environmental monitoring in ways that go beyond commercial cost saving uses. With them we can observe ecosystems in detail, monitor movements of wildlife and measure deforestation rates. Another major application is infrastructure inspection, where drones can do real-time assessments of bridges, wind turbines and pipelines to reduce the need for dangerous manual check-ups. Finally, drones are instrumental in public safety – monitoring crowds and traffic areas. Drones will also play an increasingly important role in the development of smart cities, autonomous transportation networks and sustainable resource management as their technological capabilities improve (Himeur et al., 2022).

CV and ML represent transformative technologies that empower drones to interact with their environment — visual spectrum drones to be specific. Computer Vision (CV) allows drones to process and analyze visual data in real-time for obstacle detection, path planning and object tracking. Navigating autonomously (especially in dynamic environments) is a core skill that allows drones to update their planned flight paths to avoid collisions and arrive at their objectives efficiently. Additionally, CV-based drone applications can provide high-resolution data to support tasks such as environmental monitoring and terrain mapping for agriculture, forestry, urban planning processes (Dhami, 2024).

ML on the other hand, complements CV by allowing drones to learn from data and make decisions over time. Deep learning models can, for instance, improve how a deep learning drone identifies an object and when tasked with finding its signature visual features such as vehicles or humans in highly intricate scenes. Drones learn entire sets of behaviours, e.g. flight path, structure or operation, to ensure minimisation of power consumption and successful completion of the mission through reinforcement learning by trial-and-error interactions with their environment. The integration of CV and ML enables interpretable actions with drones, facilitating operations such as real-time target detection, automatic reconnaissance, and contextual responses based on the input received during image analysis, transforming use cases in different fields. Table 1 describes different use cases of CV and ML for the advancement of drone features. It explains the combined contribution of CV and ML that enable a drone to be more intelligent, efficient, and capable pertaining towards various domains (Saha et al., 2024).

Table 1. Applications of CV and ML in drones

Application	Role of CV	Role of ML
Obstacle Avoidance	Detects obstacles and generates safe navigation paths	Predicts optimal flight paths based on prior data
Object Recognition	Identifies specific targets like vehicles or humans	Enhances accuracy through model training on labeled datasets
Terrain Mapping	Creates detailed 3D maps of landscapes	Analyzes and categorizes terrain types for better insights
Crop Monitoring	Identifies crop health issues through visual analysis	Predicts pest infestations or yield patterns
Autonomous Surveillance	Monitors large areas in real-time	Learns suspicious patterns for anomaly detection

The intention of this chapter is to identify the potential role that computer vision (CV), and machine learning (ML) technologies can play in improving drone performance by enabling functionalities such as intelligent navigation, object detection, terrain mapping, and real-time decision making. The article explores the technological infrastructures, such as edge computing that are needed to integrate these features into drones and their usage in various sectors comprising of agriculture, logistics, surveillance and disaster Management. The chapter also discusses the challenges and ethical challenges that are associated with the deployment of CV and ML in drone systems while showing pathways for innovation and development where new avenues may open up in either autonomous or swarm technologies for drones. The chapter equally contributes to academic, practitioner and industry experts due to its holistic nature of connecting CV and ML theory with practice through a topical view across all relevant fields in aerial systems.

2 FOUNDATIONAL CONCEPTS

Computer Vision (CV) is the backbone for the drones where it helps in finding and understanding visual information, making them highly intelligent and autonomous systems. Basically, cv is about making sense of images or videos of CV enables drones to perform tasks such as navigation, object detection and obstacle avoidance. Using high-resolution cameras and sensors on drones, DSP (digital signal processor) performs Feature Extraction Edge Detection Pattern Recognition of imagery. This allows drones to detect obstacles in real-time, create terrain maps, and even classify objects for accurate and efficient operation in various environments (Kiruthiga et al., 2024).

Terrain mapping and environmental monitoring is an important use of CV in drones. Surrounds its stereo cameras and depth sensors with 3D models of its environment that they assist applications such as agriculture for analyzing the health of crops, or infrastructure inspection for identifying structural anomalies. More importantly, the progress of deep learning enables CV-equipped drones to perform more complicated tasks like semantic segmentation to identify and classify objects in a scene (e.g., vehicles, pedestrians or vegetation). Coupled with other features, these massively increase the scalability of drones ensuring their use in logistics, surveillance, disaster management and more. Table 2 captures at a high-level, capabilities enabled by CV in drones. It serves as a quick reference to understand the myriad of ways that CV improves drones by bridging specific tech capabilities with their real-world uses (Zhao, 2024).

Table 2. Capabilities of computer vision in drones

Capability	Description	Application
Obstacle Detection	Identifying and avoiding obstacles in real-time	Autonomous navigation
Object Tracking	Tracking the movement of specific targets	Surveillance and logistics
3D Terrain Mapping	Generating detailed maps of the environment	Agriculture and urban planning
Semantic Segmentation	Recognizing and categorizing scene elements	Environmental monitoring and inspection
Optical Flow Analysis	Analyzing motion patterns in visual data	Path planning and dynamic navigation

Machine Learning (ML) is a game changer, capable of turning ordinary drones with little intelligence into adaptive flying devices that can learn from data and react socially to an ever-changing environment. The core of ML capabilities power drones to perform duties with precision, such as autonomous navigation and object detection, and real-time decision-making. These techniques fall under supervised learning, unsupervised learning, reinforcement learning and deep-learning category, each bringing their benefits to drone systems. In supervised learning, for example, drones can be trained on datasets that contain labeled data to recognize objects or follow certain route. Unlike supervised learning, which typically requires labelled data, unsupervised learning allows drones to discover anomalies in datalike spotting patterns of infection in crop fields or noticing structural flaws in infrastructure (Zhong, 2024).

Drones have only a limited ability to learn from other drones, and one of the most powerful ways to learn is by reinforcement learning – which fills out for corpses (behavior) across test runs and optimizes behavior with error. For instance, a drone may learn how to traverse through complex terrains or better utilize energy by maximizing reward functions. Deep learning, a sub-field of ML with major applications such as image classification and semantic segmentation, gives deep classification capabilities thus allowing drones to make sense of their visual surroundings autonomously. When combined, these techniques significantly improve the autonomy and intelligence of drones, becoming fundamental for logistics, surveillance, agriculture and to aid in disaster response (Jang and Kim, 2024). There are a number of technical, operational and ethical challenges involved in using CV and ML technologies formidably into drones. Probably the greatest technical challenge lies in the fact that drones are usually hampered by small processors and energy consumption capacity. With many CV and ML algorithms requiring large computing power, they could put pressure on the drone's onboard hardware, particularly if real-time processing is required. To make things worse, these processes consume substantial chunks of energy which limits the flight time of drones on operational missions —thus, an important concern in long-endurance protocols (Caballero-Martin et al., 2024).

However, another major hurdle of this approach is the reliance on a robust connectivity for im- mediate real-time data transfer and decision-making. Most of the CV and ML applications involve cloud-based processing or edge computing which requires stable communication networks. In re- mote/ infrastructure-poor areas, where there are connectivity issues, they can encumber a drone's ability to operate effectively. Additionally, the scalability and stability of CV and ML models in varied and dynamic ecosystems presents a unique challenge. For example, a model that performed extremely well in one environment might struggle in another and may require significant retraining efforts (Das et al., 2024).

The integration of computer vision and machine learning in drones presents ethical and societal challenges that extend beyond technical obstacles. Drones equipped with advanced surveillance technology pose a significant threat to privacy, particularly in urban areas or residential neighborhoods. Concerns regarding the potential misuse of drones for illicit surveillance or data acquisition must be addressed by comprehensive regulatory frameworks. Furthermore, biases in machine learning models may result in inaccurate or prejudiced classifications, complicating the deployment of drones in specific sensitive applications. Table 3 provides a succinct overview of the primary obstacles to implementing computer vision and machine learning in drones. It links the nature of each difficulty to its wider ramifications, enabling readers to comprehend both technical and societal dimensions at a glance (Al-kfairy et al., 2024).

Table 3. Summary of challenges

Challenge	Description	Impact
Limited Processing Power	Drones often lack the computational resources for running advanced CV and ML models.	Reduces real-time processing capability.
Energy Constraints	High energy consumption of CV and ML algorithms reduces drone flight times.	Limits operational efficiency.
Connectivity Issues	Dependency on stable networks for cloud or edge processing.	Hinders performance in remote areas.
Scalability and Robustness	Models may not generalize well to new or diverse environments.	Requires frequent retraining.
Privacy and Ethical Concerns	Surveillance and data collection raise societal and regulatory challenges.	Creates trust and acceptance issues.

3 HARDWARE AND SOFTWARE ARCHITECTURES

Drones with wide vision (CV) and machine learning (ML) hardware needs special configurations for analyzing the data. There is standard setup which includes visual camera, high resolution cameras for capturing visual data and also additional sensors like LiDAR and Infrared infrared to capture depth of the area as well as thermal information. They empower drones to perform complex tasks, such as obstacle detection, object tracking, and 3D mapping. More advanced drones utilize stereo vision systems for spatial awareness and depth perception, which are necessary for navigating complex environments (Leslie et al., 2024).

Drones are often needed to process the vast computational requirements of any CV and ML powered algorithms which can be done by adding an edge computing module with a GPU or TPU onboard (Suganya et al., 2024). These units are capable of image processing and local decision making with minimal dependence on external networks. So, small and high capacity batteries are required to maintain the high energy consumption operations of these components. Connectivity modules (Wi-Fi, 5G) for easy interaction with the cloud to process bigger model updates and logic. Together these hardware components build a balanced system that can run complex CV and ML applications while being able to operate for extended periods of time. Figure 2 effectively communicates the integration of CV and ML hardware in drones, showcasing their advanced technological features. It provides a visual representation of how drones process data, making it easier for readers to understand the hardware layout and its relevance to CV and ML tasks.

This is where the edge computing comes into play, allowing drones to perform real-time data processing and analysis for immediate results without the constant need of calling up resources from a cloud-based system (Sarkar and Sahoo, 2024). Such an approach allows computation to happen around the data source itself, employing onboard hardware such as GPUs, TPUs and edge processing modules — thus making it more ubiquitous in nature. This ability, in real-time applications like obstacle detection, can be game-changing for drones where swift moves are also required for navigation and image classification. Local data processing ensures drones are able to make autonomous decisions in real time which enhances their operational efficiency and safety, even in areas with restricted or non-existent internet access.

Edge computing in drones is also fastened by the demand for optimal energy management (Sarkar and Sahoo, 2024). In contrast to conventional cloud processing, which involves the transfer of large amounts of data from drones and other flying platforms to sequestered processors, edge computing reduces latency by lever- aging distributed architecture and conserving bandwidth so that energy can be saved within unmanned aerial vehicles (UAVs) themselves. Grouping of drones enables autonomous operation for long periods, which is especially useful in disaster response. Moreover, drones with edge-enabled intelligence can also deploy high-end machine learning models for predictive maintenance, swarm coordination and adaptive path planning thereby attaining reliable real-world performance through robust inferencing capability. Table 4 summarizes the key advantages of integrating edge computing into drone systems.

Figure 2. Advanced drone hardware for CV and ML, featuring cameras, LiDAR, edge computing, and GPUs for real-time intelligence.

Table 4. Advantages of edge computing in drones

Benefit	Description
Real-Time Processing	Enables immediate analysis and decision-making without delays from cloud dependencies.
Reduced Latency	Minimizes delays in data transfer, critical for applications like obstacle avoidance.
Energy Efficiency	Conserves power by processing data locally instead of transmitting large volumes to the cloud.
Enhanced Privacy	Keeps sensitive data onboard, reducing risks associated with cloud storage.
Autonomous Operations	Allows drones to function independently in remote or network-limited environments.

Use of highly optimized hardware such as GPUs (Grapic Processing units), TPUs (Tensor Processing Units) and AI dedicated chips improving the computation prowess of drones, allowing them to perform highly advanced computer vision (CV) and machine learning (ML) tasks (Peng et al., 2024). The GPUs play an important role in such data for processing a lot of image data in real-time and also helps in parallel computation to perform computations associated with 3D mapping, object detection, etc. GPUs can perform multiple calculations at once, so they help drone systems process visual information quickly even in difficult conditions.

TPUs, a custom-built chip for deep learning workloads, do not miss the beat of ML model performance onboard drones (Kouris et al., 2019). These units hasten the training and inference period, as a result enabling drones to execute complex functions similar to semantic segmentation, natural language comprehension needed for communicating, or adaptive path planning. Specialized AI chips, as found in edge computing modules, combine power efficiency with powerful on-site AI processing to enable more responsive real-time decision-making. Designed to help drones operate without connectivity in the field for uses like disaster relief, agriculture and security surveillance.

4 APPLICATIONS OF COMPUTER VISION IN DRONES

Image processing is essential to surveillance, logistics, and search-and-rescue missions, enabling numerous drones equipped with CV to perform object detection and tracking tasks (Abdelnabi and Rabadi, 2024). CV Power Drones To Monitor Health, Pest Infestation Analysis And Irrigation by Extracting The Images In Agriculture. Another important field of application is terrain mapping and 3d modeling which enables drones to map out construction sites, survey place hit by disaster and observe environmental changes. In addition, CV allows drones to perform autonomous navigation and obstacle avoidance thus making them suitable for safe operation in dynamic and cluttered environment (Crupi et al., 2024). They also include infrastructure inspection, which allows drones to spot anomalies such as damages in bridges, pipelines, and wind turbine structures without putting human lives at risk. As CV is increasingly integrated into drones, we see more innovation in the commercial and scientific fields.

4.1 Object Detection and Recognition

One of the most groundbreaking applications of computer vision in drones is object detection and recognition, allowing drones to identify, classify and track objects in real-time (Cazzato et al., 2020). Through sophisticated algorithms and deep learning models, drones can identify diverse objects such as vehicles, humans, animals or elements of the infrastructure – even in dynamic environments. Along with these detections are Bounding boxes, annotations and confidence scores) which can be used for different applications like in some cases surveillance, logistics, search-and-rescue missions etc.

Drone object detection enable them to maneuver around the complex space safely without collision and at the same time effectively completing tasks assigned (Sandinc et al., 2021). Abstracting one example, drones could monitor moving traffic and can track a moving vehicle in urban areas whereas Drones can detect livestock or machinery in agricultural fields. In addition, object recognition enhances these functions with the ability for the drone to distinguish between types of objects e.g. a car from a truck or identify plants down to species level. Together, detection and recognition open up possibilities across autonomous operations and precision work in a range of industries.

4.2 Terrain Mapping and Navigation

Terrain mapping and navigation represent a pivotal use of computer vision (CV) for drones, facilitating autonomous navigation and the scanning of intricate terrain, as seen in Figure 3. Drones utilize diverse sensors such as LiDAR, stereo cameras, and depth imaging technology to produce intricate 3D maps of their surroundings (Green et al., 2019). These maps illustrate color-coded elevation ranges and topographical data, used in several areas like building, agriculture, and ecology. Unmanned Aerial Vehicle By analyzing this data in real-time, they may comprehend their surroundings, identify barriers, ascertain a secure route, and navigate areas previously inaccessible to humans, such as rugged mountains or dense forests.

Using GPS coordinate systems along with mapping the terrain, drones can precisely track routes and perform autonomous missions (Nguyen et al., 2021). An example can be in disaster response situations where drones could map the impacted region, evaluate destruction and lead rescuers to zones of safety. In agriculture, they analyze the topographies of different fields to optimize irrigation and planting. In addition, by revealing safe routes and generating waypoints for effective travel, terrain mapping makes it easier to navigate. Such an ability enhances the operational efficiency of drones, while also extending their utility when deployed in remote or infrastructure sparse areas.

4.3 Obstacle Avoidance and Path Planning

It includes obstacle avoidance and path planning as shown in Figure 4, very important to feature the autonomous drones in which they can detect without human interaction an environment with complex shape. Drones detect obstacles (for instance, buildings, trees or moving objects) in real-time using computer vision (CV) and sensor fusion techniques. LiDAR, ultrasonic detectors, and stereo cameras are versatile sensors that supply depth and spatial data, which computer vision (CV) algorithms analyze to detect obstacles and determine safe distances. Such ability is crucial for applications such as delivery, search-and-rescue, and monitoring tasks

Figure 3. Terrain mapping and navigation

Path Planning allows drones to calculate the best route for its waypoints which complements obstacle avoidance. Using advanced algorithms like A* (A-star), Dijkstra, and reinforcement learning models, multiple parameters such as terrain types, obstacles on the way and other waypoints are evaluated to provide optimal paths. Real-time updates ensure that the drones adapt to the constantly changing environment, including abrupt impediments and unforeseen weather conditions. The integration of obstacle detection and adaptive path planning enhances drone agility, enabling operation in more complex and unpredictable environments.

4.4 Environmental Monitoring and Surveillance

Advanced sensors and drones equipped with computer vision (CV) are transforming environmental monitoring and extensive surveillance (Gohari et al., 2022). Drones may traverse extensive, often inaccessible areas and provide real-time data beneficial for environmental conservation, resource management, and security operations. Drones identify alterations in forest cover, areas of deforestation, and oversee rivers and lakes. The multispectral and thermal imaging cameras facilitate the assessment of vegetation health and enable wildlife tracking, so contributing to biodiversity protection.

Drones are used for surveillance, monitoring large crowds and sensitive areas, and criminal detection (Abiodun, 2020). The drones' ability to detect and track objects can be used to examine human movement, vehicles, and infrastructure giving law enforcement agencies and disaster response teams actionable intelligence. It is this diversity of capabilities and the multiplicity of applications that make drones such powerful tools for natural resource protection and public safety. Figure 5 illustrates a drone conducting environmental monitoring over a forest and river, detecting deforestation zones and water changes with real-time overlays of camera feeds and data analysis.

5 MACHINE LEARNING MODELS FOR DRONE OPTIMIZATION

By making tasks like autonomous navigation, object detection and energy-efficient flight planning possible with machine learning models, it improves drone performance (Wu et al., 2024). For example, supervised learning helps in training drones to recognize targets while reinforcement learning enhances the drone's ability to adapt and learn from unpredictable environments. At the heart of this com- plex image analysis and decision making right from within vision are deep learning models, namely convolutional neural networks (CNNs).

Figure 4. Obstacle avoidance and path planning

5.1 Supervised Learning for Navigation and Target Detection

Through supervised learning, drones can learn pinpoint maneuverability and target detection using labeled datasets of either images or sensor data (Kurunathan et al., 2023). In training, models are presented with input-output pairs — for example, images of objects (e.g., road, vehicle or pedestrian) along with their correct classifications or locations. Drones can identify patterns through this and use them in practical applications. For navigation, drones learn to identify safe paths thereby navigating their surroundings without collision whereas for target detection they classify and track objects such as vehicles or people to operate efficiently in an interactive environment. The table 5 effectively correlates supervised learning methodologies with practical drone applications, providing a transparent overview of the data, algorithms, and results involved. It underscores the adaptability of supervised learning in addressing various challenges, rendering it an invaluable resource for researchers and practitioners.

Table 5. Example applications of supervised learning in drones

Application	Input Data	Output/Action	Example Algorithms
Navigation	Images of roads, obstacles, and terrains	Safe route identification and obstacle avoidance	Decision Trees, SVM
Target Detection	Images of objects (vehicles, humans)	Bounding boxes and classifications for objects	CNNs, ResNet
Path Following	GPS coordinates and visual landmarks	Predicted waypoints for autonomous path execution	k-Nearest Neighbors (kNN)
Hazard Identification	Images of dangerous objects (e.g., debris)	Alerts for potential hazards along the path	Random Forests

5.2 Unsupervised Learning for Anomaly Detection

It entails the identification of environmental anomalies (defining normalcy versus deviation) using unsupervised learning, wherein patterns are discerned that exhibit dissimilar behavior compared to forthcoming data devoid of labeled information. Drones can identify anomalous patterns in extensive datasets using clustering and dimensionality reduction (Ahn et al., 2019). An application in agriculture involves drones utilizing unsupervised learning to transmit signals that identify clusters of anomalous vegetation patterns, which may signify sick crops or pest infestations. Similarly, in infrastructure inspection, drones analyze structural elements (such as bridges) and detect cracks or deformations using this data to facilitate predictive maintenance.

Figure 5. Environmental monitoring and surveillance

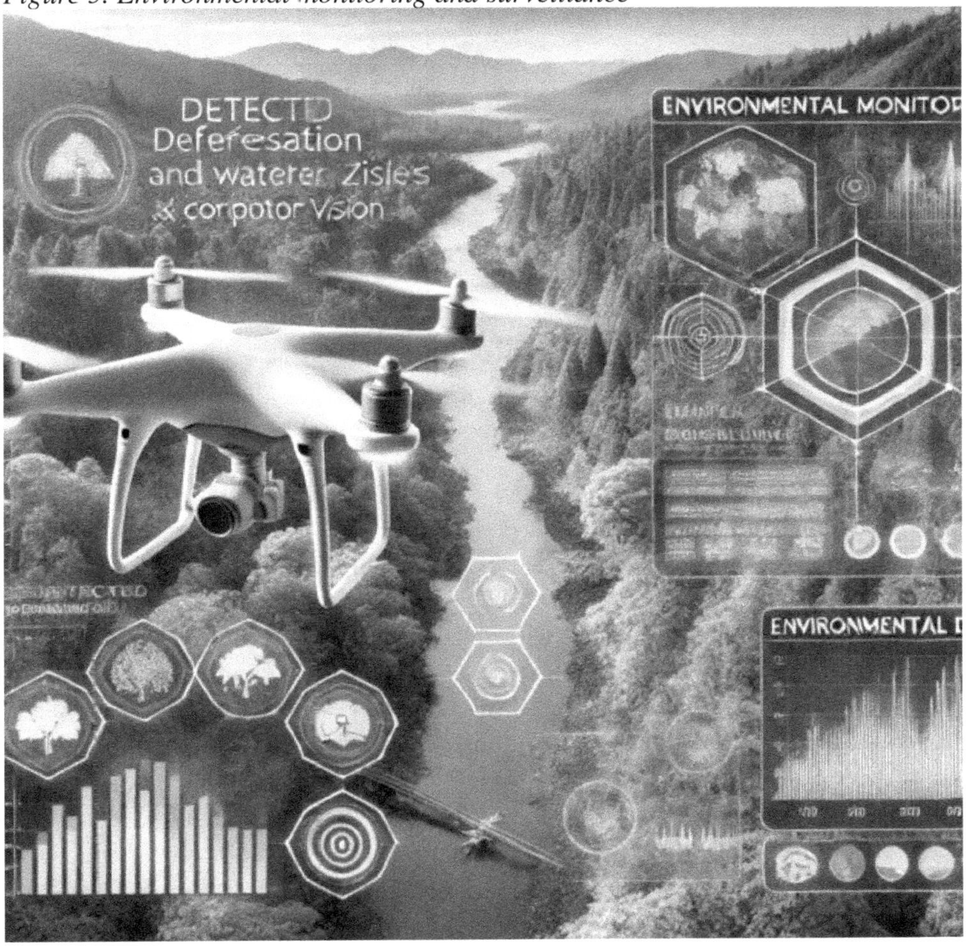

Unsupervised learning is especially appropriate for dynamic or unfamiliar contexts lacking established labels (Luber et al., 2009). These algorithms adapt based on the correlation between normal data distributions and identify outliers as potential anomalies. This allows drones to execute various tasks across multiple applications, such as environmental monitoring, security surveillance, and disaster assessment. Table 6 offers a succinct yet thorough summary of the ways in which unsupervised learning improves drones' capacity to identify abnormalities in many applications. The table provides a valuable reference for comprehending the operational function of unsupervised learning in drones by linking data kinds, techniques, and algorithms to particular use cases.

Table 6. Example applications of unsupervised learning in drones

Application	Input Data	Methodology	Output/Action	Example Algorithms
Crop Health Monitoring	Multispectral images of fields	Clustering unusual vegetation areas	Identifies unhealthy patches for action	K-Means, DBSCAN
Infrastructure Inspection	Sensor data from structures	Outlier detection in patterns	Detects cracks or structural issues	Isolation Forest
Environmental Monitoring	Thermal and visual images	Identifying deviations in heat maps	Highlights areas with unusual temperature	Gaussian Mixture Models (GMM)
Surveillance	Video feeds of monitored areas	Detecting unusual movements	Alerts for unauthorized activities	Principal Component Analysis (PCA)

5.3 Deep Learning Applications for Complex Tasks

Deep learning, characterized by neural networks with multiple layers, enhances the accuracy of com- plicated, high-level tasks done by drones (Mittal et al., 2020). Utilizing substantial processing capacity and sophisticated algorithms facilitates the execution of complex tasks, including facial recognition, semantic segmentation, real-time traffic analysis, and object tracking. Consider drones capable of recognizing individual faces in a crowd or identifying specific objects in a dynamic environment; when augmented with deep learning, they achieve a high level of accuracy. Semantic segmentation enables a drone to identify elements of a landscape, such as roads, vegetation, or water bodies, thereby offering valuable insights for applications including surveillance, environmental monitoring, and urban planning.

Deep learning models like CNNs and RNNs can process unstructured data (images, videos) with high accuracy enabling drones to leverage these capabilities (Al-lQubaydhi et al., 2024). He added that they processed data in real-time, which means these models also improve different tasks including anomaly detection, 3D mapping and autonomous navigation. With deep learning, drones can not only understand complex scenes better but also make their own decisions based on them to take a huge leap in the level of automation and intelligence. Table 7 efficiently connects specific drone tasks to the appropriate deep learning models and their outcomes. It helps readers understand the practical applications of advanced AI techniques in drone systems and serves as a quick reference for choosing the right deep learning approach for specific challenges.

Table 7. Example applications of deep learning in drones

Application	Task	Deep Learning Model	Outcome
Facial Recognition	Identifying individual faces in a crowd	Convolutional Neural Networks (CNNs)	Real-time identification for security or surveillance
Semantic Segmentation	Differentiating objects in an environment	Fully Convolutional Networks (FCNs)	Precise mapping for urban or agricultural landscapes
Traffic Analysis	Detecting and categorizing vehicles in motion	YOLO (You Only Look Once)	Real-time traffic flow and anomaly detection
3D Mapping	Constructing detailed environmental models	Generative Adversarial Networks (GANs)	Accurate 3D representation of terrains
Object Tracking	Following moving objects like vehicles or people	RNNs with Long Short-Term Memory (LSTM)	Dynamic tracking for logistics or surveillance

5.4 Reinforcement Learning for Autonomous Behavior

Drones can become autonomous by learning from experience and this is the main idea of Reinforcement Learning (RL). In contrast to supervised learning, where the model is trained on labeled datasets, RL is about maximizing cumulative rewards through exploration and interactions with the environment (Caminero et al., 2019). Specifically, in obstacle avoidance, energy-efficient navigation and path optimization tasks DRL based drones learn to adaptively improve their strategies for better performance. In other words, consider a scenario where RL can help an agent (e.g., a drone) identify the best actions to take in a potentially complex maze-like environment by tracking outcomes associated with its previous actions according to up or down reward signals (i.e., turn left no food/collision, negative reward. Turn right [positive] collision [negative]) The drone learns a policy, essentially a set of rules which govern how it behaves, and an effective way to create this policy is by practising as much as possible, even when the scenarios are not expected (Clarke and Moses, 2014). Deep reinforcement learning (RL) allows for long-term planning and adjustment to the environment, making it particularly effective in scenarios like search-and-rescue missions or through-swat swarm coordination (Hu¨ttenrauch et al., 2019). The table 8 connects RL techniques with real-world drone applications, illustrating their impact on performance and autonomy. By pairing tasks with specific algorithms and outcomes, it provides a practical understanding of how RL drives intelligent decision-making in drones.

Table 8. Applications of reinforcement learning in drones

Application	Task	RL Technique	Outcome
Obstacle Avoidance	Learning to navigate safely in cluttered areas	Q-Learning, Deep Q-Networks (DQN)	Real-time collision-free movement
Energy-Efficient Navigation	Optimizing flight paths to conserve battery	Policy Gradient Methods	Prolonged operational time
Path Optimization	Discovering the shortest route to a target	SARSA (State-Action-Reward-State-Action)	Efficient and adaptive route planning
Swarm Coordination	Synchronizing movements in drone swarms	Multi-Agent Reinforcement Learning	Coordinated, collision-free swarm operations
Search-and-Rescue	Exploring unknown areas for Survivors	Monte Carlo Methods	Faster identification of targets in complex terrains

6 INTEGRATION OF CV AND ML WITH EDGE COMPUTING

One of the greatest advantages of drones with edge computing is their ability to process data instantly on-board the drone without always requiring access to the cloud. This reduces latency, facilitating swifter decision-making about obstacle avoidance, path planning, and object detection. Edge computing diminishes bandwidth utilization, conserves energy consumption, and decreases operational and flight durations by processing data locally. This renders it a more resilient solution for remote or bandwidth-constrained areas where uninterrupted drone mobility and cooperation are desired. In addition, edge computing also improves data privacy and security by ensuring that sensitive data stays local and is not transmitted over the cloud to be processed – a crucial property for surveillance, environmental monitoring, logistics applications.

6.1 Frameworks and Platforms for Edge-Enabled CV and ML Processing

6.1.1 Edge Hardware Platforms

- **NVIDIA Jetson**: Offers GPU-accelerated modules optimized for real-time AI tasks like object detection and path planning (Hu¨ttenrauch et al., 2019).
- **Google Coral**: Provides Tensor Processing Units (TPUs) for energy-efficient ML inference (Kasioulis et al., 2024).

- **Intel Movidius Neural Compute Stick**: Compact hardware for deploying deep learning models on edge devices (Nguyen Hoang, 2019).
- **Qualcomm Snapdragon**: AI-focused processors for mobile and drone-based edge computing[1].

6.1.2 Frameworks for CV and ML

- **TensorFlow Lite**: Lightweight version of TensorFlow optimized for edge deployment.
- **PyTorch Mobile**: Enables deploying PyTorch models on mobile and embedded systems.
- **OpenCV**: A powerful open-source library for image processing and feature detection.
- **Darknet**: Lightweight framework for YOLO (You Only Look Once) object detection models.

6.1.3 Cloud-Edge Integration Platforms

- **AWS IoT Greengrass**: Synchronizes local edge processing with cloud analytics for scalable AI workflows (Shankar et al., 2020).
- **Microsoft Azure IoT Edge**:Provides tools for deploying and managing ML models on edge devices (Merenda et al., 2020).
- **Google Cloud IoT Edge**: Integrates ML models trained on Google Cloud into edge devices (Kanungo, 2019).

6.1.4 Other Tools and Libraries

- **Edge Impulse**: Platform for creating, deploying, and optimizing ML models for edge devices (Hymel et al., 2022).
- **NVIDIA DeepStream**: Optimized for video analytics and real-time CV tasks on NVIDIA Jetson (Gupta et al., 2022). [1]https://www.qualcomm.com/products/mobile/snapdragon/smartphones/mobile-ai

6.2 Case Studies

- **Case Study 1: Precision Agriculture**
  - **Problem:** Traditional agricultural methods struggle to efficiently monitor crop health across large fields (Ayaz et al., 2019).
  - **Solution:** Drones equipped with edge-enabled CV and ML using platforms like Google Coral TPU perform real-time multispectral imaging

to identify stressed crops and detect pest infestations (Reyes-Hung et al., 2024).
  - **Outcome:** Farmers receive immediate insights on affected areas, enabling targeted interventions. This reduces pesticide usage and improves yield quality while conserving resources.
- **Case Study 2: Disaster Response and Recovery**
  - **Problem:** Rapid damage assessment is critical in disaster-hit areas, where access is often restricted (ESCAP, 2017).
  - **Solution:** Drones powered by NVIDIA Jetson modules use ML models to analyze imagery for identifying damaged structures, blocked roads, and safe zones in real-time (Hern´andez Vicente, 2023).
  - **Outcome:** Emergency teams deploy resources faster and more efficiently, reducing response times and potentially saving lives.
- **Case Study 3: Urban Traffic Management**
  - **Problem:**Monitoring and analyzing traffic flow in congested urban environments is labor- intensive and inefficient (Alkaabi et al., 2024).
  - **Solution:** Drones equipped with edge computing modules and YOLO object detection models analyze vehicle and pedestrian movements, identifying bottlenecks and violations (Bakirci, 2024).
  - **Outcome:** Traffic authorities receive actionable data, enabling real-time optimization of traffic signals and improved urban mobility.
- **Case Study 4: Wildlife Conservation**
  - **Problem:** Monitoring wildlife populations in remote areas without disturbing ecosystems is challenging (Nagendra et al., 2013).
  - **Solution:** Drones with TensorFlow Lite deploy edge models for object detection and heat mapping to track animal movements and detect poaching activities (Mantau).
  - **Outcome:** Conservationists gain critical insights into animal behavior while ensuring habitat protection.
- **Case Study 5: Industrial Inspection**
  - **Problem:** Manual inspection of infrastructure like pipelines and wind turbines is hazardous and time-consuming (Dimitrova et al., 2022).
  - **Solution:** Drones using AWS IoT Greengrass perform autonomous inspections, detecting anomalies like cracks or corrosion using edge-processed CV and ML models [(Zhang et al., 2024).
  - **Outcome:** Companies achieve safer, faster, and more cost-effective maintenance workflows.

7 CHALLENGES AND LIMITATIONS

Due to its ability of processing data in real-time, Computer Vision (CV) and Machine Learning (ML) based algorithms have found plenty of applications in drones (Zhou et al., 2021). However conventional CV or ML algorithm usually require a lot of computing power to operate, making it infeasible for the drone with limited onboard sensors process data in near real-time due to their low computation capabilities. Flight time is additionally limited in terms of power, making energy efficiency an important requirement for longer missions. The scalability and robustness of AI models in diverse and dynamic environments also limits drone performance.

7.1 Processing and Power Constraints in Drones

Due to their hardware choices, a mobile drone is constrained in terms of processing and power, making conventional CV/ML difficult (e.g., real-time CV or machine learning). Unfortunately, the combi- nation of small size and weight constraints imposed on drones limits the use of high-performance hardware. Due to this, onboard processors are frequently not powerful enough to run sophisticated AI algorithms such as object detection, path planning, or environmental mapping in an efficient manner (Mengozzi et al., 2024).

Additionally, power limitations make the issue worse since drones are battery-powered: they depend on batteries which are necessarily lightweight and have low storage capacity (Apeland, 2021). Model- driven flight operations require running inherently energy-intensive computer vision (CV) and machine learning (ML) models, which quickly drain the battery and result in limited operational range or flight time. In certain cases, like cloud-based processing, the need for continuous connectivity further adds to power requirements (Markovic et al., 2013). Edge computing solutions and optimized AI models are becoming keys to unlocking further progress for drones because they hit the sweet spot — balancing computational efficiency with energy consumption (Alanhdi and Toka, 2024).

7.2 Scalability and Robustness of CV and ML Models

Computer vision (CV) and machine learning (ML) models have scalability and robustness as critical challenges of drone applications (Abadade et al., 2023). Models need to remain performant as they scale with data volumes, environments, and complexity without crumpling under pressure. So, it can be like a model that is trained for navigating cities cannot directly perform in rural or forested environments without retraining. In contrast, robustness is concerned with the consistency of models under varying or unforeseen conditions; an example would be changing lighting, weather, or especially unexpected obstacles.

To get there, scalable solutions need to be rapid to train, employ lightweight models and transfer learning so that they actually generalize from one task/environment to the next. A mission-critical operation, therefore, requires advanced-level preprocessing, complex model tuning and real-time adaptation to avoid failure which does not suit the requirements of robustness. This makes it imperative for the CV and ML community to do more work in order to build models that perform well across various situations, potentially evolving through time (Althati et al., 2024).

7.3 Ethical and Privacy Concerns in Drone Usage

The ethical and privacy issues related to drones with CV and ML are important, especially in surveil- lance and data collection applications. With drones able to take high res pictures and videos, frequently without consent, this could violate a citizen's privacy. In cities in particular, where drones could passively or actively surveil private space, this becomes even more problematic, and requires a consideration of the beneficial security expectations against encroaches to personal freedom (Al-kfairy et al., 2024).

Another important ethical problem with ML models is that these have biases which result in discrimination for tasks such as facial recognition. For example, biased or non-representative training data - setting wrong people related to the category of ML agent Apart from the safety side, with unauthorized surveillance or data harvesting in mission also drones might manipulate it that calls for strong regulatory frameworks. Transparency about how the sensor data from a drone is collected, processed and used; accompanied by strong accountability mechanisms to address misuse of that technology, addresses these challenges while also building public trust in drones.

8 FUTURE DIRECTIONS

Edge AI is making a massive change in performance tweaking concerning drones by allowing data processing, decision-making, adaptability directly on the device. Drones equipped with powerful edge hardware such as NVIDIA Jetson modules and Google Coral TPUs can perform advanced AI operations using on-board processing (without cloud connections) for tasks like obstacle avoidance, object detection, and path planning for power-efficient autonomous navigation. Flight time and operational range are augmented by lightweight, energy-efficient algorithms that further cement drones into applications ranging from surveillance to logistics to disaster response. Furthermore, advances in federated learning and compression of models enable drones to refresh and enhance AI models together, without the need to share sensitive data. All these enhancements drive the progress of drone autonomy and intelligence, providing reliable functionalities amid deviant and remote environments.

Recent capabilities in computer vision (CV), and to some extent, machine learning (ML) research for drone applications are suited towards churning out more of the drones into an autonomous, adaptable and energy-efficient ecosystem. To overcome the computational and energy limitations of drones, we develop lightweight neural networks and model compression techniques for real-time processing. Transfer learning and domain adaptation have become essential for developing scalable models to generalise across environments like urban areas, forests, or disaster zones. Moreover, increase use of reinforcement learning to develop the algorithm for path planning, swarm coordination and obstacle avoidance in dynamic environment. Drones still struggle with less optimal scenarios, but we are making good progress with the multi-modal CV systems that gather knowledge by combining data from cameras, LiDAR, and thermal sensors to build a more precise (and robust) machine in these difficult conditions. In addition, trends such as ethical AI practices including fairness and data privacy are helping to make sure that drones can function safely in increasingly sensitive applications like surveillance and public safety. All these research directions together are focused on making drones more intelligent, adaptable, and effective in a broad range of applications.

From intelligent, collaborative and large scale operations, the autonomous and swarm drone technologies offer limitless possibilities across various sectors. With proper computer vision (CV) and machine learning (ML), autonomous drones are becoming commonplace in applications ranging from precision agriculture to search-and-rescue missions, to logistics, where the systems need to react and adapt in real time. They can function without constant human intervention so they cut down on operational costs and are ideal for complex environments for them.

These technologies allow multiple drones to work in tandem—they can communicate with each other and translate natural systems, such as bird flocks or bee colonies. Each doddle can do everything a single drone can do — like environmental monitoring, disaster assessment and large-scale infrastructure inspection but more quickly through the formation of swarms. Horizon in this area can be multi-agent reinforcement learning, where drones communicate and coordinate action dynamically or otherwise. Swarm technologies also have considerable promise in other domains, including defense applications where swarms of drones can be utilized for surveillance or to transport supplies to remote locations. These technologies have transformative capabilities for industries with an increased demand for scalability, accuracy, and independence as these technologies evolve.

9 CONCLUSION

A positive understanding of the transformative roles played by computer vision (CV) and machine learning (ML) in improving the capabilities and operational efficacy of drones, including applications ready examples utilizing edge computing systems for real-time processing and good scale-able reliable AI models for autonomous operation capabilities/adaptive behavior. It deals with the main issues such as processing restrictions, energy limits, and ethical issues, applying supervised, unsupervised or reinforcement learning for applications like object detection, anomaly detection or autonomous navigation. Furthermore, this chapter also looks at the untapped potential of autonomous and swarm drone technologies to drive transformative change in key sectors such as agriculture, logistics, surveillance, and disaster response and highlights how innovations powered by artificial intelligence (AI) are vital for addressing many of the complex challenges we face today as well as creating opportunities that are yet to be realized

Drones equipped with computer vision (CV) and machine learning (ML) will be at the forefront of many industry breakthroughs as well as literally solving some of our biggest global challenges. Drones are being designed and programmed with greater autonomy, adaptability, and efficiency with continuous innovations in edge computing, scalable AI models, and collaborative swarm technologies. Such capabilities will allow drones to conduct sophisticated applications — like precision agriculture, disaster response, and nationwide infrastructure inspections — with a level of accuracy and dependability never seen before. With ethical and privacy challenges addressed through responsible AI practices and strong regulations, the incorporation of CV + ML will create new opportunities for innovation that will undoubtedly make drones viable indispensable tools in diverse applications. The future of AI- powered drones is going to be determined by how well, both, they learn and evolve from that learning but also the collaborative endeavours between ongoing human innovation which will shape a healthier world where these drones can operate in tandem with each other enabling smart cities that are safer & sustainable.

REFERENCES

Abadade, Y., Temouden, A., Bamoumen, H., Benamar, N., Chtouki, Y., & Hafid, A. S. (2023). A comprehensive survey on tinyml. *IEEE Access : Practical Innovations, Open Solutions*, *11*, 96892–96922. DOI: 10.1109/ACCESS.2023.3294111

Abdelnabi, A. A. B., & Rabadi, G. (2024). Human detection from unmanned aerial vehicles' images for search and rescue missions: A state-of-the-art review. *IEEE Access : Practical Innovations, Open Solutions*, *12*, 152009–152035. DOI: 10.1109/ACCESS.2024.3479988

Abiodun, T. F. (2020). Usage of drones or unmanned aerial vehicles (uavs) for effective aerial surveillance, mapping system and intelligence gathering in combating insecurity in nigeria. *African Journal of Social Sciences and Humanities Research*, *3*(2), 29–44.

Ahn, H., Choi, H.-L., Kang, M., & Moon, S. (2019). Learning-based anomaly detection and monitoring for swarm drone flights. *Applied Sciences (Basel, Switzerland)*, *9*(24), 5477. DOI: 10.3390/app9245477

Al-kfairy, M., Mustafa, D., Kshetri, N., Insiew, M., & Alfandi, O. (2024). Ethical challenges and solutions of generative ai: An interdisciplinary perspective. [MDPI.]. *Informatics (MDPI)*, *11*(3), 58. DOI: 10.3390/informatics11030058

Al-lQubaydhi, N., Alenezi, A., Alanazi, T., Senyor, A., Alanezi, N., Alotaibi, B., Alotaibi, M., Razaque, A., & Hariri, S. (2024). Deep learning for unmanned aerial vehicles detection: A review. *Computer Science Review*, *51*, 100614. DOI: 10.1016/j.cosrev.2023.100614

Alanhdi, A., & Toka, L. (2024). A survey on integrating edge computing with ai and blockchain in maritime domain, aerial systems, iot, and industry 4.0. *IEEE Access : Practical Innovations, Open Solutions*, *12*, 28684–28709. DOI: 10.1109/ACCESS.2024.3367118

Alkaabi, K., Raza, M., Qasemi, E., Alderei, H., Alderei, M., & Almheiri, S. (2024). Using crowd-sourced traffic data and open-source tools for urban congestion analysis. *Transportation Research Interdisciplinary Perspectives*, *28*, 101261. DOI: 10.1016/j.trip.2024.101261

Althati, C., Tomar, M., & Malaiyappan, J. N. A. (2024). Scalable machine learning solutions for heterogeneous data in distributed data platform. *Journal of Artificial Intelli- gence General science (JAIGS) ISSN: 3006-4023*, 4(1):299–309.

Apeland, J. (2021). Use of fuel cells to extend multirotor drone endurance.

Ayaz, M., Ammad-Uddin, M., Sharif, Z., Mansour, A., & Aggoune, E.-H. M. (2019). Internet-of-things (iot)-based smart agriculture: Toward making the fields talk. *IEEE Access : Practical Innovations, Open Solutions*, *7*, 129551–129583. DOI: 10.1109/ACCESS.2019.2932609

Bakirci, M. (2024). Enhancing vehicle detection in intelligent transportation systems via autonomous uav platform and yolov8 integration. *Applied Soft Computing*, *164*, 112015. DOI: 10.1016/j.asoc.2024.112015

Caballero-Martin, D., Lopez-Guede, J. M., Estevez, J., & Gran˜a, M. (2024). Artificial intelligence applied to drone control: A state of the art. *Drones (Basel)*, *8*(7), 296. DOI: 10.3390/drones8070296

Caminero, G., Lopez-Martin, M., & Carro, B. (2019). Adversarial environ- ment reinforcement learning algorithm for intrusion detection. *Computer Networks*, *159*, 96–109. DOI: 10.1016/j.comnet.2019.05.013

Cazzato, D., Cimarelli, C., Sanchez-Lopez, J. L., Voos, H., & Leo, M. (2020). A survey of computer vision methods for 2d object detection from unmanned aerial vehicles. *Journal of Imaging*, *6*(8), 78. DOI: 10.3390/jimaging6080078 PMID: 34460693

Clarke, R., & Moses, L. B. (2014). The regulation of civilian drones' im- pacts on public safety. *Computer Law & Security Report*, *30*(3), 263–285. DOI: 10.1016/j.clsr.2014.03.007

Criollo, L., Mena-Arciniega, C., & Xing, S. (2024). Classification, military applications, and opportunities of unmanned aerial vehicles. *Aviation*, *28*(2), 115–127. DOI: 10.3846/aviation.2024.21672

Crupi, L., Butera, L., Ferrante, A., & Palossi, D. (2024). A deep learning-based pest insect monitoring system for ultra-low power pocket-sized drones. In *2024 20th International Conference on Distributed Computing in Smart Systems and the Internet of Things (DCOSS-IoT)*, pages 323–330. IEEE. DOI: 10.1109/DCOSS-IoT61029.2024.00056

Das, D., Njilla, L., Ghosh, U., Shetty, S., & Levin, E. (2024). An edge-cloud approach for dynamic resource allocation in drone communication networks. In *IEEE INFOCOM 2024-IEEE Conference on Computer Communications Workshops (INFOCOM WKSHPS)*, pages 1–6. IEEE. DOI: 10.1109/INFOCOMWKSHPS61880.2024.10620670

Dhami, H. S. (2024). *Planning and Perception for Unmanned Aerial Vehicles in Object and Environmental Monitoring*. PhD thesis, University of Maryland, College Park.

Dimitrova, M., Aminzadeh, A., Meiabadi, M. S., Sattarpanah Karganroudi, S., Taheri, H., & Ibrahim, H. (2022). A survey on non-destructive smart inspection of wind turbine blades based on industry 4.0 strategy. *Applied Mechanics*, *3*(4), 1299–1326. DOI: 10.3390/applmech3040075

ESCAP, U. (2017). Innovations in disaster rapid assessment: a framework for early recovery in asean countries.

Gohari, A., Ahmad, A. B., Rahim, R. B. A., Supa'at, A. S. M., Abd Razak, S., & Gismalla, M. S. M. (2022). Involvement of surveillance drones in smart cities: A systematic review. *IEEE Access : Practical Innovations, Open Solutions*, *10*, 56611–56628. DOI: 10.1109/ACCESS.2022.3177904

Green, D. R., Hagon, J. J , G'omez, C., & Gregory, B. J. (2019). Using low-cost uavs for environmental monitoring, mapping, and modelling: Examples from the coastal zone. In *Coastal management* (pp. 465–501). Elsevier. DOI: 10.1016/B978-0-12-810473-6.00022-4

Gupta, S., Gupta, S. P., Kumar, S. A., and Wasim, M. (2022). Robust multi- sensor facial recognition in real time using nvidia deepstream.

Hern'andez Vicente, D. (2023). *Analysis design and implementation of ar- tificial intelligence techniques in edge computing environments*. PhD thesis, Universitat Polit`ecnica de Val`encia.

Himeur, Y., Rimal, B., Tiwary, A., & Amira, A. (2022). Using artificial intelli- gence and data fusion for environmental monitoring: A review and future perspectives. *Information Fusion*, *86*, 44–75. DOI: 10.1016/j.inffus.2022.06.003

Hüttenrauch, M., Šošić, A., & Neumann, G. (2019). Deep reinforcement learning for swarm systems. *Journal of Machine Learning Research*, *20*(54), 1–31.

Hymel, S., Banbury, C., Situnayake, D., Elium, A., Ward, C., & Kelcey, M. (2022). Edge impulse: An mlops platform for tiny machine learning. *arXiv preprint arXiv:2212.03332*.

Jang, S., & Kim, H.-I. (2024). Efficient deep reinforcement learning under task variations via knowledge transfer for drone control. *ICT Express*, *10*(3), 576–582. DOI: 10.1016/j.icte.2024.04.002

Kanungo, S. (2019). Edge-to-cloud intelligence: Enhancing iot devices with machine learning and cloud computing. *International Peer-Reviewed Journal*, *2*(12), 238–245.

Kasioulis, M., Symeonides, M., Ioannou, G., Pallis, G., & Dikaiakos, M. D. (2024). Energy modeling of inference workloads with ai accelerators at the edge: A benchmarking study. In *2024 IEEE International Conference on Cloud Engineering (IC2E)*, pages 189–196. IEEE. DOI: 10.1109/IC2E61754.2024.00028

Kiruthiga, R., Nithya, B., et al. (2024). Advancing uav multi-object tracking: Integrating yolov8, nano instance segmentation, and dueling double deep q-network.

Kouris, A., Venieris, S. I., & Bouganis, C.-S. (2019). Towards efficient on-board deployment of dnns on intelligent autonomous systems. In *2019 IEEE Computer Society Annual Symposium on VLSI (ISVLSI)*, pages 568–573. IEEE. DOI: 10.1109/ISVLSI.2019.00107

Kurunathan, H., Huang, H., Li, K., Ni, W., & Hossain, E. (2023). Machine learning-aided operations and communications of unmanned aerial vehicles: A contemporary survey. *IEEE Communications Surveys and Tutorials*.

Leslie, P., Chau, E., Watson, T., Furxhi, O., Jacobs, E., Driggers, R., & Conroy, J. (2024). Visible, swir, and lwir drone wide area coverage sensor systems. *Optical Engineering (Redondo Beach, Calif.)*, *63*(2), 025101–025101. DOI: 10.1117/1.OE.63.2.025101

Luber, M., Arras, K. O., Plagemann, C., & Burgard, W. (2009). Classifying dynamic objects: An unsupervised learning approach. *Autonomous Robots*, *26*(2-3), 141–151. DOI: 10.1007/s10514-009-9112-4

Mantau, A. B. J. Human observing for wide area surveillance using autonomous unmanned aerial vehicles (uavs).

Markovic, D. S., Zivkovic, D., Branovic, I., Popovic, R., & Cvetkovic, D. (2013). Smart power grid and cloud computing. *Renewable & Sustainable Energy Reviews*, *24*, 566–577. DOI: 10.1016/j.rser.2013.03.068

Mengozzi, S., Zanatta, L., Barchi, F., Bartolini, A., & Acquaviva, A. (2024). Towards nano-drones agile flight using deep reinforcement learning. In *2024 IEEE International Conference on Omni-layer Intelligent Systems (COINS)*, pages 1–6. IEEE. DOI: 10.1109/COINS61597.2024.10622558

Merenda, M., Porcaro, C., & Iero, D. (2020). Edge machine learning for ai-enabled iot devices: A review. *Sensors (Basel)*, *20*(9), 2533. DOI: 10.3390/s20092533 PMID: 32365645

Mittal, P., Singh, R., & Sharma, A. (2020). Deep learning-based object detection in low-altitude uav datasets: A survey. *Image and Vision Computing*, *104*, 104046. DOI: 10.1016/j.imavis.2020.104046

Nagendra, H., Lucas, R., Honrado, J. P., Jongman, R. H., Tarantino, C., Adamo, M., & Mairota, P. (2013). Remote sensing for conservation monitoring: Assessing pro- tected areas, habitat extent, habitat condition, species diversity, and threats. *Ecological Indicators*, *33*, 45–59. DOI: 10.1016/j.ecolind.2012.09.014

Nguyen, D. D., Rohacs, J., & Rohacs, D. (2021). Autonomous flight trajectory control system for drones in smart city traffic management. *ISPRS International Journal of Geo-Information*, *10*(5), 338. DOI: 10.3390/ijgi10050338

Nguyen Hoang, A. M. (2019). Computer vision deployment on edge devices.

Peng, H., Ding, C., Geng, T., Choudhury, S., Barker, K., & Li, A. (2024). Evaluating emerging ai/ml accelerators: Ipu, rdu, and nvidia/amd gpus. In *Companion of the 15th ACM/SPEC International Conference on Performance Engineering*, pages 14–20.

Reyes-Hung, L., Soto, I., & Majumdar, A. K. (2024). Neural network- based stress detection in crop multispectral imagery for precision agriculture. In *2024 14th International Symposium on Communication Systems, Networks and Digital Signal Processing (CSNDSP)*, pages 551–556. IEEE.

Saha, B., Kunze, S., & Poeschl, R. (2024). Comparative study of deep learning model architectures for drone detection and classification. In *2024 IEEE International Mediterranean Conference on Communications and Networking (MeditCom)*, pages 167–172. IEEE. DOI: 10.1109/MeditCom61057.2024.10621214

Sandino, J., Maire, F., Caccetta, P., Sanderson, C., & Gonzalez, F. (2021). Drone-based autonomous motion planning system for outdoor environments under object detection uncertainty. *Remote Sensing (Basel)*, *13*(21), 4481. DOI: 10.3390/rs13214481

Sarkar, M., & Sahoo, P. K. (2024). Leveraging edge computing for video data streaming in uav-based emergency response systems. *Sensors (Basel)*, *24*(15), 5076. DOI: 10.3390/s24155076 PMID: 39124122

Shankar, K., Wang, P., Xu, R., Mahgoub, A., & Chaterji, S. (2020). Janus: Benchmarking commercial and open-source cloud and edge platforms for object and anomaly detection workloads. In *2020 IEEE 13th International Conference on Cloud Computing (CLOUD)*, pages 590–599. IEEE.

Suganya, B., Gopi, R., Kumar, A. R., & Singh, G. (2024). Dynamic task offloading edge-aware optimization framework for enhanced uav operations on edge computing plat- form. *Scientific Reports*, *14*(1), 16383. DOI: 10.1038/s41598-024-67285-2 PMID: 39013972

Wu, J., Ye, Y., & Du, J. (2024). Autonomous drones in urban navigation: Autoencoder learning fusion for aerodynamics. *Journal of Construction Engineering and Management*, *150*(7), 04024067. DOI: 10.1061/JCEMD4.COENG-14787

Zhang, T., Qin, L., Zou, Q., Zhang, L., Wang, R., & Zhang, H. (2024). Crackscopenet: A lightweight neural network for rapid crack detection on resource-constrained drone platforms. *Drones (Basel)*, *8*(9), 417. DOI: 10.3390/drones8090417

Zhao, J. (2024). Drone technology for precision agriculture: Advancements and optimization strategies. *Highlights in Science. Engineering and Technology*, *111*, 185–191.

Zhong, Q. (2024). The analysis of drone track planning based on machine-visual. *High- lights in Science. Engineering and Technology*, *103*, 178–183.

Zhou, S., Ng, S. T., Yang, Y., & Xu, J. F. (2021). Integrating computer vision and traffic modeling for near-real-time signal timing optimization of multiple intersections. *Sustainable Cities and Society*, *68*, 102775. DOI: 10.1016/j.scs.2021.102775

Chapter 3
Drones in Manufacturing and Industrial Applications

Damodharan Palaniappan
https://orcid.org/0009-0003-0721-3068
Marwadi University, India

T. Premavathi
https://orcid.org/0009-0003-0172-2021
Marwadi University, India

Rituraj Jain
https://orcid.org/0000-0002-5532-1245
Marwadi University, India

Kumar Parmar
https://orcid.org/0000-0002-2502-5680
Marwadi University, India

ABSTRACT

Computer vision enhances the environment and hence enhances the use of the drones. Most computations are local and are thus is the reason why edge computing reduces drone latency. The capability of fast data processing allows the drones to operate in safety-critical industrial missions on their own. Self-driven AI drones are better than those that are manually controlled. In the industrial application, LAD precision enhances the security, the agility and the effectiveness. The drones that are controlled by artificial intelligence can protect industrial businesses from being hacked. When implementing decentralization to computers the Robotics will monitor the networks, analyze weaknesses, and prevent predicted threats in industrial

DOI: 10.4018/979-8-3693-8497-8.ch003

facilities. UAV relativism is an environment that integrates with the surroundings as soon as the UAV is deployed using artificial intelligence and edge computing. To support normal drone operations, edge computing is applied for the purpose of processing as well as providing AI analysis.

1. INTRODUCTION

The use of Unmanned Aerial Vehicles (UAVs) or the more popularly known drones have become common in many industries and processes. Bearing this in mind the adoption of drones in the industry 4. 0 framework has provided various avenues that have made their application common in production and supply chain domain (Awasthi et al., 2022). The inherent characteristics of drones including their portability and flexibility and ability to support a variety of sensors are key aspects that makes them suitable for numerous industrial applications. The tasks comprise monitoring of ongoing production processes, product inspection programs, tracking of inventory, and material and component supply all done by intelligent robots. As the fundamental technology of drones remains to grow especially in aspect like computer vision, edge computing and artificial intelligence, the future opportunities of drone application in the industrial segment are expected to grow to greater limits.

More to the point of the present and future use of drones in manufacturing and industry, this chapter focuses on the key technological developments driving these trends. It explores how artificial intelligence and computer vision assist in enhancing drones and how edge computing is enabling these to take a much bigger role in industrial enabling and automation. The upgrades in computer vision and edge computing allow drones to handle situational awareness and reactivity more effectively, while the implementation of artificial intelligence algorithm helps drones to become learners, decision-makers in making decisions to optimize industrial logistics and improve efficiency and productivity of overall industries.

The increased accuracy of computer vision and the developments of edge computing increase the drones' awareness and reactivity about the environment, the inclusion of AI algorithms makes them learn from experience and act accordingly improving the overall productivity and efficiency of industrial processes. These development in drone technology are revolutionizing the manufacturing and industrial setup thus enhancing the future expected capabilities of automation, precision and optimization. If the interest continues to grow, new drone technologies that will combine AI, computer vision, and edge computing will be an essential component of modern manufacturing industries and factorization to allow change and innovation in the industry.

2. ADVANCEMENTS IN THE DRONE SECTOR

This outreach has especially been pioneered in several fields such as computer vision and edge computing over the last few years. This technological development has helped to improve the features of drones making them more efficient, more independent, and more responsive to the industrial use.

2.1 Computer Vision in Drones

The other major technological enhancement is in algorithmic of computer vision of the drones. Computer vision allows the drones to acquire enhanced perception of the environment helping them comprehend what is going on around them. Today's drones are equipped with even higher grades of sensors for instance, high-resolution cameras and LiDAR systems which allow the drones to capture better quality visuals and process it in Realtime to sense the objects and track them, or detect any abnormality, and manoeuvre through the environment (Hamdi et al., 2021) (Dilshad et al., 2020).

Computer vision implemented into drones has something revolutionized the industrial usage of the technology. Automated drones that are built with computer vision capabilities will offer services such as surveillance of manufacturing, assessment of machines and facilities, and stock counting with great accuracy and within a short span of time. Through the aid of computer vision, drones can identify defects, estimate dimensions, and foresee possible problems that may arise in an organization hence quickly rectify them thereby improve industrial processes.

Thanks to CV, drones can now pay attention to details when observing production processes, inspecting equipment and infrastructures or counting inventory. Precise detection of the defects, measurement of necessary dimensions, or identification of the possible problems in the production process offer an opportunity to correct the situation and gain the most efficient results for industrial processes and increased production rate. In addition, the high-detail imagery provided by such drones can be further examined towards the development of patterns and trends, then used for the decision-making processes which form the foundation of constant optimisation of solutions within the manufacturing and industrial domains.

Computer vision enables the unmanned aerial vehicles to have sight and analyse the images in the same way that an observer would. Figure. 1 refers what are technologies relies on:

Figure 1. Enabling drones to "see" and interpret surroundings

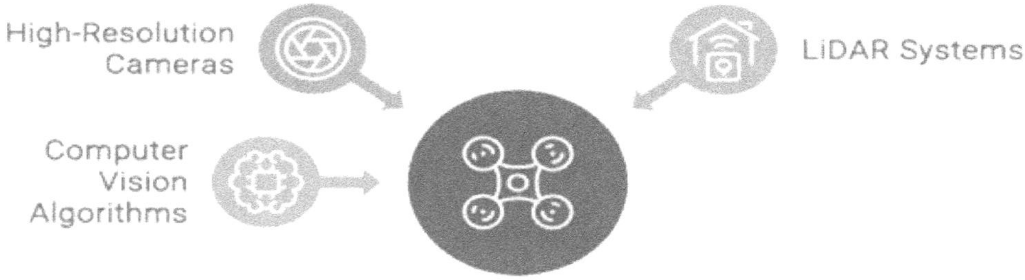

High-Resolution Cameras: Take different photos of the surrounding area. LiDAR Systems: Laser pulse to determine distance and build a topographic map. Computer Vision Algorithms: Alter visual data so that it is suitable for, putting an Object recognition system for identification and tracking of objects, detect abnormalities (e. g., flaw in structure), navigate complex environments. Applications of Computer Vision in Drones: Monitoring Production Processes: Monitor the flow of goods, find out the constraints. Inspecting Equipment and Infrastructure: Identify cracks, rot, wear Tears in upholstery and carpeting, and other measure of deterioration. Conducting Inventory Audits: Check the status of the inventory, improve the flow of the warehouse. Apart from the computer vision, there has also been the integration of edge computing that has been a great achievement in drones.

2.2 Brief on Edge Computing in Drones

In addition to the progress made in the field of computer vision the use and incorporation of edge computing has been a key trend in the development of the drone industry. As opposed to performing computations on a remote cloud site, edge computing takes place on the drone itself. This approach helps to keep the drones free of latency and can make decisions, and perform the necessary safety-critical functions, promptly. This high-speed data processing enables the efficiency that is needed for several manufacturing operations where use of drones may be in dangerous or in time-sensitive conditions. This approach is reliable and efficient in the industrial application of drones because it makes the drones independent of their current operating environment to process any data and make subsequent decisions in real-time. Furthermore, the use of edge computing in drones also helps in efficiency improvement of the industrial processes through real-time processing of data and decision making. Edge computing also enhances the performance of drone applications by enabling drones to collect and analyze data, to identify patterns that can be useful in areas such as production planning, maintenance schedules and the like. Use of AI in manufacturing of drones is one more advancement that is evident and has revolutionised the manufacturing and industrial systems. Reconnaissance drones that are propelled by artificial intelligence can adapt to the data they gather in the air so the performance of identifying and following the target increases. Edge computing is the extension of data processing to devices that are closest to the data source that in the case of drones is the drone itself. In contrast to the method used in most conventional platforms where data is transferred to a server or cloud for analysis, edge computing enable drones to process the information on their own.

3. ADOPTION OF COMPUTER VISION AND EDGE COMPUTING

The following factors have initiated the utilization of the computer vision and edge computing technologies in the drone industry. Figure 2 describes what are the benefits of computer vision and edge computing in drones. Improved Efficiency and Productivity: Drones, through use of computer vision, can reduce cycles while identifying irregularities; improve product output, efficiency and therefore, profitability for all manufacturers and operational industrial units. Enhanced Safety and Risk Mitigation: Self-driving drones designed with computer vision and edge computing capability can go and work in combative or difficult to access areas and thereby sparing human worker's risky work. Increased Accuracy and Precision: That is because advanced sensing and data processing offered by computer vision in drones lead to more precise inspections, measurement, and inventory tracking which is very necessary for meeting quality standards customer needs.

In the same sense, the incorporation of edge computing in drones further improves the device's ability to act quickly, be more dependable, and operate independently in real-time safety-related applications in industries. Edge computing also allows drones to make decisions faster, function in networks with limited access, and relieve the load of principal systems by processing data at the edge. These advancements in computer vision and edge computing have led to a growing adoption of drones in a wide range of industrial applications, including - Monitoring and Inspection: Drones also can overview equipment, facilities, and production processes, evaluate the problems, and contribute to predictive maintenance. Inventory Management: Using drones, inventories and audit of the stocks can be done automatically, meaning that it can also increase the efficiency of the warehousing systems. Aerial Surveying and Mapping: Civil use encompasses aerial photography or videography, inspection of industrial structures and infrastructure, surveying and mapping of construction sites among others through drones fitted with cameras or sensors.

Figure 2. Key benefits of computer vision and edge computing in drones

There is much potential to utilizing drones for manufacturing or industrial, but one must also introduce the limitation and factors that go along with its use which consist of the regulation in manufacturing systems to ensure compliance in use, privacy issues where it relates to the public and how it can sync with current systems in manufacturing and the industries involved. Using the computer vision and edge computing technologies, drones are steadily gaining importance as helpful tools for the manufacturers and operators of industrial facilities, helping them boost the performance and generate fewer risks (Hamdi et al., 2021). The uses of drones in manufacturing and industrial applications are likely to be of high impact though there some potentials which require a close look such as the laws regulating them, ownership, handling, and privacy issues and how they can be integrated into their follow-up systems. With computer vision and edge computing, drones are turning into essential tools for manufacturers and industrial operators, as the methodology of their effective use helps the enhancement of effectiveness and safety as well as the optimization of processes.

However, its application of drones in industries comes with several factors that need to be taken into consideration in the industrial settings. Different regulating authorities has set down rules and regulation concerning the use of drones such as flight space prohibition, use of drone for privacy violation and accident responsibility. Types of permits that are involved are meant to guide manufacturers and industrial operators on the appropriate ways to use drones. Furthermore, the interfacing of the drones with other systems and processes may pose technical and operational problems. Transmitting and receiving of data, protection of data and proper communication protocols are paramount in the deployment of drones in the industrial space. They also show that managing change and training people are also important to make the use of technology, in this case drone, and fully realise as its potential.

However, based on the growing technological enhancements in the fields of the computer vision, edge computing and artificial intelligence, there is a promising prospect in the employment of drones in manufacturing and industrial sectors. With progression in these technologies and extension of accessibility, it is evident that the advantages arising from improved efficiency, safety as well as superior decision making will always outweigh these difficulties, hence acting as the catalyst towards industrial revolution in different fields (Hamdi et al., 2021).

4. UTILIZATION OF UAVS IN MONITORING, MAPPING, INSPECTION, AND INVENTORY CONTROL

It is widely known that utilization of unmanned aerial vehicles referred to as drones for manufacturing and industrial purposes has seen a rapid rise in recent years. Built on new body and wing designs, these sophisticated airborne systems with a variety of chemical and electric sensors, as well as digital vision systems, are disrupting traditional industries' ways of performing tasks and projects including surveillance, surveying, assessment, and stock checking. Based on the literature, the use of drones in the construction industry is for progress tracking, inspection or analysis with many advantages and prospects for contractors. These are as follows: the first one is the ability of taking aerial images and other relevant data, which improve the monitoring of construction sites, and help in the identification of possible problems or new opportunities for learning.

Figure 3. Utilization of unmanned aerial vehicles (UAVs)

Figure 3 concerns the use of UAV's. Thus, the Unmanned Aerial Vehicles, commonly referred to as the drones are revolutionising how industries go about in monitoring, mapping, inspection and inventory. Drones are endowed with enhanced sensors, high-definition cameras, and progressively improving software leading to combined capability of aerial access, data acquisition, and better cost management. Monitoring - By using drones, industrial facilities and production processes, structures and infrastructure can be constantly 'spied' to capture useful data required for maintenance, safety evaluation, and operations efficiency. Construction Sites: Real-time monitoring of the progress of the construction, pointing out risks that may be encountered, and enhance the organization of the site are just some of the advantages of using drones (Regona et al., 2022). Infrastructure: Performing evaluation or assessment of damages or defects on bridges, power lines, pipelines as well as wind turbine systems increases in safety, rate and effectiveness. This is on the advice given by other scholars as noted by Giordan et al. in their recent studies. Agriculture: Maintain records of crop condition, detect problems with irrigation and assess the returns on investment through details gathered through precision agriculture (Zinno et al., 2022). Environmental Monitoring: Keep record of the removal of forests, record the status of animal species, and review the effects of disasters (Smeesters et al., 2023).

Drone mounted with high end sensors and photogrammetry software can generate ortho image maps with excellent ground resolution and detailed 3D maps of the large structures such as industrial complexes and construction sites. Creating Detailed Maps: Produce accurate and detailedinki 20 2-D and n-dimensional maps including terrain, buildings and construction sites (Guimarães et al., 2020). Surveying and Topography: Carry out aerial reconnaissance for land use purposes, planning, and natural resources prospecting among others (Giordan et al., 2020). Supporting Disaster Relief: Instance: For quick response, it helps in mapping of disaster areas to facilitate search and rescue as well as identification of affected regions' level of damage. Drones can routinely patrol areas which are difficult to access or potentially hostile, and thus, avoid human exposure. Activities such as inspection of the civil works structures and infrastructures dense areas such as the bridges, electrical power lines, pipelines and wind farms, become safer, faster and more efficient since drones can determine and do the work without at all having to involve human beings to risk their lives for the same. Visual Inspections: Checking on the structures, pipelines, equipment and other tangible fixtures for flaws, rusting or destruction in businesses including oil and gas, energy and transport among others. Thermal Imaging: Identify dangerous or potentially troublesome hot spots in equipment, structures or other systems. Supporting Predictive Maintenance: Machines do not have to be stopped frequently due to possible problems hence saving on the amount of money that may be spent on maintaining the machine. It is now possible to track and even manage inventory using drones possessing the most sophisticated sensors and computer vision at their disposable. Warehouse Management: Automate inventory checking and counting, manage inventory resources and organise the warehouse. Based on the related literature According to the study conducted by (Guimarães et al., 2020). Outdoor Stockpiles: Control large yard stocks of materials used in construction such as timber, stones, or mining industries products. Improving Efficiency and Accuracy: The manual and repetitive process of doing inventory counts tires the employees and tends to be erroneous at times, while drones eliminate such problems. The increased adoption of drones in manufacturing and industrial applications is driven by several key advantages - Increased Safety: Reaching areas that can be dangerous for employees or areas that are otherwise hard to navigate. Cost-Effectiveness: Lesser amount of expenses compared to manned aircraft or ground operating methods that are usually employed. Improved Efficiency and Speed: They are also faster in data collection hence providing a quicker way to decision making. High-Quality Data: Take photographs, thermographic surveys, and LiDAR surveys to get relevant information. Application of UAV in monitoring, mapping, and inventory, inspection is revolutionizing industries through enhancing work completion safety, optimizing resources' usage and technically informed decision making. Observers of the drone industry anticipate that with the development of new technologies the application of drones within fields of different industries will further expand.

5. ARTIFICIAL INTELLIGENCE AND MACHINE LEARNING ALGORITHMS IN DRONES

AI, Machine learning algorithms as well as other advanced technologies have been another reason for the improving and enhancing of drones as well as their application in manufacturing industries. With the use of AI, and more particularly of ML, drones can now operate more independently, accurately and efficiently across a broad range of activities. AI progress in promotion of computer vision helps drones to have better perception and awareness of what surrounds it, which is important for an efficient and safe operation in industrial environment. Edge computing also increases drone's interactivity by processing data and making decisions onboard the drone, thus improving the general performance (Awasthi et al., 2022). The incorporation of AI with machine learning algorithms in drones as significant enhancements in the possibility of the several areas of uses in UAVs.

5.1 How AI and ML Enhance Drone Capabilities?

Constant advancements in fields of AI and ML help drones in their tasks to perform functions like object detection, image recognition and decision making with higher degree of precision (Qu et al., 2021). This allows drones to automate and optimize a variety of industrial applications: For instance, drones controlled by AI can inspect structures, plants and equipment, and machinery, defects, irregularities and possible problems more efficiently than human beings. Furthermore, several algorithms that are implemented to machine learning will allow the drones to have intelligent learning and decision making in aspects like navigation, way from obstacles and executing tasks. The use of artificial intelligence and machine learning to drones has greatly improved with the use of unmanned aerial vehicles in manufacturing industries. These novelties have provided drones with higher levels of freedom, accuracy and effectiveness for numerous fields along with altering how different sectors conduct observation, survey, inspection and stock checking procedures.

Enhanced Autonomy: They have an ability to decide and control their movement by their own without the real time intervention of operators. This is especially of great significance where the environment under consideration is contentious or quite risky (Telli et al., 2023). Improved Object Detection and Tracking: The use of AI in computer vision allows drones to detect and follow objects most accurately especially when they are surrounded by other objects. This is essential especially in surveillance, inspection and search and rescues operations among other similar uses (Lamberti et al. 2023). Real-time Data Analysis: It (edge computing with ML) enables drones to perform analytics onboard and take decisions on the fly; critical operations such as avoiding an obstacle or determining flight path (Jagannath et al., 2020). Predictive Maintenance: Relying on data from the sensors of drones, one can teach the ML algorithms to foresee cases of equipment failure, thus avoiding them in advance (Cheng et al., 2023). Optimized Route Planning: The paths that the drones must take can also be decided with the help of AI algorithms in view of such factors as the weather occurrence, the airspace regulations, and energy consumption (Raivi et al., 2023). Further development of the AI & ML is expected in future; there will be more unique possibilities and improved performance of drones in manufacturing and industrial industries resulting higher safety and cost-effectiveness.

5.2 Use Case of AI and ML in Drones

AI and machine learning algorithms impact drones in that these have helped the drones to improve their performance to undertake many industrial and manufacturing applications. Figure 4 refers some of the examples include - Automated Inspection and Monitoring: Computer vision and AI incorporated in drones can help in the effective examination of infrastructure, equipment, and facilities for defects and anomalies or possible problems than taking physical tours (Nelson et al., 2023).

Figure 4. AI and ML applications in drones

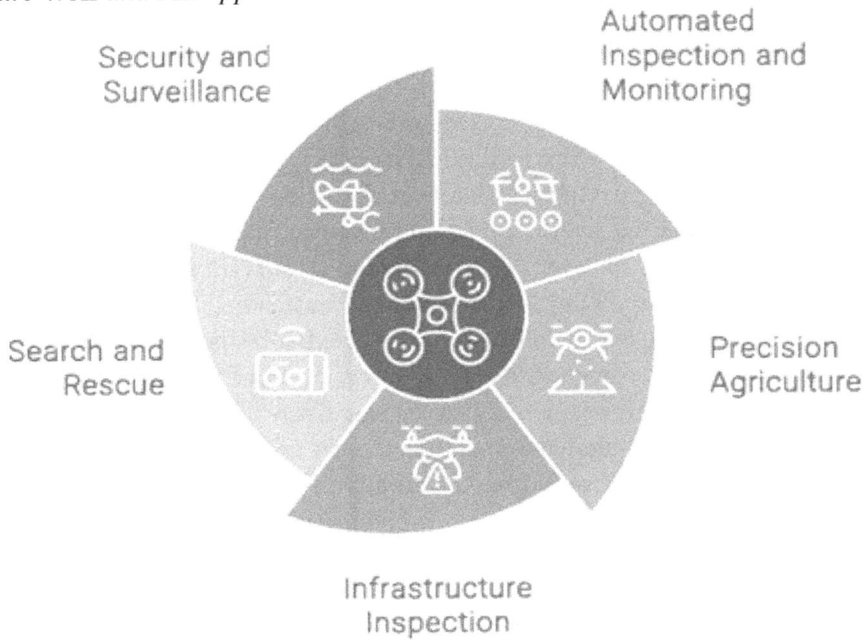

Precision Agriculture: They apply artificial intelligence in imaging analysis of crops, pests and diseases detection, determination of quantity and time of irrigation as well as application of fertilizers (Gikunda, 2024). Infrastructure Inspection: Automated drones, by use of artificial intelligence, can effectively identify defects in bridges, power lines and pipelines within the shortest time possible and with a lot of precision (Giordan et al., 2020). Search and Rescue: AI enabled drones can search for lost individuals in complex and disaster areas resulting to quick response and rescue (Xing et al., 2023). Security and Surveillance: AI involves drones scanning large areas and following suspicious activities while at the same time providing real-time information (Merkle et al., 2023). AI with the help of machine learning algorithms has also changed the inventory management scenario in manufacturing and industrial sectors.

5.3 Challenges and Ethical Considerations

As positive this is, the use of artificial intelligence and machine learning in drones has both implications and drawbacks that need to be figured out. Among them, the stability and safety of AI operated drones and conditions for their responsibility must also be considered. Some of the drawbacks that have been raised include risks with algorithmic choices, system vulnerability and malicious use that can result to disaster. In addition, there are concerns regarding the computational need of AI and machine learning in drones which results to the negative impacts on the environment such as high energy usage (Huertas-García et al., 2023). Some key ethical considerations regarding the integration of AI and machine learning in drones include - Data Privacy: It is for such reasons that the use of surveillance through drones with Artificial Intelligence is a cause for alarm because it may lead to the violation of data privacy and misused the information collected. Job Displacement: There has been discussions and debates regarding the effect of drone technology on human beings especially near future where the drones are likely to replace people in some industries. Algorithmic Bias: This means that there is need for rigorous measures to be taken to avoid cases that would see the AI algorithms within drones have some negative bias that would lead to certain prejudiced conclusion being arrived at. The adoption of AI and ML technology has been instrumental in the advancements seen in the drone industry especially in relation to safer, efficient and possibly more autonomous operations. The ongoing advancement of these technologies gives indications of even other unique prospects and uses in disparate fields as well. But having discussed the effective use of AI-powered drones, it is crucial to touch upon the ethical issues and the possible concern regarding such technologies.

6. ENHANCED APPLICABILITY OF DRONES

With the contribution of Artificial Intelligence and Machine Learning, drones have been effectively incorporated and spread out in the field of manufacturing and industries. Different tasks that require higher precision and accuracy, speed and flexibility, and decision-making are now performed by drones and this have changed the way industries perform many of their operations (Nelson et al., 2023), (Baracaldo et al., 2022).

6.1 Traffic Survey and Rehabilitation

Remote controlled aerial vehicles which are installed with high end sensors as well as artificial intelligent computing algorithms enable extensive inspection of crucial structures like; bridges, power lines and pipelines. These performances can detect faults, crack, and other forms of problems in comparison to manual checking that requires workforce to go to risky and stringent terrains that are challenging to access. Drones hold the potential of delivering real time information to estimate the upcoming maintenance schedule which can detect the potential areas of failures, or which part deserves immediate repairing along with ensuring the safety of core infrastructures thereby computing with the help of vision processing, edge computing and predictive analytics. Technological developments in areas of UAVs, sensors, artificial intelligence & data analysis of UAV captured data that has made the use of drone on infrastructure inspection & monitoring much broader in practice today. The use of drones in infrastructure has been identified as safe efficient, and cheaper than the usual means of working making it a more efficient method of working. Some of the specific applications of drones are evidenced by the rising popularity of using of drones in managing inventory and Transport in production and industrial organizations.

6.2 New Areas of the Usage of Drones in Production and Commerce

Applications of drones especially in manufacturing and industrial processes are increasing at a very high rate because of improving technology and improvements in discovering the advantages of using drones. Drones are increasingly being deployed in a wide range of industrial applications, including (Nunes, 2023), (Shahmoradi et al., 2020). Drones are increasingly being deployed in a wide range of industrial applications, including (Nunes, 2023) (Shahmoradi et al., 2020). Inventory Management and Logistics: Drones have been applied in inventory monitoring and control especially in the giant stores and stockist's centres. Similarly, using barcode scanners, RFID readers, and computer vision algorithms attached to the drones; the counting of inventory can be automated, error free, and less costly than when relying on human input (Feroz & Dabous, 2021). Facility Inspection and Maintenance: From the roof, walls, pipes, machinery and other areas which are hard to access, drones mounted with high-definition cameras, thermography and other sensing instruments can make inspections on industrial facilities. This makes it possible to detect certain problems that may range from structural defects, mechanical faults, or even existence of dangerous hazards at an early stage so that necessary corrective actions are taken (Alabanc et al., 2020). Emergency Response and Disaster Management: During an event of an industrial calamity, a natural disaster, or any other form of disaster, the use of drones can shortly be dispatched to evaluate the occurrence, identify the extent of the damage and participate in alleviation or mitigation procedures. Drones offer an opportunity to carry out aerial photography, detect risks and respond to disasters by capturing critical assets, assets and samples for delivery by helping to increase awareness for emergency operations (Merkle et al., 2023).

Automated Quality Assurance and Defect Detection: The use of drones in the manufacturing sector can be used in the aspect of inspecting the quality control of products, and identifying any form of defect that may be present. Manufacturers can know about product defects, wrong size, and any other quality issues by using computers vision and machine learning algorithms in drones to take samples (Islam, 2023). The usage of computer vision and machine learning as well as uprising professional sensors allow drones to detect manufacturing defects, assess dimensional parameters of a product, and guarantee that they adhere to the standards which results in better production quality and eliminated scrap.

7. DETECTION OF ABNORMALITIES AND RISK ESTIMATION

They hypothesized that one of the major benefits that come with incorporating the drones with artificial intelligence learning is that it offers the models the ability to identify deviations and to estimate the level of risks associated with several industrial activities and processes. AI-powered drones can capture and process data from their instruments in real time and detect trends that deviate from normal or if there is an existence of threats (Nunes, 2023). For instance, in a manufacturing environment, these drones used AI to identify machines that are developing faults, business processes that are deviating or manufacturing employees who have violated set safety standards by comparing the current events with previous occurrences and set norms respectively. They can then notify the operators of such anomalies so that appropriate measures are taken to avoid evolution of bigger problems or mishaps. Further, it is educed that AI algorithms can be trained to evaluate the severity and likelihood of the repercussions linked to the recognition of anomalies. Concerning the practical applications of using drones, the capabilities mentioned above combined with other metrics such as the type of the problem, the degree of interference with the functioning of the organization etc., can serve as important guidelines that help choose among the highest-priority work-related issues (Gerstein & Leidy, 2024). This capability is mainly compelling in segments such as oil and gas, mining, and chemical processing, where circumstances are critical and the outcomes of equipment failures, and operational misfortunes are usually adverse. That is how AI-powered drones' abilities to monitor the environment in real time and assess the risks make it possible for companies to increase safety and efficiency, optimise decision-making processes, increase productivity with less downtime and, thereby, minimise risks.

8. TRANSFORMING OPERATIONAL DECISION-MAKING PROCESS

Hence, the application of drones, artificial intelligence, and edge computing into manufacturing and industrial environments is not limited to data collection and analytics. These disruptive technologies therefore present a great opportunity to catalyse operational improvements and provide valuable support to strategic decisions that define the future of these industries. Smart drones with state-of-the-art sensors and computer vision technologies capture data's which AI can mine and harness to offer valuable information and optimized solutions to operations managers and the company executives. For example, unmanned aerial vehicles charged with artificial intelligence can predict when the vital machines require upkeep to schedule them airtight and undertaking maintenance that is often cost-effective and timely in addition to extending the useful lives of precisely the equipment. Besides, these AI systems can assist drones in proactively identifying most of the probable equipment failures that may happen soon to reduce their impacts on the productivity, safety, and operational continuity of the drones. The application of drones in manufacturing and industrial environments has improved decisions made in operations by a notch. Real-time data, automation, photo and video capabilities, and aerial viewpoints improve decision-making in multiple areas of organisational functioning.

8.1 Data-Driven Insight for Improved Planning and Control

Due to the large quantity of data acquired by drones and the opportunities of analyzing it using AI, the firms can get the depth of knowledge that will radically change the decision-making process. From streamlining production processes as well as inventory management, to enhancing the maintenance of facilities and one's safety, AI drones provide a revolutionary approach to operational decision making. The use of drones n manufacturing and industrial applications has defined several aspects of capabilities that are changing operational decision making. Real-time Data Collection: Sensors, cameras and high-end imaging technologies (including LiDAR) in drones enable the capture of massive data on facilities, equipment and processes (Gonçalves et al., 2023). Visualizing Complex Environments: Drones used in aerial imaging and surveying allow the generation of 2D and 3D maps of industrial complexes; layouts facilitate easy layout comprehension and modeling of bottlenecks in layouts, allowing enhanced modeling of industrial operations (Nunes,

2023). Monitoring and Tracking Operations: Drones follow the flow of product, observe assembly line and equipment to gather information for performance analysis and for optimization of the process based on specific concerns detected (Behroozi & Ma, 2023). The additional features of predictive maintenance capabilities improve the decision-making process since they help to perform maintenance in advance.

8.2 Expanding Operations and Becoming More Efficiency

That is why taking the advantages of the use of drones by various companies will ensure high efficiency in their operations. It is very time consuming and labour intensifying to undertake physical reconnaissance of large areas of land; drones can do all this in some few hours. They can also help in facility management particularly in the field of mapping and surveying of industrial complexes where high-definition maps as well as 3D models of buildings and structures in relation to the comprehensive layout of the place comes as an important source of information in determining the allocation of resources, the infrastructural development and control of industrial process and flow. In addition, there are drones with different sensors, for instance, LiDAR, hyper-spectral cameras which are highly valuable for acquiring very accurate data for analysis and decision making. They are also increasingly being deployed in indoor manufacturing facilities for instance in delivering goods to different floors in a building. Nonetheless, drones are being employed to conduct scouting, material delivery, and our accident-free deliveries, which tremendously enhance productivity and safety in operations performed in enclosures (Günal, 2019). Some key ways in which drones are transforming manufacturing and industrial operations: Automating Inspections: Drones do regular scans of structures, assets and stocks in construction, industries and stores more efficiently than the traditional approaches, thus saving time which human beings would otherwise use to undertake such work (Aragon et al., 2020). Optimizing Logistics and Supply Chain: They facilitate movement of goods within an organization's premises, help in inventory monitoring and tracking of supply chain movements hence increasing efficiency and decreasing on time (Abdullah et al., 2024). Supporting Predictive Maintenance: Drone data along with the AI and Machine Learning technology is used to predict the failure points of the equipment in advance and hence reduces the downtime and the maintenance cost (Bousdekis et al., 2021). Moreover, the use of drones in manufacturing and industrial environment generally improves safety and reduce risk.

8.3 Enhancing the Probabilistic and Tactical Problem Solving

Therefore, with the help of such elements as drones, artificial intelligence, and edge computing, businesses raise levels of operational decision-making to a higher level – predictive. Machine learning and other forms of artificial intelligence can learn from the masses of data gathered by drone technologies and use such data to make accurate predictions and offer well-coordinated, focused, and effective solutions that help executives and operations managers make the best strategic decisions that can have long term impacts on the organisations. Improved Risk Management: The ability to pinpoint areas in potential risks, violation of safety procedures as well as other security threats is made easier through UAV inspection thus enhancing safety in the working places (Ali et al., 2023). Informed Resource Allocation: The tracking and monitoring of production processes, inventory and machinery provides a real time view of field events to allocate resources more efficiently, man power and materials (Alsayed et al., 2021). Data-Driven Process Optimization: The data received through the drones enables the organizations to recognize the discrepancies, problem areas and constraints within the manufacturing and industries, which in compound, increases productivity while decreasing expenses (Kurbanzade et al., 2023). Of the three, emerging technologies like drones, artificial intelligence, and edge computing, the most significant shift is being witnessed in the manufacturing and industrial industries – enabling organisations to obtain improved 'big data' insights and decision-making for better, operationally-efficient manufacturing.

8.4 Allowing for Agile and Responsive Functioning

It also makes the use of drones, AI as well as edge computing more suitable and versatile for quick change in market condition, customer behaviours, and other company needs. Real-time data on the production, supply chain and the facility can also be made available by drones to help the managers make quick decisions. Also, with the processing of data on the drone, using edge computing, it makes it easier for the drones to perform tasks on their own without having to wait for the main hub's response time. Rapid Response to Emergencies: The use of drones during emergencies gives a bird's eye view hence achieves faster response, better coordination, and excellent decisions making during disasters. Adapting to Dynamic Environments: Real-time information collected from drones enables organizations to make quick decisions and adapt the processes when the conditions change: equipment breakdowns, lack of materials, or slowdowns. This is supported by recent studies (Xu et al., 2020). Manufacturing and Industrial organizations must take advantage of drone technology as it opens vast opportunities for better performance, safety, and business advantage in today's complex business environment. The application of drones, artificial intelligence, and edge computing in manufacturing and industrial environments is already promising to revolutionize operations, safety, and decision-making later in factories thus opening doors for future, sustainable growth and competitive advantage in a complex environment to all manufacturers globally.

9. MAINTAINING ACCURACY, SECURITY, SPEED, AND EFFECTIVENESS IN INDUSTRIAL USAGE

The existing skills of using AI drones are accuracy, speed and security and these skills will make the drones valuable assets in so many sectors economically. Integrated application of computer vision and machine learning improves the ability of drones to accomplish their tasks more accurately and stably than if inspections were done by hand or with previous generations of automated machinery. For instance, drones may be used to analyse product dimensions and check for defects hence guaranteeing product quality and Low product wastage. Moreover, since the drones expand their functionality by incorporating AI, the need to involve people to manage or oversee their activities reduced considerably since the drones are self-sufficient and can work independently without the possibility of human interference and its associated potentially fatal, errors. The high-speed data processing and decision-making ability provided by edge computing of these drones makes it possible for real time responses to dynamic situation to be made so that important operations are done fast. Despite these benefits of using drones especially in the manufacturing and industrial sectors, several factors must be put into consideration, implemented and followed to ensure the accuracy, security, speedy and effective use of drones.

9.1 Accuracy

Another advantage of using ai driven drones in industries is the kind of job they can do with a lot of efficiency. This is mainly realised using computer vision and machine learning where drones are controlled to undertake observation and recognition of several aspects in the environment. For instance, high-resolution cameras as well as different sensors can be used in drones for inspecting the industrial equipment, structures, and products with the noticeable defects or unevenness. It able to guarantee that the products that being manufactured are of the right quality, reduce on wastages and to be able to detect some problems which are likely to occur and which are of larger magnitude if not detected early. That said, there are several measures that can be taken to guarantee AI powered drone precision in an industrial environment. Sensor Calibration and Maintenance: Drones should be calibrated and maintained frequently especially the sensors such as cameras, LiDAR, and thermal sensors. Strong calibration measures need to be in place; ground control points would help to check the accuracy of the results. Data Processing and Analysis: Ensure that there are fewer errors through proper analysis of data using advanced methods in data processing as well as using the appropriate software in analyzing raw data collected by drones. For data validation, low quality control measures should be implemented. Skilled Operators and Pilots: Ensure that those operating the drone have adequate training on how the flight is to be conducted, data acquisition processes as well as safety measures to be followed in a bid to reduce chances of human interference that may lead to provision of inaccurate data. Hence by properly deploying and utilizing the UAVs that come with AI systems, industrial corporations stand to benefit from better results in their processes, quality products and less strain since the decision-making process will be greatly facilitated by the systems.

9.2 Security

The use of technology such as drones, AI and edge computing in industries however, presents some concerns on security. However, as these technologies increase in usage within an organization, it is imperative to provide strong security measures to against acts of cybercrimes, data leakage and illegitimate access to a firm's systems. Key security measures for AI-powered drones in industrial applications include - Data Encryption and Storage: Use strong methods of securing data transmission and data storage to minimize the risk of data leakage and guarantee data accuracy. Encrypt your documents on Cloud or on web servers with filtered access. Drone Security Features: Go for drones that have basic security measures such as password protection, geographical restrictions and capability to shut down remotely in case of theft. Compliance with Regulations: Comply to every codified regulation of the aviation sector, every privacy policy, and other specific standards of the industry regarding drones' usage, data gathering, and cybersecurity. Continuous Monitoring and Incident Response: They also must use monitoring mechanisms to discover security breaches or unusual activities as well as possessing a good response plan. With proper consideration to security in the deployment of the AI-powered drones in industrial organizations, security challenges such as data breaches, cyber-attacks and other threats can be reduced hence allowing proper and safe functioning of the AI technologies.

9.3 Speed

This is due to the incorporation of edge computing especially when used together with artificial intelligence various drones in industrial premises. Drones can use computations and decision-making at their edge instead of comparative cloud infrastructure and they can adapt on-the-go because they do not have to wait for the data to be transmitted to a data center and processed afterward. Optimized Flight Planning: Design the routes of the flights and use software that would consider different factors such as weather conditions, airspace restrictions and the battery to ensure that the flight durations are minimal while at the same time the effort to collect data is optimized. Automated Data Collection: Exploring how to apply automation to the collection of data and how the drones can fit into the system, for instance, working in conjunction with assets' management databases to cut down on time when transferring and processing data. Edge Computing Capabilities: Invest in drones capable of performing edge computing where data is processed locally and analyzed on the spot thus avoiding data transfer to remote servers may be advantageous. Focusing on the speed and response time, industrial organisations can get the most of the use of AI drones for increasing operations efficiency and reducing time spent on troubleshooting and decision making.

9.4 Effectiveness

Use of AI in the drones is a factor that has encouraged the uptake of drones in industries. Application of these technologies results to gain in productivity and dealing cost, minimizing the discrepancy between operation and output. Some of the ways in which AI-powered drones can enhance the effectiveness of industrial operations include - Clearly Defined Objectives: Introduce definite targets and key performance indicators of drones' usage to evaluate the success and relevance of the implementation to definite objectives. Monitor and measure performance frequently and make revisions as necessary. Integration with Existing Systems: Automate the feeding of drone data into current business systems to enhance informed decision-making to various departments to boost the worth of drones to the operational systems. Collaboration and Training: Coordinate data usage among the drone groups and the data analysis teams as well as the operational users of the data. One way that organizations can help their teams remain fresh is to give the regular training to ensure that they remain up to par with the organizational change, emerging strategies as well as existing best practices. Thus, it is possible to locate the major opportunities for business and use drones in manufacturing and industrial safety to the maximum without declining accuracy, security, speed, and efficacy.

10. FUTURE ADVANCEMENTS OF DRONES IN MANUFACTURING AND INDUSTRIAL APPLICATIONS

It is therefore expected that with time there will be enhancements as to how drones can be utilized intensively in manufacturing and industrial applications. One area that is currently receiving more attention is how to increase the levels of decision-making regarding the drones. In the future, new AI algorithms will allow drones to function more independently and to address new problems in real time; edge computing will also play a major role in drones' future development. Moreover, 5G connectivity along with other prospective communication technologies will enhance the capability of active drones to send and receive information as well as suction feedbacks with low latency and high capacity therefore enhancing the reactivity of drones and easing their integration with other industrial components. Some more features include; increased efficiency and first as well as improved duration of the drone platforms. With values such as the battery life expanded, aerodynamics enhanced and advanced propulsion systems incorporated in the future model of drones, they will access the operational periods without regular recharging or refuelling hence increasing the operational flexibility in industries. Furthermore, the on-going innovation and advancement in drone hardware leading to the incorporation of miniatures sensors will create the foundation for lightweight drones. Such drones can manoeuvre in small and confined spaces, survey areas that are difficult to access and accomplish very accurate operations. As these developments occur, various sectors will observe a rising use of drones in manufacturing and industrial applications with the resulting outcome that lead to higher efficiency, safety, and potential decision-making abilities.

Manufacturing and industrial uses of drone are still expanding, but a surge of new prospects offering even more innovations in manufacturing and other sectors is soon to come. Here are some key areas where we can expect significant progress - Enhanced Autonomy and Intelligence: The future drones are expected to exhibit added self-governance to perform chores with little interference from the human beings. This will be made possible by unleashing the potential use of Artificial Intelligence, Computer vision and sensor fusion systems. Furthermore, the drones will make complex decisions which will be an integration of an AI and machine learning capability to enable the drones to take and even change operational decisions without the need to consult any human. Moreover, multiple drone swarms will be coordinated during their operations to perform large-scaled tasks and surveys, inspections, mapping, or construction with better results and within a shorter amount of time. Advanced Communication and Connectivity: Telecommunication enhancement including 5G technology and satellite communication will foster efficient data transfer, real-time control and extension of drone operation in decisions geographical areas. Further, edge computing capacity will enable drones to provide data processing on the fly; faster analysis and even real-time decisions without much dependence on the cloud.

Improved Energy Efficiency and Endurance: Future versions of drones will find improvements in battery technology, other power potentials, and optimal designing of the drone body that will further augment the energy density and the endurance capacity of drones. Longer battery life will enable operations for a longer period as well as larger area scanned at a given time before the batteries need to be recharged. New impressive sources of power like power in hydrogen fuel cells and power from solar panels will also increase the endurance of drones and allow for constant use of the drones with little or no charging time. Further developments in the drone design technology with enhanced aerodynamics and better propulsion systems will lead to enhanced energy efficiency thus more hours in the air and lower operating expenses. Specialized Sensors and Payloads: Miniaturization will lead to the increased use of even smaller sensors on drones which shall help them gather a broader array of information with much accuracy and precision. Also, the more use of hyperspectral, multispectral, and thermal imaging will lead to further penetration into infrastructure status, environmental conditions, and asset productivity. Moreover, the incremental concept of using modular payload systems will also facilitate tool and sensor installation for various types of work such as fabrication, transportation, or inspection tasks such as 3D printing, handling of material or non-destructive examination. Enhanced Safety and Regulation: Propulsion through AI and computer vision along with enhanced and integrated drones and sensors will prevent collisions and make drone navigation safe. Advanced forms of autopilots that are accompanied by backup safety mechanisms shall decrease the probability of an error occurring and hence improve on the safety of the flight. Better rules and guidelines along with maturing of advanced Unmanned Traffic Management solutions will provide proper collision-free integration of drones into the pre-existing airspace. The future of drones to be integrated in manufacturing and Industrial application is very bright due to the kind of technological improvement that is being adopted in the market. And as these technologies advance and become more embedded, we expect more drastic improvements on operational productivity, safety and automation that would change how industries operate and thrive in the future.

CONCLUSION

The application of drones into manufacturing, and industrial processes is not just a technological development, it correlates with the revolution of operations, safety and decision-making processes of organizations. Admittedly, drones have become choice tools that have automated routine operations and offered real-time data that can be used for timely maintenance among other things, reshaping industries' capabilities and competitive advantages. So, as we have seen the development of computer vision, edge computing, and artificial intelligence, we will find that the capabilities of drones will also only increase. Advanced quoting ratio, communication media, processes and special sensing technologies shall enhance the versatility of drones in performing their functions in various roles such as surveillance and inspection, material and 3D printing. In addition, the optimization of drone formation flying and Incorporation of drones into Unmanned traffic management systems will add another safety dimension in managing drone operations in complex industrial settings. The progress of UAV and drones in manufacturing and industrial sectors hold the predisposition of revolutionizing soon. By integrating the gains from such technologies, industries can attain higher standards of operating efficiency, flexibility and creativity so sorely required to secure their strategies in today's more global and complex environment. It will become even bigger in the future as the four key areas of autonomy, connectivity, energy consumption and sensor will continue to improve. It will be possible to expect further development concerning many intelligent drones that would interact effectively within intricate environments and provide enormous data streams to analyse processes enhancing safety and promoting new solutions. This pattern of penetration of drone in evolving industries suggests a future of more usage of aerial workhorses, changing the course of operations in the future industrial world.

REFERENCES

Abdullah, A. S., AB, A. A., & Geetha, S. (2024). DecentraliDrone: A decentralized, fully autonomous drone delivery system for reliable, efficient transport of goods. *Alexandria Engineering Journal*, *88*, 1–30.

Albano, M., Ferreira, L L., Orio, G D., Maló, P., Webers, G., Jantunen, E., Gabilondo, I., Viguera, M., & Papa, G. (2020, July 21). Advanced sensor-based maintenance in real-world exemplary cases. Taylor & Francis, 61(4), 537-553. DOI: 10.1080/00051144.2020.1794192

Ali, S S., Kaur, R., & Khan, S. (2023, January 21). Identification of innovative technology enablers and drone technology determinants adoption: a graph theory matrix analysis framework. Springer Science+Business Media, 16(2), 830-852. DOI: 10.1007/s12063-023-00346-3

Alsayed, A., Yunusa-Kaltungo, A., Quinn, M. K., Arvin, F., & Nabawy, M. R. A. (2021, August 24). Drone-Assisted Confined Space Inspection and Stockpile Volume Estimation. *Remote Sensing (Basel)*, *13*(17), 3356–3356. DOI: 10.3390/rs13173356

Aragon, R. V., Castaño, C A R., & Cerón, A. (2020, September 30). Impact and Technological Innovation of UAS/Drones in the World Economy: *VI International Conference on Innovation and Trends in Engineering (CONIITI 2020)*. DOI: 10.1109/CONIITI51147.2020.9240228

Awasthi, S., Gramse, N., Reining, C., & Roidl, M. (2022, January 1). UAVs for Industries and Supply Chain Management. Cornell University. https://doi.org//arxiv.2212.03346DOI: 10.48550

Baracaldo, N., Anwar, A., Purcell, M., Rawat, A., Sinn, M., Altakrouri, B., Balta, D., Sellami, M., Kühn, P., Schöpp, U., & Buchinger, M. (2022, January 1). Towards an Accountable and Reproducible Federated Learning: A FactSheets Approach. Cornell University. https://doi.org//arxiv.2202.12443DOI: 10.48550

Behroozi, M., & Ma, D. (2023, January 1). Last Mile Delivery with Drones and Sharing Economy. Cornell University. https://doi.org//arxiv.2308.16408DOI: 10.48550

Bousdekis, A., Lepenioti, K., Apostolou, D., & Mentzas, G. (2021, March 31). A Review of Data-Driven Decision-Making Methods for Industry 4.0 Maintenance Applications. *Electronics (Basel)*, *10*(7), 828–828. DOI: 10.3390/electronics10070828

Cheng, N., Wu, S., Wang, X., Yin, Z., Li, C., Chen, W., & Chen, F. (2023, August 15). AI for UAV-Assisted IoT Applications: A Comprehensive Review. *IEEE Internet of Things Journal*, *10*(16), 14438–14461. DOI: 10.1109/JIOT.2023.3268316

Dilshad, N., Hwang, J., Song, J., & Sung, N. (2020, October 21). Applications and Challenges in Video Surveillance via Drone: A Brief Survey. DOI: 10.1109/ICTC49870.2020.9289536

Feroz, S., & Dabous, S. A. (2021, May 6). UAV-Based Remote Sensing Applications for Bridge Condition Assessment. *Remote Sensing (Basel)*, *13*(9), 1809–1809. DOI: 10.3390/rs13091809

Gerstein, D M., & Leidy, E. (2024, January 1). Emerging Technology and Risk Analysis: Unmanned Aerial Systems Intelligent Swarm Technology. DOI: 10.7249/RRA2380-1

Gikunda, K. (2024, January 1). Harnessing Artificial Intelligence for Sustainable Agricultural Development in Africa: Opportunities, Challenges, and Impact. Cornell University. https://doi.org/DOI: 10.48550/arXiv.2401

Giordan, D., Adams, M., Aicardi, I., Alicandro, M., Allasia, P., Baldo, M., Berardinis, P D., Dominici, D., Godone, D., Hobbs, P., Lechner, V., Niedzielski, T., Piras, M., Rotilio, M., Salvini, R., Segor, V., Sotier, B., & Troilo, F. (2020, April 1). The use of unmanned aerial vehicles (UAVs) for engineering geology applications. Springer Science+Business Media, 79(7), 3437-3481

Giordan, D., Adams, M., Aicardi, I., Alicandro, M., Allasia, P., Baldo, M., Berardinis, P D., Dominici, D., Godone, D., Hobbs, P., Lechner, V., Niedzielski, T., Piras, M., Rotilio, M., Salvini, R., Segor, V., Sotier, B., & Troilo, F. (2020, April 1). The use of unmanned aerial vehicles (UAVs) for engineering geology applications. Springer Science+Business Media, 79(7), 3437-3481. DOI: 10.1007/s10064-020-01766-2

Gonçalves, R. S., Oliveira, M. D., Rocioli, M., Souza, F. D., Gallo, C. A., Sudbrack, D. E. T., Trautmann, P. V., Clasen, B., & Homma, R. Z. (2023, June 13). Drone–Robot to Clean Power Line Insulators. *Sensors (Basel)*, *23*(12), 5529–5529. DOI: 10.3390/s23125529 PMID: 37420696

Guimarães, N., Pádua, L., Marques, P., Silva, N., Peres, E., & Sousa, J. J. (2020, March 24). Forestry Remote Sensing from Unmanned Aerial Vehicles: A Review Focusing on the Data, Processing and Potentialities. *Remote Sensing (Basel)*, *12*(6), 1046–1046. DOI: 10.3390/rs12061046

Günal, M M. (2019, January 1). Data Collection Inside Industrial Facilities with Autonomous Drones. Springer International Publishing, 141-151. DOI: 10.1007/978-3-030-04137-3_9

Hamdi, A., Salim, F. D., Kim, D. Y., Neiat, A. G., & Bouguettaya, A. (2021, January 1). Drone-as-a-Service Composition Under Uncertainty. Cornell University. https://doi.org//arxiv.2103.06513DOI: 10.48550

Huertas-García, Á., Martí-González, C., Maezo, R. G., & Rey, A. E. (2023, January 1). A Comparative Study of Machine Learning Algorithms for Anomaly Detection in Industrial Environments: Performance and Environmental Impact. Cornell University. https://doi.org//arxiv.2307.00361DOI: 10.48550

Islam, S. M. R. (2023, January 1). Drones on the Rise: Exploring the Current and Future Potential of UAVs. Cornell University. https://doi.org//arxiv.2304.13702DOI: 10.48550

Jagannath, J., Jagannath, A., Furman, S., & Gwin, T. (2020, January 1). Deep Learning and Reinforcement Learning for Autonomous Unmanned Aerial Systems: Roadmap for Theory to Deployment. Cornell University. https://doi.org//arxiv.2009.03349DOI: 10.48550

Kurbanzade, A. K., Baig, A. M., & Mehrotra, S. (2023, January 1). Elevating Industries with Unmanned Aerial Vehicles: Integrating Sustainability and Operational Innovation. Cornell University. https://doi.org//arxiv.2312.13551DOI: 10.48550

Lamberti, L., Bompani, L., Kartsch, V., Rusci, M., Palossi, D., & Benini, L. (2023, January 1). Bio-inspired Autonomous Exploration Policies with CNN-based Object Detection on Nano-drones. Cornell University. https://doi.org//arxiv.2301.12175DOI: 10.48550

Merkle, N., Bahmanyar, R., Henry, C., Azimi, S. M., Yuan, X., Schopferer, S., Gstaiger, V., Auer, S., Schneibel, A., Wieland, M., & Kraft, T. (2023, January 1). Drones4Good: Supporting Disaster Relief Through Remote Sensing and AI. Cornell University. https://doi.org/DOI: 10.48550/arXiv.2308

Merkle, N., Bahmanyar, R., Henry, C., Azimi, S. M., Yuan, X., Schopferer, S., Gstaiger, V., Auer, S., Schneibel, A., Wieland, M., & Kraft, T. (2023, January 1). Drones4Good: Supporting Disaster Relief Through Remote Sensing and AI. Cornell University. https://doi.org//arxiv.2308.05074DOI: 10.48550

Nelson, J. P., Biddle, J. B., & Shapira, P. (2023, January 1). Applications and Societal Implications of Artificial Intelligence in Manufacturing: A Systematic Review. Cornell University. https://doi.org//arxiv.2308.02025DOI: 10.48550

Nunes, E. C. (2023, January 1). Employing Drones in Agriculture: An Exploration of Various Drone Types and Key Advantages. Cornell University. https://doi.org//arxiv.2307.04037DOI: 10.48550

Qu, Y., Dai, H., Zhuang, Y., Chen, J., Dong, C., Wu, F., & Guo, S. (2021, January 1). Decentralized Federated Learning for UAV Networks: Architecture, Challenges, and Opportunities. Cornell University. https://doi.org//arxiv.2104.07557DOI: 10.48550

Raivi, A. M., Huda, S. M. A., Alam, M. M., & Moh, S. (2023, January 28). Drone Routing for Drone-Based Delivery Systems: A Review of Trajectory Planning, Charging, and Security. *Sensors (Basel)*, *23*(3), 1463–1463. DOI: 10.3390/s23031463 PMID: 36772502

Regona, M., Yiğitcanlar, T., Xia, B., & Li, R Y M. (2022, March 1). Opportunities and Adoption Challenges of AI in the Construction Industry: A PRISMA Review. Springer Science+Business Media, 8(1), 45-45. DOI: 10.3390/joitmc8010045

Shahmoradi, J., Talebi, E., Roghanchi, P., & Hassanalian, M. (2020, July 15). A Comprehensive Review of Applications of Drone Technology in the Mining Industry. *Drones (Basel)*, *4*(3), 34–34. DOI: 10.3390/drones4030034

Smeesters, L., Verbaenen, J., Schifano, L., Vervaeke, M., Thienpont, H., Teti, G., Forconi, A., & Lulli, F. (2023, February 23). Wide-Field-of-View Multispectral Camera Design for Continuous Turfgrass Monitoring. *Sensors (Basel)*, *23*(5), 2470–2470. DOI: 10.3390/s23052470 PMID: 36904674

Telli, K., Kraa, O., Himeur, Y., Ouamane, A., Boumehraz, M., Atalla, S., & Mansoor, W. (2023, January 1). A Comprehensive Review of Recent Research Trends on UAVs. Cornell University. https://doi.org/DOI: 10.48550/arXiv.2307

Xing, J., Cioffi, G., Hidalgo-Carrió, J., & Scaramuzza, D. (2023, January 1). Autonomous Power Line Inspection with Drones via Perception-Aware MPC. Cornell University. https://doi.org//arxiv.2304.00959DOI: 10.48550

Xu, K., Li, Y., Liu, C., Liu, X., Hao, X., Gao, J., & Maropoulos, P. (2020, May 25). Advanced Data Collection and Analysis in Data-Driven Manufacturing Process. China Machine Press, 33(1). DOI: 10.1186/s10033-020-00459-x

Zinno, R., Haghshenas, S. S., Guido, G., & Vitale, A. (2022, January 1). Artificial Intelligence and Structural Health Monitoring of Bridges: A Review of the State-of-the-Art. *IEEE Access : Practical Innovations, Open Solutions*, *10*, 88058–88078. DOI: 10.1109/ACCESS.2022.3199443

Chapter 4
Edge Computing, Emerging Trends, and Drone Technology for the Business Industry

Imdad Ali Shah
https://orcid.org/0000-0003-2015-1028
Faculty of Engineering Science and Technology, Iqra University, Pakistan

Manzoor Ahmed
https://orcid.org/0009-0007-3551-2195
Government of Sindh, Pakistan

ABSTRACT

The primary objective of this chapter is to focus on drone technology used in the business industry. The use of drone technology is causing a revolutionary change in business assiduity, adding a functional edge. Drone advancements in Artificial Intelligence (AI) and Machine literacy (ML) allow independent operations and real-time data analysis. Drone use is getting safer and more common because of legislative developments between tech businesses and regulations recent developments in drone technology, their uses in the marketable sector, and the possible obstacles and chances for expansion they offer. Keeping up with these new developments is essential for sustaining a competitive edge and promoting sustainable invention as companies continue to use drone technology.

DOI: 10.4018/979-8-3693-8497-8.ch004

INTRODUCTION

This preface intends to give a general overview of the new progressions in drone technology and how they are revolutionizing the marketable region. We may gain a better understanding of how drones can transfigure business missions, and goad unborn excrescency by appearing at the country of drone operations, technology improvements, and nonsupervisory revise. Drone technology has become a ground-breaking invention in recent times, with significant marketable assiduity ramifications. Unmanned upstanding instruments (UAVs), appertained to as drones, have evolved from their initial martial usage to become essential instruments in a variety of marketable disciplines(Al-Fuqaha et al., 2015, Lin et al., 2017, Vailshery, 2021). Rapid progressions in tackle, software, and legit fabrics are driving this specialized excrescency and perfecting the capabilities and availability of drones for marketable operations. Figureure 1 presents the drone's activities.

Figure 1. Overview of drone activities

Drone integration into commercial operations is opening new openings and changing established procedures. Drones are being used by sectors like construction, real estate, logistics, husbandry, and security to increase productivity, save charges, and gain a competitive edge drone(Wang et al., 2020, Luong et al., 2019, Nguyen et al., 2021). The development of independent drones that can make complicated opinions and analysis data is being backed by rising trends that show an adding confluence of drone technology with artificial intelligence (AI) and Machine literacy (ML). Drone integration into corporate operations is opening new opportunities and changing established procedures(Lin et al., 2020, Yang et al., 2019). Drones are being used by sectors like construction, real estate, logistics, agriculture, and security to increase productivity, save expenses, and obtain a competitive edge drone. The development of autonomous drones that can make complicated decisions and analysis data is being aided by emerging trends that show an increasing convergence of drone technology with artificial intelligence (AI) and Machine Learning (ML) (Fitwi et al., 2019, Guo et al., 2019, Shah et al., 2024a). These intelligent drones can complete complex jobs with little assistance from humans, creating new opportunities for operational excellence. Furthermore, the legislative environment is changing to support the growing several commercial drone uses. To facilitate the wider deployment of drone technology, government and regulatory agencies around the world are striving to establish frameworks that guarantee safety, privacy, and effective airspace management. Figureure 2 Overview of growth of technology in various business sector 2020-24.

Figure 2. Overview of growth of technology in various business sector 2020-24

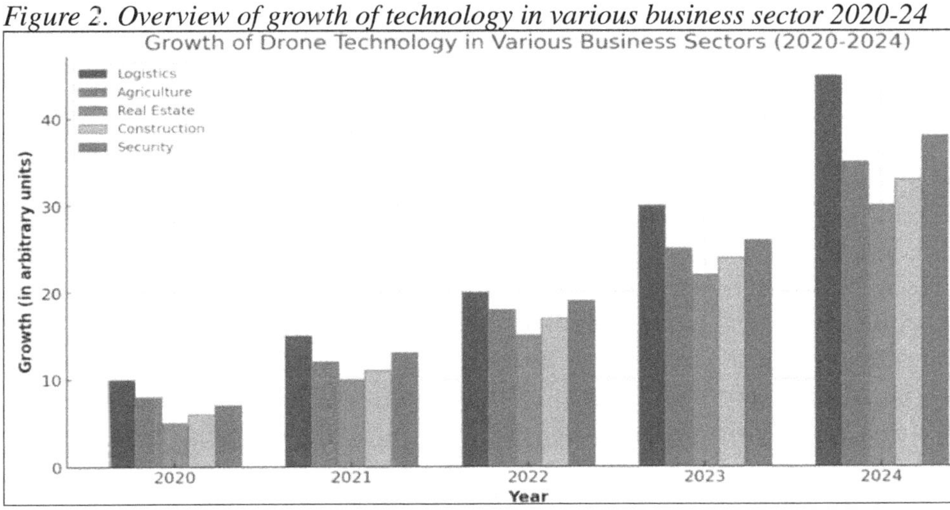

The device edge consists of edge devices; the local edge, which consists of the application and network layer; and the cloud edge. But we're just getting started. In our next piece in this series, we will go into greater detail about the different layers and technologies needed by developers to build an edge computing architecture. Due to the rapid development and deployment of unmanned aerial vehicles (UAVs), the drone industry has experienced exponential growth in recent years(Shah and Jhanjhi, 2024, Guo et al., 2021). These multifunctional instruments are increasingly necessary for a wide range of uses, including agricultural monitoring, emergency response, and aerial photography in logistics. Using technologies that make use of computer offers not available in traditional or pall data centers, edge computing shifts the cargo closer to the point of data generation and makes it possible to respond to data analysis. By making use of and managing the processing power set up in remote areas, such as distribution centers, artificial shops, retail stores, storages, and hospices. Using technologies that make use of computer resources not available in traditional or cloud data centers, edge computing shifts the load closer to the site of data generation and makes it possible to respond to data analysis(Shi et al., 2016, Khan et al., 2019, Varghese et al., 2016). By making use of and managing the processing power found in remote areas, such as distribution centers, industrial plants, retail stores, warehouses, and hotels.

LITERATURE REVIEW

In recent times, there have been notable developments in incorporating drone technology into the marketable sector. Unmanned Aerial instruments (UAVs), usually called drones, have revolutionized multitudinous industriousness by offering innovational effects that meliorate data delicacy, security, and forcefulness(Carvalho et al., 2021, Jiao et al., 2013). This literature review examines drone technology's recent progressions and uses in the merchantable region. Drone technology and bite computing are two snappily developing areas that have a monumental jolt on the business region. The purpose of this literature review is to examine the new progressions in bite computing and how they relate to drone technology, featuring how they affect nonidentical business firmaments. Processing data near where it's generated, as opposed to depending on a centralized box- predicated system, is known as bite computing(Satyanarayanan, 2017, Bilal et al., 2018). This system boosts real-time data processing, lowers quiescence, and increases system forcefulness altogether. To take data more effectively and efficiently, bite computing has had to be developed in reaction to the appended measure of data created by IoT bias.

In recent times, there have been noble progressions in incorporating drone technology into the marketable region. Unmanned Aerial instruments (UAVs), generally called drones, have revolutionized numerous diligences by offering innovational results that ameliorate data delicacy, security, and effectiveness. This literature review examines drone technology's rearmost progressions and uses in the marketable region(Xia et al., 2016a, Yi et al., 2015, Shah et al., 2024b). Drone technology and bite computing are two snappily developing areas that have a monumental jolt on the business region. The purpose of this literature review is to examine the new progressions in bite computing and how they relate to drone technology, featuring how they affect nonidentical business disciplines. Processing data near where it's generated, as opposed to depending on a centralized pall-grounded system, is known as bite computing. This system boosts real-time data processing, lowers quiescence, and increases system effectiveness altogether. To take data more effectively and efficiently, bite computing has had to be developed in reaction to the adding measure of data created by IoT bias. Figureure 3 Overview taxonomy of literature review.

Figure 3. Overview taxonomy of literature review

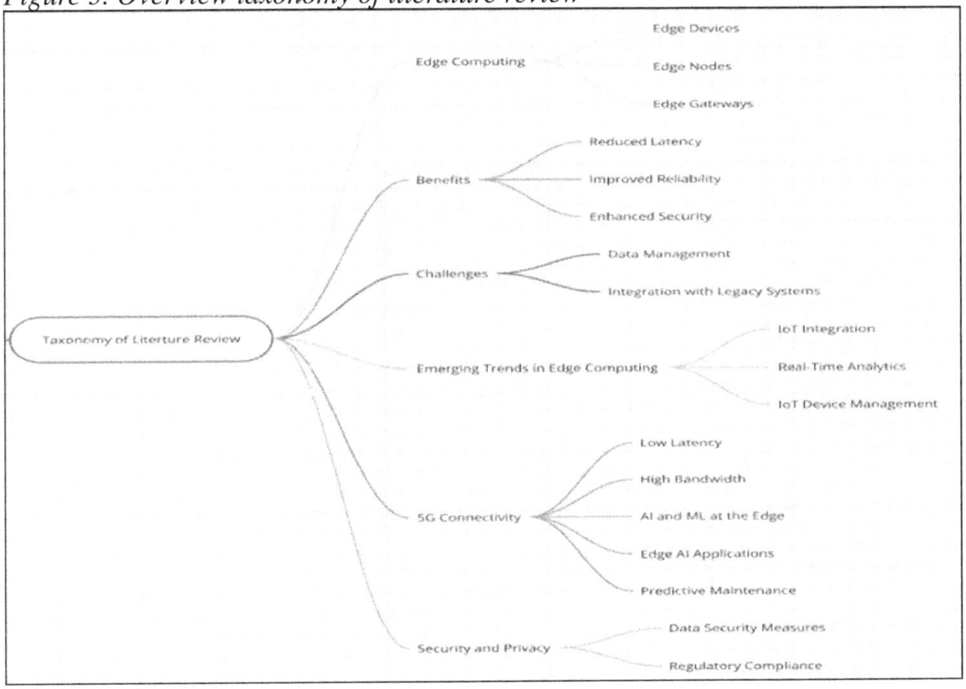

One of the most important tasks in UAV operation, especially in combined airspace, is detecting and precluding concussions with manned aircraft. Both manned and unmanned aircraft must be suitable to identify other aircraft to operate safely. Advanced systems that can identify and track aircraft in real- time utilizing a variety of detectors and optic cameras have been the seat of recent exploration in the field of aircraft discovery for UAVs. still, it can be delicate to exercise usual radar systems to describe fragile aircraft, similar to general aeronautics aircraft(Othman et al., 2015, Nithya et al., 2020, Wu et al., 2021). The drone assiduity's relationship with pall waiters has steered into a new period of technological invention and functional effectiveness. This community is revolutionizing the use of drones in a batch of diligence, involving husbandry, guidance, environmental monitoring, structure examination, and logistics. pall computing offers storehouse capacities and reliable data operation systems for managing the massive volumes of data produced by drones(Luo et al., 2021, Xia et al., 2019). This allows for bettered real-time processing, sophisticated dissection, and flawless messages. At the edge of a network, where data processing must happen, powerful computers called "edge servers" are placed. The physical location of an edge server is often close to the applications or systems that produce the data that the edge servers consume or store. This contrasts with a more traditional server conFigureuration, which frequently stores data in a data center or the cloud(Chen et al., 2018, Hong et al., 2019). At the bite of a network, where data processing must be, important computers called" bite waiters" are placed. Figure 4 Overview of the issues & challenges of edge computing and drone technology.

Figure 4. Overview of the issues & challenges of edge computing and drone technology

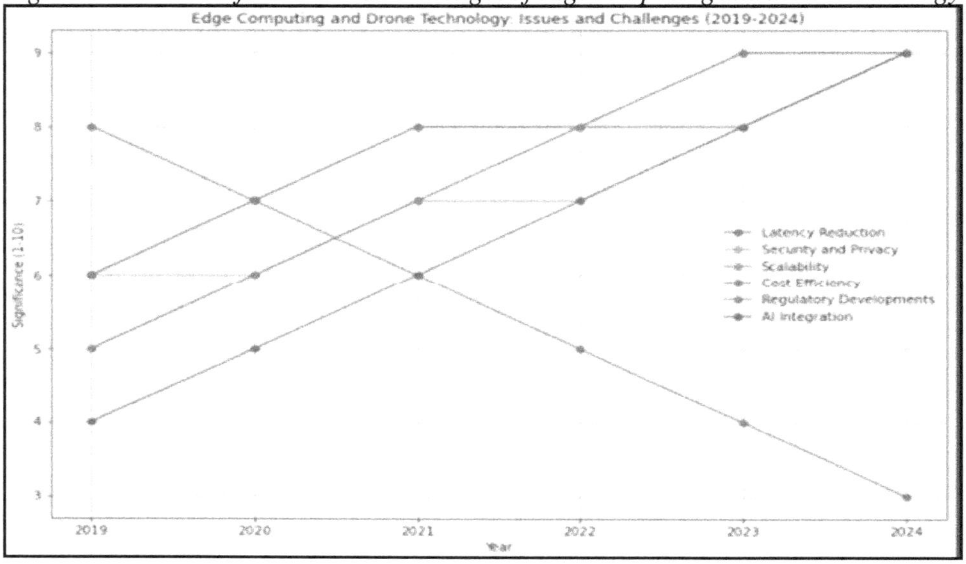

The physical position of a bite garçon is frequently close to the operations or systems that produce the data that the bite waiters consume or store. This contrasts with a more traditional garçon conFigureuration, which constantly stores data in a data locus or the pall. Edge servers are an essential part of edge computing, a computer paradigm that moves processing and data storage closer to the point of demand, improving reaction times and saving bandwidth(Wei et al., 2021, Cisco, 2020). Implementing edge AI with UAVs presents several difficulties, chief among them being the creation of distributed training algorithms, privacy and security concerns, resource allocation issues, and real-time demands. When it comes to the practical difficulty of "developing distributed training algorithms," instances that must be secure and mission-critical call for online distributed training algorithms where on-device constraints and communication bottlenecks are significant.

Whenever it comes to the implementation challenges of "security and privacy," some attacks on UAV systems are limited to specific software/hardware combinations, and implementing well-researched countermeasures of other communication systems can present major deployment challenges(Mocnej et al., 2018, Su et al., 2022). Furthermore, distinguishing between typical and unexpected network traffic requirements might be challenging. Additionally, edge AI lowers energy consumption by facilitating preprocessing at the edge, which minimizes the amount of data that must be transmitted to the distant cloud and, consequently, the energy consumption related to data transport. For applications like precision agriculture and civil infrastructure inspection, this lower energy consumption may enable longer searches, coverage, and inspections before requiring recharging. Cloud computing's centralization makes it challenging to meet the requirements of some time-sensitive applications that require cheap cost, great performance, and ultra-low latency(Cao et al., 2020, Siriwardhana et al., 2021). In this regard, edge computing is suggested as a viable solution to cloud computing's drawbacks. Edge computing, which is thought to be the key enabling technology for the 5G future, uses an open platform with fundamental network capabilities, processing resources, and data storage to offer customers nearest-end services. Edge computing, which is perfect for sophisticated processing and real-time data analysis, is an addition to cloud computing rather than a replacement for it. The era known as the Internet of Everything (IoE) has emerged in recent years. Value is continuously created by people, processes, data, and objects that are intelligently connected to the Internet. The intrinsic limits of the current computing paradigms, particularly edge computing, are exacerbated by IoE. Certain aspects of edge computing architecture, such as service migration, privacy and security protection, and edge node deployment, are challenging to effectively handle using traditional techniques(Kong et al., 2021b, Husain and Askar, 2021). These difficulties may be lessened by innovative solutions that have surfaced because of the quick development of emerging technology. Regarding service migration, the Q-learning algorithm is used by reinforcement learning (RL) to optimize the return of migration and minimize the cost of user equipment migration and communication. By uploading only parameters, like weights, federated learning effectively avoids privacy issues by allowing users to jointly train an algorithm while retaining local data samples on the device(Gao et al., 2021, Ranaweera et al., 2021). By preventing nodes from disclosing metadata during edge coordination, blockchain technology protects users' changeable keys and guarantees the security and privacy of edge networks. Microservices may integrate the different components of IoT system design layers to enable distributed software development, which solves the problem of the application package's distribution about node deployment concerns. The field of edge computing research is quite busy because of the abundance of research opportunities at the nexus between the computer model and other disciplines. Apart from resolving the issues, the combination of edge computing and certain technologies—such as blockchain, 6G, artificial intelligence (AI), and digital twins powered by other technologies—can fully realize their potential and gain from one another(Xia et al., 2016b, Ye et al., 2020). The reciprocal advantages of integrating several fields need more investigation and study.

OVERVIEW EDGE COMPUTING DRONE TECHNOLOGY

Drone and edge computing technologies are revolutionizing several industries by improving operational efficiency, providing creative solutions, and facilitating real-time data processing. A distributed computing paradigm known as edge computing is closer to the data sources. By serving this, quiescence is dropped, recycling pets are swelled, and data handling responsibility is bettered(Ye et al., 2020, Kong et al., 2021a). Unmanned aerials (UVs) are adaptable instruments employed in a variety of spots for jobs taking upstanding guidance, data collection, and quittance services. Figureure 4 Overview of computing & drone technology.

Figure 5. Overview of computing & drone technology

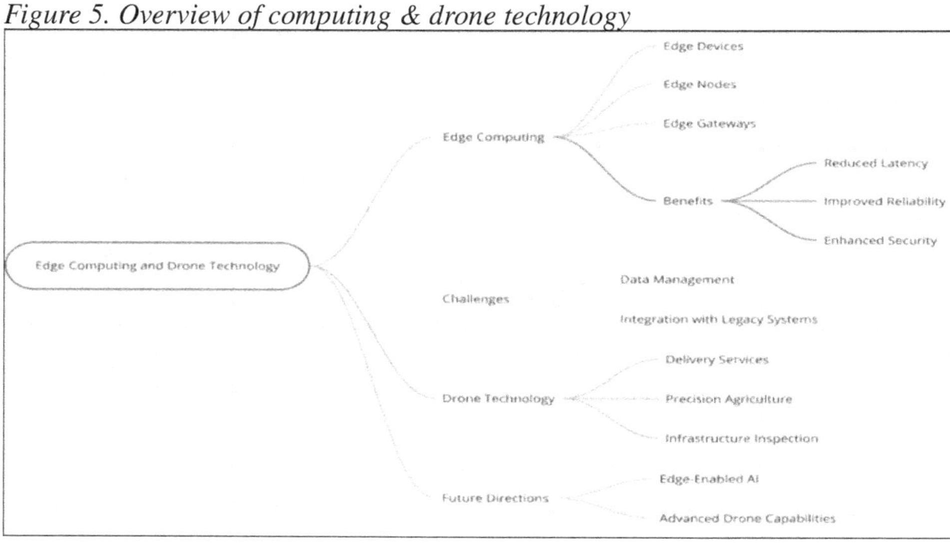

Drone and edge computing technologies are causing major shifts in several diligence. Strong results for real- time data processing, swelled operation forcefulness, and innovational business models are handed by their integration. This technology's implicit to revise spots will only boost with further evolution, opening new profitable openings for companies around the world.

Overview Edge Computing and Drone Technology for the Business Industry

Drone technology and bite computing are making crucial progressions in the commercial region, furnishing real-time data processing innovational operations, and significant earnings in functional effectiveness. By moving data recycling near the data source, bite computing improves the screen and lowers quiescence(Khan et al., 2021, Zhou et al., 2022). Drones' technology contemporaneously offers adjustable results for structure examination, husbandry, and quittance services. Business experience, and prospects of bite computing and drone technology, featuring how they could revolutionise the marketable region. Figureure 6 Overview of Drone technology in the business industry

Figure 5. Overview of drone technology in the bossiness industry

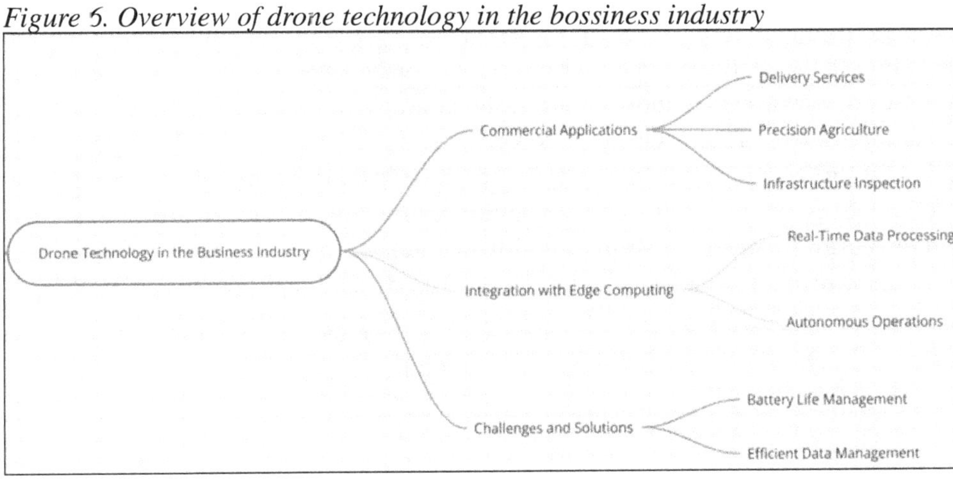

The commercial geography is invariably changing due to the rapid-fire elaboration of technology, and two of the most significant progressions in recent technology. Data recycling and storehouse are shifted to lower quiescence, pets up reaction moments and resolution-making capabilities which are essential in the whirlwind-paced world of the moment by recycling data locally. multitudinous diligence has witnessed a noble suge in the exercise and expansion of drone technology. For jobs taking upstanding data collecting, covering, and delivering services, drones have special advantages in fields where usual ways could be less operative or more precious, such as logistics, husbandry, and structure examination(Li et al., 2021, Lim et al., 2020, Firdaus and Rhee, 2021). Drone technology and bite computing combine to produce a potent community that enhances technology vantages. Real-time data processing and dissection by drones with bite computing capabilities enable independent missions and prompt resolution- timber. Companies now have further options because this connectivity helps them ameliorate customer experience cut charges and streamline processes.

CONCLUSION

The use of drone technology is causing a revolutionary change in business assiduity, adding a functional edge. Drone advancements in Artificial Intelligence (AI) and Machine literacy (ML) allow independent operations and real-time data analysis. Drone use is getting safer and more common because of legislative developments between tech businesses and regulations recent developments in drone technology, their uses in the marketable sector, and the possible obstacles and chances for expansion they offer. Keeping up with these new developments is essential for sustaining a competitive edge and promoting sustainable invention as companies continue to use drone technology. Drone technology has endured unknown growth, transitioning from military origins to wide mercenary and marketable operations. This will further reduce the demand for centralized data processing and ameliorate bite device responsiveness by allowing indeed more sophisticated real-time analytics and resolution-making capabilities. With an emphasis on boosting capabilities and autonomy, drone technology will continue to advance. unborn drones will have further sophisticated detectors, longer battery life, and better navigation systems, which will boost their utility for a range of marketable usages.

REFERENCES

Al-Fuqaha, A., Guizani, M., Mohammadi, M., Aledhari, M., & Ayyash, M. (2015). Internet of things: A survey on enabling technologies, protocols, and applications. *IEEE Communications Surveys and Tutorials*, *17*(4), 2347–2376. DOI: 10.1109/COMST.2015.2444095

Bilal, K., Khalid, O., Erbad, A., & Khan, S. U. (2018). Potentials, trends, and prospects in edge technologies: Fog, cloudlet, mobile edge, and micro data centers. *Computer Networks*, *130*, 94–120. DOI: 10.1016/j.comnet.2017.10.002

Cao, K., Liu, Y., Meng, G., & Sun, Q. (2020). An overview on edge computing research. *IEEE Access : Practical Innovations, Open Solutions*, *8*, 85714–85728. DOI: 10.1109/ACCESS.2020.2991734

Carvalho, G., Cabral, B., Pereira, V., & Bernardino, J. (2021). Edge computing: Current trends, research challenges and future directions. *Computing*, *103*(5), 993–1023. DOI: 10.1007/s00607-020-00896-5

Chen, C., Jiao, S., Zhang, S., Liu, W., Feng, L., & Wang, Y. (2018). TripImputor: Real-time imputing taxi trip purpose leveraging multi-sourced urban data. *IEEE Transactions on Intelligent Transportation Systems*, *19*(10), 3292–3304. DOI: 10.1109/TITS.2017.2771231

Cisco, U. 2020. Cisco annual internet report (2018–2023) white paper, March 2020.

Firdaus, M., & Rhee, K.-H. (2021). On blockchain-enhanced secure data storage and sharing in vehicular edge computing networks. *Applied Sciences (Basel, Switzerland)*, *11*(1), 414. DOI: 10.3390/app11010414

Fitwi, A., Chen, Y., & Zhu, S. A lightweight blockchain-based privacy protection for smart surveillance at the edge. *2019 IEEE International Conference on Blockchain (Blockchain)*, 2019. IEEE, 552-555. DOI: 10.1109/Blockchain.2019.00080

Gao, L., Luan, T. H., Gu, B., Qu, Y., & Xiang, Y. (2021). *Privacy-preserving in edge computing*. Springer. DOI: 10.1007/978-981-16-2199-4

Guo, F., Tang, B., Tang, M., Zhao, H., & Liang, W. (2021, June). Microservice selection in edge-cloud collaborative environment: A deep reinforcement learning approach. In 2021 8th IEEE International Conference on Cyber Security and Cloud Computing (CSCloud)/2021 7th IEEE International Conference on Edge Computing and Scalable Cloud (EdgeCom) (pp. 24-29). IEEE.

Guo, H., Li, W., Nejad, M., & Shen, C.-C. (2019). *Access control for electronic health records with hybrid blockchain-edge architecture. 2019 IEEE international conference on blockchain (Blockchain).* IEEE.

Hong, Z., Chen, W., Huang, H., Guo, S., & Zheng, Z. (2019). Multi-hop cooperative computation offloading for industrial IoT–edge–cloud computing environments. *IEEE Transactions on Parallel and Distributed Systems*, *30*(12), 2759–2774. DOI: 10.1109/TPDS.2019.2926979

Husain, B. H., & Askar, S. (2021). Survey on edge computing security. *International Journal of Science and Business*, *5*, 52–60.

Jiao, L., Friedman, R., Fu, X., Secci, S., Smoreda, Z., & Tschofenig, H. (2013). Cloud-based computation offloading for mobile devices: State of the art, challenges and opportunities. 2013 Future Network & Mobile Summit, 1-11.

Khan, L. U., Saad, W., Han, Z., Hossain, E., & Hong, C. S. (2021). Federated learning for internet of things: Recent advances, taxonomy, and open challenges. *IEEE Communications Surveys and Tutorials*, *23*(3), 1759–1799. DOI: 10.1109/COMST.2021.3090430

Khan, W. Z., Ahmed, E., Hakak, S., Yaqoob, I., & Ahmed, A. (2019). Edge computing: A survey. *Future Generation Computer Systems*, *97*, 219–235. DOI: 10.1016/j.future.2019.02.050

Kong, X., Wang, K., Hou, M., Hao, X., Shen, G., Chen, X., & Xia, F. (2021a). A federated learning-based license plate recognition scheme for 5G-enabled internet of vehicles. *IEEE Transactions on Industrial Informatics*, *17*(12), 8523–8530. DOI: 10.1109/TII.2021.3067324

Kong, X., Zhu, B., Shen, G., Workneh, T. C., Ji, Z., Chen, Y., & Liu, Z. (2021b). Spatial-temporal-cost combination based taxi driving fraud detection for collaborative internet of vehicles. *IEEE Transactions on Industrial Informatics*, *18*(5), 3426–3436. DOI: 10.1109/TII.2021.3111536

Li, A., Sun, J., Wang, B., Duan, L., Li, S., Chen, Y., & Li, H. (2021, December). Lotteryfl: Empower edge intelligence with personalized and communication-efficient federated learning. In *2021 IEEE/ACM Symposium on Edge Computing (SEC)* (pp. 68-79). IEEE.

Lim, W. Y. B., Luong, N. C., Hoang, D. T., Jiao, Y., Liang, Y.-C., Yang, Q., Niyato, D., & Miao, C. (2020). Federated learning in mobile edge networks: A comprehensive survey. *IEEE Communications Surveys and Tutorials*, *22*(3), 2031–2063. DOI: 10.1109/COMST.2020.2986024

Lin, J., Yu, W., Zhang, N., Yang, X., Zhang, H., & Zhao, W. (2017). A survey on internet of things: Architecture, enabling technologies, security and privacy, and applications. *IEEE Internet of Things Journal*, *4*(5), 1125–1142. DOI: 10.1109/JIOT.2017.2683200

Lin, S., Yang, G., & Zhang, J. (2020, November). A collaborative learning framework via federated meta-learning. In 2020 IEEE 40th International Conference on Distributed Computing Systems (ICDCS) (pp. 289-299). IEEE.

Luo, Q., Hu, S., Li, C., Li, G., & Shi, W. (2021). Resource scheduling in edge computing: A survey. *IEEE Communications Surveys and Tutorials*, *23*(4), 2131–2165. DOI: 10.1109/COMST.2021.3106401

Luong, N. C., Hoang, D. T., Gong, S., Niyato, D., Wang, P., Liang, Y.-C., & Kim, D. I. (2019). Applications of deep reinforcement learning in communications and networking: A survey. *IEEE Communications Surveys and Tutorials*, *21*(4), 3133–3174. DOI: 10.1109/COMST.2019.2916583

Mocnej, J., Miškuf, M., Papcun, P., & Zolotová, I. (2018). Impact of edge computing paradigm on energy consumption in IoT. *IFAC-PapersOnLine*, *51*(6), 162–167. DOI: 10.1016/j.ifacol.2018.07.147

Nguyen, D. C., Ding, M., Pham, Q.-V., Pathirana, P. N., Le, L. B., Seneviratne, A., Li, J., Niyato, D., & Poor, H. V. (2021). Federated learning meets blockchain in edge computing: Opportunities and challenges. *IEEE Internet of Things Journal*, *8*(16), 12806–12825. DOI: 10.1109/JIOT.2021.3072611

Nithya, S., Sangeetha, M., Prethi, K. A., Sahoo, K. S., Panda, S. K., & Gandomi, A. H. (2020). SDCF: A software-defined cyber foraging framework for cloudlet environment. *IEEE Transactions on Network and Service Management*, *17*(4), 2423–2435. DOI: 10.1109/TNSM.2020.3015657

Othman, M., Xia, F., & Khan, A. N. (2015). Context-aware mobile cloud computing and its challenges. *IEEE Cloud Computing*, *2*(3), 42–49. DOI: 10.1109/MCC.2015.62

Ranaweera, P., Jurcut, A. D., & Liyanage, M. (2021). Survey on multi-access edge computing security and privacy. *IEEE Communications Surveys and Tutorials*, *23*(2), 1078–1124. DOI: 10.1109/COMST.2021.3062546

Satyanarayanan, M. (2017). The emergence of edge computing. *Computer*, *50*(1), 30–39. DOI: 10.1109/MC.2017.9

Shah, I., Jhanjhi, N., & Rajper, S. (2024a). Use of Deep Learning Applications for Drone Technology. Cybersecurity Issues and Challenges in the Drone Industry. *IGI Global.DOI*, *10*, 979–8.

Shah, I. A., & Jhanjhi, N. Z. (2024). Cybersecurity Issues and Challenges in the Drone Industry.

Shah, I. A., Jhanjhi, N. Z., & Ujjan, R. M. A. (2024b). *Use of AI applications for the drone industry. Cybersecurity Issues and Challenges in the Drone Industry.* IGI Global.

Shi, W., Cao, J., Zhang, Q., Li, Y., & Xu, L. (2016). Edge computing: Vision and challenges. *IEEE Internet of Things Journal*, *3*(5), 637–646. DOI: 10.1109/JIOT.2016.2579198

Siriwardhana, Y., Porambage, P., Liyanage, M., & Ylianttila, M. (2021). *A Survey on Mobile Augmented Reality with 5G Mobile Edge Computing: Architectures.* Applications, and Technical Aspects.

Su, X., Xue, S., Liu, F., Wu, J., Yang, J., Zhou, C., Hu, W., Paris, C., Nepal, S., & Jin, D. (2022). A comprehensive survey on community detection with deep learning. *IEEE Transactions on Neural Networks and Learning Systems*. PMID: 35263257

Vailshery, L. S. (2021). Internet of Things (IoT) and non-IoT active device connections worldwide from 2010 to 2025. *Online,https://bit. ly/2SwKVuB*

Varghese, B., Wang, N., Barbhuiya, S., Kilpatrick, P., & Nikolopoulos, D. S. (2016). *Challenges and opportunities in edge computing. 2016 IEEE international conference on smart cloud (SmartCloud).* IEEE.

Wang, X., Han, Y., Leung, V. C., Niyato, D., Yan, X., & Chen, X. (2020). Convergence of edge computing and deep learning: A comprehensive survey. *IEEE Communications Surveys and Tutorials*, *22*(2), 869–904. DOI: 10.1109/COMST.2020.2970550

Wei, Y., Zhou, S., Leng, S., Maharjan, S., & Zhang, Y. (2021). Federated learning empowered end-edge-cloud cooperation for 5G HetNet security. *IEEE Network*, *35*(2), 88–94. DOI: 10.1109/MNET.011.2000340

Wu, Y., Zhang, K., & Zhang, Y. (2021). Digital twin networks: A survey. *IEEE Internet of Things Journal*, *8*(18), 13789–13804. DOI: 10.1109/JIOT.2021.3079510

Xia, F., Jedari, B., Yang, L. T., Ma, J., & Huang, R. (2016a). A signaling game for uncertain data delivery in selfish mobile social networks. *IEEE Transactions on Computational Social Systems*, *3*(2), 100–112. DOI: 10.1109/TCSS.2016.2584103

Xia, F., Liu, L., Jedari, B., & Das, S. K. (2016b). PIS: A multi-dimensional routing protocol for socially-aware networking. *IEEE Transactions on Mobile Computing*, *15*(11), 2825–2836. DOI: 10.1109/TMC.2016.2517649

Xia, F., Wang, J., Kong, X., Zhang, D., & Wang, Z. (2019). Ranking station importance with human mobility patterns using subway network datasets. *IEEE Transactions on Intelligent Transportation Systems*, *21*(7), 2840–2852. DOI: 10.1109/TITS.2019.2920962

Yang, R., Yu, F. R., Si, P., Yang, Z., & Zhang, Y. (2019). Integrated blockchain and edge computing systems: A survey, some research issues and challenges. *IEEE Communications Surveys and Tutorials*, *21*(2), 1508–1532. DOI: 10.1109/COMST.2019.2894727

Ye, D., Yu, R., Pan, M., & Han, Z. (2020). Federated learning in vehicular edge computing: A selective model aggregation approach. *IEEE Access : Practical Innovations, Open Solutions*, *8*, 23920–23935. DOI: 10.1109/ACCESS.2020.2968399

Yi, S., Li, C., & Li, Q. (2015, June). A survey of fog computing: concepts, applications and issues. In *Proceedings of the 2015 workshop on mobile big data* (pp. 37-42).

Zhou, X., Liang, W., Ma, J., Yan, Z., Kevin, I., & Wang, K. (2022). 2D federated learning for personalized human activity recognition in cyber-physical-social systems. *IEEE Transactions on Network Science and Engineering*, *9*(6), 3934–3944. DOI: 10.1109/TNSE.2022.3144699

Chapter 5
Enhancing Security and Surveillance With AI-Driven Drones

Muhammad Usman Tariq
https://orcid.org/0000-0002-7605-3040
Abu Dhabi University, UAE & University College Cork, Ireland

ABSTRACT

This chapter focuses on the possibility of improving security and surveillance through the adoption of AI driven drones across different fields. It continues with the illustration of deployment and use of unmanned aerial vehicles (UAVs) with integration of artificial intelligence to respond to security issues and to help in processing real time information and surveillance. In the chapter, a number of application areas are discussed, such as military, law enforcement, private business ventures and disasters. Drones, due to Artificial Intelligence, have brought the improvements in surveillance and intelligence in the defence operations in present warfare. Police with drones that contain artificial intelligence is making policing and subsequently, public safety possible through surveillance and monitoring. The private sector has also incorporated the use of the AI-operated drones in offering security to the corporations in their observation of the physical properties.

INTRODUCTION

Drone technology and applications have grown dramatically over the past years, moving from military use only to being incorporated across many fields such as surveillance and security. Drones have been in existence since the mid-20th century but the first time they were used was in military with special reference to reconnaissance

DOI: 10.4018/979-8-3693-8497-8.ch005

was during the Second World War. The first drones developed were stationary and called ''Radioplane'' targets for use in exercises (Maguluri et al., 2023). During the several decades, the technology experienced constant refinement in terms of design and construction material to fundamentally enhanced aspects of drones during the last century and the early years of the current century. Increased freedom of movement, increased flight duration and capabilities of UAV sensors became a turning point. Over time, drones have been taken and used for other purposes different from the military; for instance, photographic and aerial surveying, growing crops, and surveillance (Veprytska & Kharchenko, 2024). One of the more recent innovations within the drone category is the integration of artificial intelligence as the driving force behind the drone's performance in security-surveillance missions. AI has been incorporated into drone systems to offer autonomy, meaning that drones can accomplish most of their tasks independently. AI capabilities enable the drones to work with large volumes of data, to understand them and to make the decisions in real time (Maguluri et al., 2023). Such independence has changed the capacity of the drones in accomplishing challenging surveillance operations including facial acknowledgment, object identification, and movement tracking among others. Accordingly, armed drones have revolutionized the method and practice of security in ways that have hardly been noticed in-efficiency, precision, and immediacy owing to the advances in Artificial Intelligence (Tlili et al., 2024).

The use of high levels of security has over the recent years become more important due to the following challenges in the world. The advancement in security risks, whether physical or virtual, have continued to evolve such that it requires enhancement of better instruments and equipment to protect people, structures and enterprises (Veprytska & Kharchenko, 2024). More specifically, corresponding to the increased activity by terrorists and organized criminals as well as the growing flow of immigrants and refugees it has become imperative to improve surveillance. Point security surveillance approaches like surface patrols and stationary cameras have been found to be wanting as they can hardly meet today's security demands especially where there is large insecurity or insecurity in large provinces. AI than implemented on drones is more flexible and can be used to monitor large areas and supply data to security personnel in real-time (Tlili et al., 2024). Applications of drones in areas like criminal investigation or policing or at the border control, AI controlled drones are now compulsory. These drones can work on their own flying over vast terrain, they can provide constant video feed and security check for any intrusion in the compound. For instance, border control authorities across the globe have commonly used drones to watch borders for breach, drug smuggling and human trafficking activities (Mallikarjunaradhya et al., 2023). Only with the help of AI algorithms can drones recognize the movement in the video stream, notice if there is something suspicious and alert security services. The fact that large and hard to access regions can be surveilled for long durations without human intervention has brought about a drastic shift in the manner border guards are accomplished (Veprytska & Kharchenko, 2024; Tariq, 2025).

Such organizations as those which control the critical infrastructure and different corporate assets also experience a growing need for enhanced security solutions. Security automatons are perfect for supervision of large structures like power stations or chemical plants, or simply enormous manufacturing complexes; this means constant patrolling that would take a lot of manpower otherwise. These drones have three or more different types of sensors such as thermal, to identify intruders or unauthorized access in regions that are difficult for people to monitor (Mallikarjunaradhya et al., 2023). In addition, the reconnaissance drone allows checking condition of fences, gates, and other infrastructures that might be damaged or worn out without any immediate obvious signs of it. The application of AI improves the drones' capability of recognizing threat or failure indications and respond to them to avoid incidents that would be more expensive (Tlili et al., 2024). Drones have also been deployed more often by urban centers for surveillance with the use of artificial intelligence. Today urban areas are more highly developed and populated than before hence a big challenge to security with physical structures. Due to distinct characteristics, drones with Artificial Intelligence can fly along city streets, oversee squares, and help police officers regulate the crowd or check an occasion. These drones can effectively scan faces to detect people that are on the watch list or follow the motion of a specific person of interest. Drones are used to increase effectiveness in large urban areas because monitoring large groups of people and large open spaces is often difficult (Veprytska & Kharchenko, 2024; Tariq, 2024).

The increasing need for intelligent drones in security and surveillance can also be explained by the necessity of cheap and scalable technology. Physical security systems, which include security cameras, security guards, rovers, etc. can be very costly to implement and manage especially within huge or hard to access areas (Mallikarjunaradhya et al., 2023).Drones can also be of more advantages than the manned aircraft as they cover large areas using little resources, and few operators are needed. The scalability in drone system also reflects another advantage of this technology; it is flexibility in deployment as a response to new threats because deployment of security agencies and resources can be equally scalable (Tlili et al., 2024). With regards to security threats, development of enhancing sophisticated technology including the use of drone with artificial Intelligence are expected to form a more central part of current surveillance. That is why drones today are an indispensable tool for responding to the modern security challenges of the world – they can collect data in real time, make decisions on their own, and cover large territories. From police work, boosting border security to protecting company's assets, AI drones hold the potential to become the key answer to the ever-growing need for more effective, intelligent surveillance systems. Besides improving the functionality of the security teams, the combination of AI with drone technology is revolutionizing surveillance in numerous sectors (Tariq, 2024).

BACKGROUND

Artificial intelligence (AI) soaring with drone technology has revolutionized the notion of security-slash-surveillance with progressive leaps in the ways that security can be managed and threats in diverse settings can be detected and addressed. Drones also referred to as UAVs or Unmanned Aerial Systems have been around for more than 50 years, starting with their use in military operations to conduct reconnaissance. But as technology develops drones are slowly introduced to a range of civil applications such as agricultural, cinematography, delivering goods and services and security among others. Whereas the usage was basically in Para policing and aerial reconnaissance, it has proven to be a highly developed smart watching system, which can enhance security in real-time mode in a wide range of fields and industries (Obiuto et al., 2024). The development of UAVs over the years has been immense in terms of its payload, its endurance and the sensors incorporated in the unit, making it possible for the device to execute tasks beyond human thinking. The initial drones are rudimentary and could perform a few tasks and were engineered to support fewer uses; they were mainly useful in the military to test munitions or gather intelligence. Drones eventually had the better sensors, cameras, and better navigation; that made them applicable in the commercial markets and businesses. The step moved to AI, especially the ML and CV, as the next big progression of drones had to progress from mere remote-controlled tools to systems that can make decisions on their own and work in real-time (Farsath et al., 2024).

These enhancements have enabled drones to form a significant technique of today's security systems that provide innovative solutions to surveillance large areas and identify vulnerabilities and threats that could lead to other security breaches. The desire to implement higher levels of protection has increased side by side with an increase in the risks in the world today. In the process of rapid urbanization, industrialization and infrastructural development the old-style security system, which included fixed CCTV cameras based on building constructions and ground mobile patrolling vehicles, have become inadequate to secure the facilities and habitats (Jain et al., 2024). Security teams are today required to manage extensive geographical space, ranging from sparsely developed rural regions, expansive factories, and complex urban cities, with inadequate resources. However, the increase in threats such as terrorism, organized crime, and other threats make the use of surveillance technologies that can quickly pick threats that exist and provide intelligence to security personnel more necessary. In this context the drones that are incorporated with the AI technologies have proved to be a valuable tool. This makes them ideal in places where one cannot set up normal security systems, or they end up being inefficient due to the vast coverage areas they can secure (Obiuto et al., 2024; Tariq, 2025).

AI helps in increasing the functionality of drones from mere surveillance and monitoring tools besides other operations. Some of the benefits that result from the application of drones include machine learning, computer vision, and autonomous navigation which means that a drone can analyze data on the go. Artificial intelligence lets drones to memorize encounters and behave according to them, while computer vision lets drones recognize objects, faces and movements including in crowded and in low-light scenarios. Such high ADA makes drones capable of identifying threats sooner than human operators or even other forms of CCTV surveillance. In addition, AI enables drones to function independently without human supervision and intervene in new security threats quickly since they are constantly flying (Farsath et al., 2024). Another advantage that drones with artificial intelligence should be talked about is the possibility of their growth and flexibility. It may be used for secured places within the city or for specific sites in remote or dangerous territories, which are difficult to protect by security teams, while reaching larger territories much quicker. The fact that they can patrol and monitor on their own as well as in real time reduces greatly the amount of people and money that must be invested in security. For instance, drones mounted with thermal sensors are capable of tracing heat at night or during poor light conditions, which affords a plus against conventional surveillance cameras (Jain et al., 2024).

Moreover, since AI can be linked to other security systems like motion detectors, alarm systems or others, it covers an enhanced overall security system. The next new opportunities in this situation are using the AI-powered drones for security purposes, which can go beyond the traditional environment. For instance, drones are gradually being employed in the surveillance of large tracks of land or large areas of coastline for border violations and other unlawful activities. For instance, drones can be evaluated for a variety of critical infrastructure segments including energy or transport since they can cover vast grounds ensuring for instance, presence of suspicious individuals or detecting system failure. It also has its usage in policing in tracking of criminals, crowd policing, or managing large crowds during events. AI drones are not only beneficial in big scenarios but also help in even city surveillance and help in finding lost or endangered persons in a crowd (Obiuto et al., 2024). Self-organizing drones will play a more significant part in security and surveillance as they progress in the future. The constant development in AI combined with improvement in drone hardware means even better prospects in offering increased object recognition facility, better navigation, and increased range of operations it can perform. This chapter provides a discussion of increasing adoption of

AI integration with security applications of drones and their ability to improve the efficiency and effectiveness of security operations. As more security applications rely on artificial intelligence to drive autonomy, it can be expected that the future of security will involve even more entwining of drones into security and surveillance, becoming essential to forces of security all around the world (Jain et al., 2024).

Artificial Intelligence as an Innovative Tool

Artificial Intelligence AI is an innovative tool that has made varied transformations to many sectors and improved functionality of diverse technologies. The use of drones is another area where AI is widely incorporated, and security and surveillance are perhaps the most popular ways to use drones (Mohamed & Pasam, 2020). Drones have been deployed in various fields for several years now across the agriculture, entertainment, etc., but the integration of AI in drones led to an outstanding innovation of performance, efficiency, self-driven, and advanced drones that can enhance the surveillance beyond expectations. This provides the drones not only with the capability of gathering information, but also to process that information and even make decisions based upon that information in real time, which makes for a powerful tool when it comes to surveillance and security of assets, structures, and people (Goruganthu et al., 2024). AI as applied to drones can be described as the employment of machine learning; computer vision; and autonomous navigation systems through which drones can make decisions based on gathered data. These capabilities take drone from just being aerial vehicles to intelligent systems, which can perform multiple surveillance operations independently. Combining AI within drones has revamped how security operates with real-time decision making and no longer require operators. Drones can learn about threats on their own using their AI systems and then respond to threats using pre-programmed protocols. This gives a level of sophistication and end user control that greatly improves the effectiveness of surveillance systems (Joshi et al., 2024).

Machine learning belongs to the major AI technologies that drone uses. This technology makes sure that drones can use the information acquired from large data sets and keeps them getting better on their performance over time. Because of machine learning algorithms, drone can analyze the data collected by the different sensors such as those of camera, thermal imaging, LiDAR and more to reveal patterns and make predictions (Cadet et al., 2024). For instance, a drone with the ability to employ machine learning can detect the one behaving abnormally, like a person who enters a prohibited zone or follows random movements of cars in an extensive region. As time goes on, the accuracy and effectiveness of the machine learning model enhances, by recognizing potential security risks more evidently. The kind of industrial learning that drones allow makes them superior to conventional surveillance systems where analysts would have to pore through data and make sense of it (Cadet et al., 2024).

Another important AI tech applied to drones include Computer vision, which means the drone can "see" or interpret what it "sees" in a way like how humans do. With the help of computer vision drones, facial recognition or license plates, and identifying suspicious behavior in their environment. This technology is very suitable when it comes to security personnel and surveillance where the drone improves its chances of finding specific targets (Mohamed & Pasam, 2020). For example, drones can be incorporated with facial detection technology to detect different people in large groups or compel trespassers to turn back into secure areas. In addition to facilitating surveillance, this capability will also shorten the amount of time and personnel required to conduct visual analysis on videos. Computer vision system integrated drones are independent and able to make decisions in real-time to determine who or what to follow (Ramadan et al., 2024). Self-navigation of the drone is another vital feature of the drones under AI technology. Historically, drones needed people to dictate how they are to maneuver, and so their deployment was rigid and had a fixed scale. AI in drones makes it possible to manage drones with autonomous navigation and control systems for their flying and operations. These systems make use of GPS, laser scanning and detection, also known as LiDAR, and cameras to chart out the space around the vehicle and avoid other objects in the road. It also allows drones to navigate areas that are crowded or are constantly changing, for instance urban centers or industrial estates in an accurate manner (Goruganthu et al., 2024).

The ability to navigate on its own means that drones can perform extended patrols over large territories and react to threats as soon as they appear without constant direct control. Also, these drones are flexible and can work in worst conditions like unfavorable weather or hinderances along the line to provide uninterrupted surveillance (Ramadan et al., 2024). Self-navigation with the help of numerous algorithms and computer vision, and machine learning make AI drones have certain privileges compared to traditional monitoring. That was one of the most essential advantages – the information can be analyzed in real-time. Standard security solutions compel guards to spend countless hours monitoring the recorded video feeds or other sensory data feed which in turn gives rise to considerable time delay to discern a threat. Drones controlled by the AI system can go through the data on the move, pointing out the suspicious activities as these are being conducted and alerting the security personnel on the same immediately. It dramatically increases the speed at which the organization can respond to new information and increases cognitive range visually. For example, drones can capture unauthorized persons on the secure area ahead of time and inform security officers of the breach (Cadet et al., 2024).

One of the advantages of drone solution implemented using AI is that the drones can follow the terrain and are able to collect information in large and inaccessible regions. Conventional security measures including use of stationary cameras or ground patrols may be inconvenient by the physical barriers or size of the area under surveillance. While about the monitoring of large territories and areas distant from populations, constant surveillance of vast territories, and territories with limited access, namely roofs and intensive forests, drones are irreplaceable. This flexibility enables them to survey large industrial facilities, borders, or cities with little or no requirements for infrastructure and manpower curb. That they can operate autonomously through such regions implies that environments can be scanned perpetually without much human intercession. AI-based drones also tend to be more scalable and less expensive than other known surveillance approaches (Ramadan et al., 2024). The initial cost of acquiring drone technology may be high, but once deployed, the drones can cost very little to maintain and they can patrol large areas at comparatively cheaper price to employ security personnel. Also, drones are flexible in the needs of security elaboration; they can be easily redirected to other points of protection when necessary. CCTV cameras for instance, are fixed larger networks of cameras and people to monitor and maintain them, which are expensive to run. Drones on the other hand can fly on their own around the area and process data on the field in real time thereby cutting down on the need to deploy people to supervise the process and enhancing the efficiency of the surveillance (Cadet et al., 2024).

In addition, AI technology in drones can be combined with other technologies to improve its functionality. For example, drones can be integrated with large data analytical systems, which collect data from many sources to improve surveillance. They can also be linked to other security contexts as we have seen; for example, they can be combined with ambient motion detectors or with alarm systems to form a comprehensive security web. Such integration helps in strengthening the security system since all the components of its framework are closely linked and well-coordinated (Ramadan et al., 2024). Therefore, with the introduction of AI technology in the drone's aircraft, security and surveillance have enhanced. For drones, machine learning, computer vision, and autonomous navigation systems provide capabilities for drones to fly independently, process the provided data, and do not require assistance to find the optimal way in different environments. Many advantages in comparison with traditional methods of surveillance: better visibility, operability, speed, and coverage of large or inaccessible areas. With the advancement of AI, the chance of AI bolstering alarmed drones with idealistic security solutions or possibly better, efficient, reliable surveillance than manned security will increase (Goruganthu et al., 2024).

Applications for AI-Powered Drones in Security and Surveillance

AI doubled drones have been used widely to increase security and surveillance system in different areas of industries. It has made them to be valuable instrument when it comes to enhancing safety and efficiency in different setting by autonomously collecting and analyzing data in real time. The foremost area where drones enforced with artificial intelligence are widely used is in the military and security sector. Drone surveillance is used to guard borders, and this is especially frequent where the territory is large or difficult for humans to cover (Illiashenko et al., 2023). These drones with inbuilt AI camera and sensors have capabilities to track and identify intruder or any unauthorized vehicle or person trying to cross borders. Since drones do not require human control, it means that they can search for specific patterns, for instance movement and heat on their own, hence they are able to inform the authorities when there is a breach, thus a fast action is taken (Mohamed et al., 2023). Moreover, the use of drones with machine learning can enhance the technology's functionality to recognize threats in real-time and therefore, grow constantly in the capability to detect unauthorized intrusions. This capability is especially relevant where large and unmanned territories complicate conventional monitoring techniques, including cameras or patrols (Joshi et al., 2024).

Besides border surveillance, drones also apply AI for urban security to increase safety in cities. As the level of the complexity created in cities rises the protection of the citizens and public areas is becoming a more difficult problem. With computer vision, and other AI technologies, drones can constantly and effectively monitor vast regions comprising city roads, events, or key installations. These drones are capable of pinpointing abnormal such as people staying in prohibited zones or car that move along the wrong direction in a one-way street and notify law enforcement to take necessary actions (Mohamed et al., 2023). The drones can also patrol specific areas without supervision, which also enhances surveillance while at the same time cutting on supervision costs. Thus, the UAVs equipped with artificial intelligence are used for traffic management, and traffic flow tracking, as well as for identifying events, such as an accident or congestion, which may need further investigation (Bekmurzaeva et al., 2024). Feeds and data from drones enable reconstructions of environments thereby enhancing better traffic flow and reaction time making the existing urban set up safer and efficient. Drones also assist in public security during occasions such as politically charged events with an eye on population density and its behavior. The AI algorithms can monitor patterns and recognize risks as well as signal the security teams to act to prevent dangerous incidents (Illiashenko et al., 2023).

Security in Industries has also received the advantage of using AI or Intelligent drones, especially in the protection of key assets. In energy, manufacturing, and transportation industries, protecting extensive structures, pipelines, or power plants always remains a big concern. Security forces have started deploying unmanned air vehicles over these large tracts of land and these aircrafts scan for incidences of any breach of security which may include intrusion or failure of equipment. For example, drones can be used to optimize exploration and production activities in the oil and gas industry because they can be fitted with thermal sensors and machine learning algorithms to scan pipelines for gas leakage or fire risks in advance of an accident or fire occurs (Salama et al., 2023). Likewise, it used the use of drones in identification of irregularities, structural flaws as well as assessment of damage at power plants independently in real time as opposed to the manual approaches used. Drones can afford large size and frequently inaccessible spaces to be investigated without requiring man personnel to be presence on the site and are more secure, and operationally more efficient thus allowing extended surveillance and maintenance activities with less (Illiashenko et al., 2023).

Other important application of intelligent drones is in the area of disaster management or response services. During calamities like disasters, drone is an essential equipment in the search and rescue missions. By using AI-driven drones, it is possible to make fast aerial observations of areas affected by the disaster and collect both photographic images and thermal images to identify victims or evaluate the extent of the destruction (Bekmurzaeva et al., 2024). These drones can survey large regions at much the same speed as human crews on the ground, zeroing in on the areas of interest and allow the emergency services to concentrate on the critical parts of the disaster. During rescue operations, drones for example can fly into areas that could have been rendered inaccessible such as buildings that have collapsed, or areas surrounded by forests or floods. The opportunity to send and receive actual time data – GPS coordinates, live video stream – helps to make the right choices and act more quickly. Also, it is stated that drones can perform sign of life with the help of their AI technologies and identify heat signatures of the people alive trapped under the rubble, which enables the faster process of the rescue (Salama et al., 2023).

AI-enabled drones are also essential for capture and reporting of the status on ground during disaster like hurricanes, wildfire or even earthquakes. In the above situations, the emergency responders should have timely, if not real-time information to help them comprehend the scale of the disaster and to mobilize resources (Nguyen et al., 2024). AI crops sensitive equipment including sensors and cameras on Unmanned Aerial Vehicles UAVs that can survey disaster affected areas for better understanding. They can detect situation like damage to structures, water levels or fire outbreaks and feedback such details to command posts in real sense (Bekmurzaeva et al., 2024). This data is especially valuable for stream making decisions about exactly where to provide the relief and from where, how to evacuate populations and where to deliver supplies and in what quantity. Many attendees found that the capability of drones in providing frequent updates and working in unfavorable environment like a room full of smoke or high-rise building, make them a valuable asset as far as disaster response is concerned (Salama et al., 2023).

All in all, the new technology where Artificial Intelligence is in command of the drones is transforming the security and surveillance industry into many sectors and uses. Be it the improvement of the security along the borders, to dab + the safety of cities and towns, to + the protection of sensitive structures and in case of disaster support, these drones serve a numerous application (Nguyen et al., 2024). Drones can carry out complicated tasks on their own by combining AI technologies like machine learning, computer vision, and autonomous navigation. This allows for real-time insights and solutions to a range of security issues. The capabilities of AI-powered drones will surely be further improved with additional research and deployment, making them even more crucial instruments in the continuous efforts to protect infrastructure, property, and human life (Joshi et al., 2024).

Technical Components of AI-Powered Drones

Drones with AI capabilities are outfitted with a variety of technological components that allow them to carry out complex surveillance operations with a high level of efficiency and autonomy. These elements, which vary from sophisticated sensors and cameras to potent data processing systems, cooperate to offer danger detection, real-time monitoring, and decision-making capabilities (Nalini, 2024). What sets AI-powered drones apart from conventional drones is the integration of these elements, which allows them to offer useful security insights in addition to basic aerial observation (Kumar & Altalbe, 2024). The sensors and cameras that AI-powered drones employ to collect data while in flight are among their most important technological elements. Depending on the surveillance requirements, these drones can include a variety of sensors, but some of the more popular ones are thermal, infrared, and LiDAR sensors. Drones can identify warm objects or people even in total darkness or low visibility situations thanks to thermal sensors, which can identify heat signatures. This feature is very helpful for spotting illegal activity, including people crossing borders at night or trespassers on important infrastructure locations (Kumar & Altalbe, 2024). Like this, infrared sensors detect temperature changes, but they are especially good at seeing heat signatures through smoke, fog, and other environmental barriers. In contrast, LiDAR (Light Detection and Ranging) sensors use laser pulses to detect distances and produce intricate three-dimensional maps of their surroundings. LiDAR is particularly helpful for keeping an eye on sizable industrial sites and scanning regions for potential security risks or changes in the landscape. Even in difficult or complicated surroundings, these sensors provide drones with a situational awareness level that is essential for spotting any threats (Zhou et al., 2024).

High-resolution cameras are another essential part of AI-powered drones, in addition to specialized sensors. These cameras enable drones to take sharp pictures and videos, which are necessary for in-depth surveillance operations (Nalini, 2024). Drones with high-resolution cameras may capture sharper, more detailed photos than conventional cameras or security systems because they can track moving targets or zoom in on far-off things. Real-time processing of the video recorded by these cameras enables security staff to act quickly on the information they receive. For example, the camera can give live feeds to assist identify possible disruptions or track the movement of individuals during crowd surveillance or public events. High-resolution cameras can recognize people, cars, and certain behaviors when paired with AI-driven object identification and tracking algorithms, which further improves the drone's high degree of accuracy in identifying security concerns (Arya et al., 2023). Real-time data processing is necessary once the drone collects data to guarantee prompt decision-making. Drones with AI capabilities have onboard computers that can process enormous volumes of data produced by their cameras and sensors. Drones can respond nearly instantly to possible threats or changing conditions thanks to these processors, which enable them to assess data as it is acquired. Real-time data processing is especially crucial for surveillance applications where prompt reactions might be vital. For instance, in the context of border protection, a drone might instantly notify security staff of an unlawful entry or an odd movement pattern, enabling them to respond promptly. Drones may function more effectively, and only pertinent, processed data is provided when real-time data processing is used to minimize the quantity of data that must be sent to a central control system (Zhou et al., 2024).

Drones' efficiency in security and surveillance activities is further increased by AI-driven analytics. These analytics algorithms look for trends in the data that might point to a possible danger using machine learning models. For example, the AI may be able to identify suspicious activity, such as driving strangely or someone lingering in a prohibited area. Over time, the AI system can become more adept at spotting possible dangers by continuously learning from the data it gathers and adjusting to new trends and shifting circumstances. Drones can proactively scan high-risk locations and foresee possible security breaches before they happen thanks to AI-driven analytics that can also be used to predict risks based on past data. Security teams' capacity to react to problems more swiftly and efficiently is greatly improved by these predictive capabilities (Shah et al., 2024).

The ability of AI-powered drones to navigate and make decisions on their own is another essential feature. Artificial intelligence (AI)-enabled drones can fly on their own without continual human supervision. Drones can explore complicated terrain and avoid obstacles thanks to sophisticated path-finding algorithms, which enable this autonomy. To determine the optimum route to its destination, pathfinding algorithms consider variables such the drone's present position, potential obstructions in its flight path, and the surrounding environment (Zhou et al., 2024). This is particularly crucial in settings like construction sites, woodlands, or densely populated metropolitan regions where the landscape is always shifting or where unforeseen impediments may arise. Drones use obstacle avoidance technologies in addition to pathfinding to make sure they don't run into anything while in flight. The drone may take evasive action if needed thanks to these technologies, which employ sensors and cameras to identify obstructions in its route. To ensure safe and effective navigation, a drone can autonomously alter its flight route to avoid obstacles like trees or buildings. Obstacle avoidance is especially important in industrial security, where drones must maneuver through big, crowded facilities, and urban surveillance, where they must function in intricate, densely populated situations (Shah et al., 2024). Drones with AI capabilities also make judgments on their own using their navigation systems and the data they gather. Drones may change their behavior in real time based on the situation thanks to their decision-making capabilities. For instance, a drone may autonomously alter its flight route to follow an intruder and record more video if it finds them in a prohibited area. Like this, a drone can automatically change its route or return to a safe area if it comes across hazardous environmental conditions like high winds or a possible collision risk (Arya et al., 2023).

AI-powered drones are an effective tool for improving security and surveillance in dynamic and unpredictable situations because of their capacity to make judgments in real time. To sum up, the technological elements of AI-powered drones—such as powerful sensors and cameras, real-time data processing, and autonomous navigation—combine to allow drones to carry out complex security and surveillance missions. These drones are transforming the security industry by processing data, identifying dangers, and navigating complicated situations on their own. Compared to conventional approaches, they provide faster, more accurate, and more efficient solutions (Kumar & Altalbe, 2024).

CHALLENGES AND LIMITATIONS

AI powered drones have slowly been adopted in different sectors, doing away with traditional methods in security, surveillance and response systems the world over. In military application, drones governed by artificial intelligence have been central in the current warfare and defense frameworks. The following are blessed with advanced technologies that include sensing, analyzing data and threatening, and the operational ability to work independently. For example, an AI system can study a landscape and recognize irregularities or attacks by the enemy, which enhances storage and organizational performance while minimizing danger to people. A prime example is the application of machine intelligence aloft the unmanned aerial vehicles, specifically by the military forces of the United States with an operational theater in the Middle East. These drones can move around independently, identify and follow the movement of the foe and deliver this vital information to commanders without much interference by human personnel. AI inclusion leads to quick decision making since the appropriate data is fed into the system as a way of boosting the military capability and avoiding human mistake (Kumar et al., 2024).

Autonomous drones have been deployed in the policing services, and the results have also shown that artificial intelligence has helped to improve the policing situation and make the services better. For instance, application of drones for managing the crowds and monitoring people during occasions such as sporting activities, festivals, among others, has today become almost rampant (Ahmed & Echi, 2021). Police forces in cities, especially Dubai, utilize drones to cover crowded places, follow criminals and respond to an incident in the same area in real-time. Some of the drones include an AI-based facial recognition system and motion detectors to capture and follow suspects and relieve the police departments' reaction time. In the US, drones have been used for search and rescue operations in a way that AI aids in the enhancement of aerial pictures in search of a missing person. In some of the zones, the AI drones are also used for traffic police related issues such as implementation of traffic rules and regulation, traffic accident indications, and congested traffic zones. This policing using AI is not only efficient but also more effective in maintaining the safety of the public since the AI powers the real-time surveillance (Kumar et al., 2024).

The private sector also sees that application of the concept of artificial intelligence infused drones is indispensable for corporate security and surveillance. Some industries, including oil and gas platforms, power stations, and vast industrial facilities, now use drones to watch and secure their property. For example, autonomous drones in industries are employed to monitor expansive areas of manufacturing units through constantly monitoring the scenes to capture security threats (Ahmed & Echi, 2021).

Drones are used in oil and gas industries involving pipeline infrastructure for the purpose of naturally assessing their conditions for corrosion, leakage, or even intrusion. This area explains that, with the aid of AI, imaging data pulled from drones' cameras and sensors allows organizations to address maintenance concerns before they escalate (Suryanarayana, 2023). In retail sectors, these robots have been used for surveillance of large distribution centers to control stock and prevent acts of theft. AI's adoption also maintains a secure environment for employees and significant corporate properties while lowering costs since operations such as, surveillance consume a lot of manpower. Enterprises can extend the constant security surveillance while minimizing the human factors leading to more effective security management (Ahmed & Echi, 2021).

Other uses of self-flying drones include the management of disaster incidents where service is immensely offered to the needy at appropriate interval. Thermal imaging cameras, infrared cameras and drones that also can receive real time data have been used in searching for people trapped in areas affected by disasters such as earthquakes, hurricanes, and wildfires. After the natural disasters that took place in Puerto Rico in 2017 in the form of hurricanes, drones were applied much in the doing of the assessment of the destruction, infrastructure and this which needed humanitarian aid (Suryanarayana, 2023). The data arriving from drones was analyzed by AI algorithms, which helped to quickly determine the location of potential victims under the rubble, as well as find out which facilities are critical and damaged, and which regions are the most relevant for rescuers to work in. In California, for instance, artificial intelligence-based drones conducted surveillance on fire fronts and how the flames responded to changes in wind directions that are critical in planning evacuation and in managing aerial assaults on the fires. The real-time data processing procedures of the AI drones applied to the assessment of disaster scenes have had significant impacts on rescue activities since it enables emergency response teams to make the right decisions within the shortest time possible (Ahmed & Echi, 2021).

The following case studies are a representation of versatility and potential of using AI-powered drones in different fields. In the military, they improve surveillance and reconnaissance functions, giving real-time information with little danger to human personnel. In policing, there is the benefit of using drones for crime prevention and ways in which AI can act proactively by processing events and developing countermeasures while they occur. We have seen the private sector using ai drones in enhancing corporate security and guarding of key installations/property and this also at reduced cost. Finally, in disaster management, AI-specific drones should participate in the search-and-rescue operation to quickly identify unique and important data necessary to minimize the effects of natural disasters and save people's lives. These real-life examples show that AI integrated drones would be a necessity in improving the dynamics of security, surveillance and rescue operations throughout the world (Suryanarayana, 2023).

CASE STUDIES AND REAL-WORLD EXAMPLES

Smart drones have become significant assets in different sectors, helping the way security and surveillance as well as responses are conducted. Use of the AI drone in military is progressive as is evident from various military forces and strategies all over the world. These are surveillance and reconnaissance drones, used also for tactical and combat support, collecting data in a live fight without endangering the human lives (Adil et al., 2023). An example of the AI application is the use of military drones by the United States military in the Middle Eastern theater. These drones can self-fly through such terrains, pursue enemy formations, then send useful data back to headquarters. In addition to being controlled via AI, the drones can also process data and give instant recommendations on threats. The adoption of AI technology in the military-use drones has revolutionized the fast and accurate decision that has made operational effectiveness to improve, and finally less human input or loss of lives (Laghari et al., 2024).

Currently in the police force, AI driven drones have been successfully utilized in the enhancement of police and security work. These types of drones are specifically useful when it comes to surveillance of large groups of people, and security during events such as protests or sporting activities. For instance, the latest policing technologies reveal that police in such cities as Dubai uses drones that apply artificial neural network-based face recognition to track suspects and or criminals in real time (Adil et al., 2023).

The drones can fly on their own, process videos and sound alarms when there is an unusual movement detected. Also, drones are used for search and rescue and AI aids in the analysis of Search and Rescue aerial imagery to find lost persons. They are also used to monitor traffic, and to enable police departments to have real time information about accidents, traffic offences and congestion. This preventive measure increases security by having officers ready to act as necessary when an event occurs. This has been adopted by the private sector too especially in corporate space and security a surveillance measures (Laghari et al., 2024). Some examples of the use of the drones are Security; Large companies and industrial facilities, power plants among others have opted to use drones to inspect their properties and ensure that they are secure. They can move through large areas on their own and capture activity and security violations at construction sites. For instance, in the energy industry, the drones are used to scan the pipeline and the power lines for any signs of wear out or intrusion (Adil et al., 2023).

The footage recorded by the drone includes illustrations of dysfunctional parts that need preservation; through the analysis of cameras and sensors, AI algorithms demonstrate the presence of problems and their development as a way not to bring down drones on account of malfunctioning and jeopardizing equipment and people's lives. In the retailing and logistics, drone is utilized in the management of stocks, in addition to monitoring the fencing of the distribution centers, thus, improving the security of the property, in addition to increasing efficiency in their functioning (Laghari et al., 2024). Thus, the drones are also used in managing disasters, especially those that occurred in the recent past, with the support of artificial intelligence technology. These are unmanned aerial vehicles that have been employed where there are disasters; search and rescue involve supplying information to those on the ground. Hurricane, earthquakes, tsunamis, and major ship accidents that covered and destroyed several districts may also be explored by remote-controlled helicopter or airplanes with cameras, thermographic scanners or infrared of higher definition to detect survivors or to evaluate the damage. For instance, in the case of the hurricane Maria in Puerto Rico, drones were used to survey the affected regions, check the status of the roads for the need of help, or other such purposes. Likewise in California where wildfires have been ramparted, AI drones have been applied in following the pattern of fire, locating hotspot and weather conditions to help the firefighters make the right decisions. The application of autonomous drones in compiling and analyzing data in real-time has been a major boon to disaster response teams and following swift and accurate actions (Adil et al., 2023). These cases demonstrate how AI-drone technology can be adopted across the different fields. Used in military, police work, private security, and natural disaster response, drones with AI applications are changing the face of security and surveillance industries. Consequently, with passing time, it is possible to avail new and improved technologies in utilization of AI operated drones for elaborate enhancement of these industries at a larger scale with increased results (Laghari et al., 2024).

FUTURE TRENDS AND INNOVATIONS

Drones that are being operated by artificial intelligence has a tremendous prospect in the future due to the ever-velocity enhancements being made in the artificial intelligence and drone. New and more sophisticated levels of learning and improved computer vision systems are expected to greatly improve the capabilities of drones. These technologies assist drones to improve their abilities to evaluate situations, conflicts and risks that they meet and make decision making without the help of humans. For example, AI models are gaining increased capabilities at detecting patterns on the data to allow drones to detect and evaluate the probabilities of an event such as movement or heat with increased accuracy (Tallapaneni et al., 2024).

This evolution will soon open the way of completely autonomous surveillance drones that would not require further human intervention. In the future such drones may be used for long-term, risky missions like surveillance along borders, surveillance of cities, surveillance and response to disasters, etc., all while carrying out tasks that can be very risky for human life or require too many resources (LNC Prakash et al., 2023). Future development is expected to reveal the integration of drones with other technologies that are also developing now. There is much to explore in creating synergy between drones and the IoT, 5G connectivity options, and smart cities. The IoT makes it possible for drone to connect and convey information to other devices, leading to an interconnectivity and actionable environment. For instance, drones could gather and send data gathered from the environment to environmental Internet of Things devices in smart cities to give real information on traffic, air quality, and other components influencing urban daily existence (Tallapaneni et al., 2024). Moreover, the use of fifth generation technology will foster the effectiveness of drones through improvement of the rate in transmitting information in their operations, especially in surveillance and decision-making. Being associated with low latency, 5G enables AI drones to share data with control centers instantly so that it is possible to provide an instant response to security or emergency situations. It will mean that drones will be able to connect to several sensors and systems, thus making urban and industrial spaces far smarter, improving their efficiency (Nalini, 2024).

Another potential is the creation of the groups of the drones AI controlled, working together. The last idea is essentially the union of several drones performing certain tasks in groups and squadrons. These drones are preprogrammed to employ artificial intelligence algorithms that allow them to communicate and coordinate effectively to improve functional performance characteristics. For instance, a set of drones might be employed to scout a vast manufacturing compound, and each drone would control a specific quadrant to verify security incursion or assert an efficient surveillance area (LNC Prakash et al., 2023). Large swarms could also be deployed in Cases whereby surveillance of large areas or crowds is needed more effectively than would be done by a single drone. To add to this, collaborative drones can also relay their information in real time thus enhancing the overall competence of displaying dynamic analysis of events. This swarm intelligence is one of the many innovations today in the AI drone that is promising as it aims to provide better operation with better redundancy, scalability and efficiency (Tallapaneni et al., 2024). Nevertheless, the position of this technology in the growth of ethical and social concerns will grow with the growth of technology. The inclusion of drones in the public and private security national security systems raises concerns on privacy among the public. Steady surveillance by drones poses future threats to privacy and civil liberties mainly because drones can capture massive data from people (Nalini, 2024).

There will always be disagreements on privacy with drones becoming a common commodity in both city and rural areas while questioning the difference between public security and privacy. The use of drones at protests or in people's compounds may be seen as an invasion of privacy if some privacy concerns are not addressed to the public's positive reception. This will hence call for proper regulation and clear polices to enable security needs just like privacy rights of the citizens (LNC Prakash et al., 2023). Moreover, we find that the ethics of AI are expected to play a more significant part in the use of drones as more of the systems turn into drones, including them taking sole responsibility of decision making. Ethical consideration in the deployment of AI controlled drones includes issues to do with authority and disclosure of activities. For example, if an unmanned vehicle with AI being the controller takes wrong decisions or becomes destructive, who is held liable for such a decision? In what ways are AI systems going to be capable of bringing fairness into the sorting of the data and avoiding biases in security and surveillance? The improved advancement in artificial intelligence will be mandatory to establish such ethical standards to reign in the use of drones in the most appropriate manner. The questions about how drones should be built and operated ethically will follow AI and continue to contribute to the defining contours of a society capable of leveraging AI for the optimization of drone capacities in a responsible manner (Nalini, 2024).

CONCLUSION

Thus, in this chapter, we have subjected the use of AI-powered drones in security and surveillance with a view to assessing its impact. The coupling of AI with drones has greatly improved drone operations whereby the drones can work independently, and even analyze big data at real time and make smart decisions with little human input. Starting their journey as tools during wars and conflicts, AI drones seem to gain increasingly significant importance in modern security frameworks shaping polices and responding to various threats ranging from natural disasters to security threats in cities and industries. In this work, we have described how these drones make use of innovative technologies like machine learning, computer vision and autonomous navigation to increase efficiency and effectiveness, providing real time monitoring and surveillance, and fast reaction to the growth of threats.

We have also discussed many potential applications such as the border security, surveillance of city, monitoring the industry and in emergencies. It introduces AI drones as approaches to security within these sectors that not only are speedier but are also more precise and smear in terms of scale than conventional methods. In border control for instance, drones are capable of patrolling extensive and physically challenging terrains while independently detecting and following cross-border incursions. In urban environments, civil AI drones thus improve on security providing real-time surveillance in addition to traffic control, crowd supervision, as well as combating crime. The private sector of this industry has applied drones for security purposes including guard 24/7 of the core infrastructure, surveillance and security of industrial premises among others. Second, in disaster response, AI drones act in search and rescue and provide valuable information to manage natural disasters effectively.

The actual technical attributes of AI enabled drones such as high precision sensors, real-time data analysis, and self-navigation have been crucial in making such systems a success. They have deemed these drones fitted with thermal, infrared and LiDAR sensors able to capture high quality imagery while following the contours of the land and identify other strange features. Their capabilities are complemented further with the help of AI analytics as drones are also capable of processing data independently. The technologies of self-organized movement and avoidance of the obstacles enable these drones to operate in rather flexible manners in different security and surveillance situations and circumstances.

However, as with many things that involve leading edge technologies involving Artificial Intelligence, these advantages come with several drawbacks and limitations. Privacy related and some ethical issues are obvious and still prevail especially as the drone monitoring increases in civil spaces. This issue of a trade-off between security and the right to privacy is a constant debate and policy relevant problem. Still, technically, drones have several constraints such as battery capacity, data protection issues or susceptibility to cyber threats. Entrance and legal barriers constitute another challenge that must be met to enable the proper application of drones, airspace regulations for example and compliance with the laws of other nations.

Therefore, AI incorporated drones are the most progressive increase in the way security and surveillance are done. The autonomous mobility as well as the ways they can gather data and assess the situation in real time while responding to threats quickly make them capable of revolutionizing several industries starting from military and police forces up to the rescue services and corporate security. In this regard, it is essential that the application of the technology subsequent and ethical while paying particular attention to privacy concerns while in use. It is necessary and possible to find a balance between progress and safety to make AI drones an important part of security further development.

REFERENCES

Adil, M., Song, H., Mastorakis, S., Abulkasim, H., Farouk, A., & Jin, Z. (2023). UAV-assisted IoT applications, cybersecurity threats, AI-enabled solutions, open challenges with future research directions. *IEEE Transactions on Intelligent Vehicles*, *9*(4), 4583–4605. DOI: 10.1109/TIV.2023.3309548

Ahmed, A. A., & Echi, M. (2021). Hawk-eye: An AI-powered threat detector for intelligent surveillance cameras. *IEEE Access : Practical Innovations, Open Solutions*, *9*, 63283–63293. DOI: 10.1109/ACCESS.2021.3074319

Arya, L., Sharma, Y. K., Padmanaban, H., Devi, S., & Kumar, R. (2023, October). Eyes in the sky: Safeguarding borders security with AI-powered aerial monitoring and IoT integration. In *The International Conference on Recent Innovations in Computing* (pp. 863-873). Springer Nature Singapore.

Bekmurzaeva, R., Kalimullin, R., & Aguzarova, F. (2024). Application of drones and artificial intelligence to monitor and protect natural ecosystems. In *BIO Web of Conferences* (Vol. 140, p. 01008). EDP Sciences. DOI: 10.1051/bioconf/202414001008

Cadet, E., Osundare, O. S., Ekpobimi, H. O., Samira, Z., & Wondaferew, Y. (2024). AI-powered threat detection in surveillance systems: A real-time data processing framework.

Farsath, K. R., Jitha, K., Marwan, V. M., Jouhar, A. M. A., Farseen, K. M., & Musrifa, K. A. (2024, March). AI-enhanced unmanned aerial vehicles for search and rescue operations. In *2024 5th International Conference on Innovative Trends in Information Technology (ICITIIT)* (pp. 1-10). IEEE.

Goruganthu, T. S., Marepalli, H., Bheemineni, S., Bobba, R. C., Wang, D., Richhariya, S., & Liu, J. (2024). Building an AI-based surveillance drone cloud platform. In *Proc. of the IEEE International Workshop on Future Intelligent Technologies for Young Researchers (FITYR)* (pp. 1-10). DOI: 10.1109/FITYR63263.2024.00008

Illiashenko, O., Kharchenko, V., Babeshko, I., Fesenko, H., & Di Giandomenico, F. (2023). Security-informed safety analysis of autonomous transport systems considering AI-powered cyberattacks and protection. *Entropy (Basel, Switzerland)*, *25*(8), 1123. DOI: 10.3390/e25081123 PMID: 37628153

Jain, N., Gambhir, A., & Pandey, M. (2024). Unmanned aerial networks—UAVs and AI. In *Recent Trends in Artificial Intelligence Towards a Smart World: Applications in Industries and Sectors* (pp. 321-351). Springer Nature Singapore.

Joshi, A., Spilbergs, A., & Miķelsone, E. (2024, June). AI-enabled drone autonomous navigation and decision-making for defence security. In *ENVIRONMENT. TECHNOLOGIES. RESOURCES.Proceedings of the International Scientific and Practical Conference* (Vol. 4, pp. 138-143). DOI: 10.17770/etr2024vol4.8237

Kumar, A., Tallapaneni, K., Yadav, A., & Gautam, H. (2024). Enhancing home security with AI-driven drone surveillance. *J Sen Net Data Comm, 4*(2), 01-10.

Kumar, G., & Altalbe, A. (2024). Artificial intelligence (AI) advancements for transportation security: In-depth insights into electric and aerial vehicle systems. *Environment, Development and Sustainability*, •••, 1–51. DOI: 10.1007/s10668-023-04127-7

Laghari, A. A., Jumani, A. K., Laghari, R. A., Li, H., Karim, S., & Khan, A. A. (2024). *Unmanned aerial vehicles advances in object detection and communication security review*. Cognitive Robotics. DOI: 10.1016/j.cogr.2024.07.002

Maguluri, L. P., Arularasan, A. N., & Boopathi, S. (2023). Assessing security concerns for AI-based drones in smart cities. In *Effective AI, Blockchain, and E-Governance Applications for Knowledge Discovery and Management* (pp. 27-47). IGI Global. DOI: 10.4018/978-1-6684-9151-5.ch002

Mallikarjunaradhya, V., Pothukuchi, A. S., & Kota, L. V. (2023). An overview of the strategic advantages of AI-powered threat intelligence in the cloud. *Journal of Science and Technology, 4*(4), 1–12. DOI: 10.55662/JST.2023.4401

Mandloi, D., Arya, R., & Verma, A. K. (2024). Internet of drones. In *Recent Trends in Artificial Intelligence Towards a Smart World: Applications in Industries and Sectors* (pp. 353-373). Springer Nature Singapore. DOI: 10.1007/978-981-97-6790-8_13

Mohamed, B. (2024). AI-empowered flying ad-hoc networks for dynamic connectivity. *International Journal of Computing and Digital Systems, 15*(1), 167–177. DOI: 10.12785/ijcds/150114

Mohamed, N., Ahmed, A. A., Alsharif, A., & ElKhozondar, H. J. (2023, November). Employing AI-driven drones and advanced cyber penetration tools for breakthrough criminal network surveillance. In *2023 IEEE 9th International Women in Engineering (WIE)Conference on Electrical and Computer Engineering (WIECON-ECE)* (pp. 1-6). IEEE. DOI: 10.1109/WIECON-ECE60392.2023.10456514

Nalini, M. (2024). AI-powered drones for healthcare applications. In *Artificial Intelligence for Autonomous Vehicles* (pp. 131-149). DOI: 10.1002/9781119847656.ch6

Nguyen, V. T., Zhou, J., Dong, C., Cui, G., An, Q., & Vinnakota, S. (2024, June). Enhancing house inspections: UAVs integrated with LLMs for efficient AI-powered surveillance. In *2024 International Joint Conference on Neural Networks (IJCNN)* (pp. 1-8). IEEE. DOI: 10.1109/IJCNN60899.2024.10650902

Obiuto, N. C., Festus-Ikhuoria, I. C., Olajiga, O. K., & Adebayo, R. A. (2024). Reviewing the role of AI in drone technology and applications. *Computer Science & IT Research Journal*, *5*(4), 741–756. DOI: 10.51594/csitrj.v5i4.1019

Ramadan, M. N., Basmaji, T., Gad, A., Hamdan, H., Akgün, B. T., Ali, M. A., Alkhedher, M., & Ghazal, M. (2024). Towards early forest fire detection and prevention using AI-powered drones and the IoT. *Internet of Things : Engineering Cyber Physical Human Systems*, *27*, 101248. DOI: 10.1016/j.iot.2024.101248

Salama, R., Al-Turjman, F., & Culmone, R. (2023, March). AI-powered drone to address smart city security issues. In *International Conference on Advanced Information Networking and Applications* (pp. 292-300). Springer International Publishing. DOI: 10.1007/978-3-031-28694-0_27

Shah, I. A., Jhanjhi, N. Z., & Brohi, S. N. (2024). Use of AI-based drones in smart cities. In *Cybersecurity Issues and Challenges in the Drone Industry* (pp. 362–380). IGI Global. DOI: 10.4018/979-8-3693-0774-8.ch015

Suryanarayana, G. (2023). AI applications of drones. In *Drone Technology* (p. 153). Future Trends and Practical Applications.

Tallapaneni, K., Kumar, A., Gautam, H., Yadav, A., Kumar, A., & Yadav, H. (2024). Enhancing home security with AI-driven drone surveillance. *Authorea Preprints*. DOI: 10.36227/techrxiv.171665618.82092467/v1

Tariq, M. U. (2024). Harnessing Persuasive Technologies for Enhanced Learner Engagement and Motivation. In M. Sanmugam, D. Lim, N. Mohd Barkhaya, W. Wan Yahaya, & Z. Khlaif (Eds.), *Power of Persuasive Educational Technologies in Enhancing Learning* (pp. 30-62). IGI Global. https://doi.org/DOI: 10.4018/979-8-3693-6397-3.ch002

Tariq, M. U. (2024). Smart Transportation Systems: Paving the Way for Sustainable Urban Mobility. In Munuhwa, S. (Ed.), *Contemporary Solutions for Sustainable Transportation Practices* (pp. 254–283). IGI Global., DOI: 10.4018/979-8-3693-3755-4.ch010

Tariq, M. U. (2025). Navigating the Ethical Frontier-Human Oversight in AI-Driven Decision-Making System. In Hai-Jew, S. (Ed.), *Enhancing Automated Decision-Making Through AI* (pp. 425–448). IGI Global Scientific Publishing., DOI: 10.4018/979-8-3693-6230-3.ch013

Tariq, M. U. (2025). Enhancing Pandemic Preparedness With AI-Driven Risk Prediction Models. In Shah, I., Sial, Q., & Fateh, S. (Eds.), *Generative AI Techniques for Sustainability in Healthcare Security* (pp. 77–94). IGI Global Scientific Publishing., DOI: 10.4018/979-8-3693-6577-9.ch005

Tariq, M. U., & Sergio, R. P. (2025). Ethical Considerations and Social Impact of AI and Computational Intelligence in Society 5.0. In Pandey, R., Srivastava, N., Prasad, R., Prasad, J., & Garcia, M. (Eds.), *Open AI and Computational Intelligence for Society 5.0* (pp. 225–248). IGI Global Scientific Publishing., DOI: 10.4018/979-8-3693-4326-5.ch010

Tlili, F., Ayed, S., & Fourati, L. C. (2024). Advancing UAV security with artificial intelligence: A comprehensive survey of techniques and future directions. *Internet of Things : Engineering Cyber Physical Human Systems*, *27*, 101281. DOI: 10.1016/j.iot.2024.101281

Veprytska, O., & Kharchenko, V. (2024). Analysis of AI-powered attacks and protection of UAV assets: Quality model-based assessing cybersecurity of mobile systems for demining. In *IntelITSIS* (pp. 356-374).

KEY TERMS AND DEFINITIONS

Artificial Intelligence (AI): The simulation of human intelligence in machines programmed to perform tasks such as learning, reasoning, and problem-solving.

Unmanned Aerial Vehicle (UAV): A drone or aircraft that operates without a human pilot on board, controlled remotely or autonomously.

Computer Vision: A field of AI enabling machines to interpret and process visual data from the environment for tasks like object detection and recognition.

Autonomous Navigation: The ability of drones to self-direct and adapt to their environment without human intervention using AI algorithms.

IoT (Internet of Things): A network of interconnected devices that communicate and exchange data, often used to enhance drone capabilities in smart environments.

Chapter 6
Fundamentals of Drone Navigation and the Role of Computer Vision

Siva Raja Sindiramutty
https://orcid.org/0009-0006-0310-8721
Taylor's University, Malaysia

N. Z. Jhanjhi
https://orcid.org/0000-0001-8116-4733
Taylor's University, Malaysia

Wei Wei Goh
https://orcid.org/0000-0002-8577-8481
Taylor's University, Malaysia

ABSTRACT

Drone navigation is based on precise navigation for efficient and secure performance for delivery, surveillance, or rescue. Traditional navigation based on GPS, inertial measurement units, and magnetometers provides good guidance but is inefficient in conditions with weakened signals or unpredictable obstacles. Computer vision is changing this. By equipping drones to perceive and understand visual information about their surrounding space, it makes decision-making independent, allows for better navigation past obstacles, and builds real-time maps. Object detection, optical flow, and SLAM are some techniques being applied in aerial robotics today. Vision is complemented and enhanced when combined with other sensors like LiDAR and GPS, making navigation feasible in complex terrains. However, processing high volumes of visual data in real-time remains a challenge. Advances in edge computing and AI-driven perception are helping to overcome these limitations, bringing faster

DOI: 10.4018/979-8-3693-8497-8.ch006

and more efficient onboard processing.

1. INTRODUCTION

Overview of Drone Navigation and Its Significance

The drones have altered the way humans interact with the sky. Navigation is among the most vital things in making this functionality operate because they fly with purpose and precision. Whether deployed to deliver products or survey remote locations, GPS, computer vision, and artificial intelligence power their motion (Akbari et al., 2021; Dass et al., 2025). When in urban locations, drones have to manoeuvre around power lines, constructions, and other aerial objects (Foudeh et al., 2021). Sensor fusion in collaboration with machine learning helps them decide in an instant. Some years back, a drone was used to test delivering medicines in Africa (Nyaaba & Ayamga, 2021). It had to travel through unstable winds and rocky landscapes to demonstrate that advanced navigation is not just an engineering feat—it can be life or death, Farmers utilize drones in agricultural production to scout over lands to obtain just the perfect amount of water and fertilizer for each plot (Krishna, 2017; Hussain et al., 2024). Drone navigation is not only transferable to uses in aviation. However, issues remain. GPS is not reliable in tunnels or forests where canyoning is done, so drones must resort to cameras and onboard sensors (Grenier et al., 2023; Jun et al., 2024). Bad weather, particularly windy conditions, is yet another challenge. Engineers keep working to hone systems to gain better precision, safety, and reliability. Near-future drones can expect to fly by making decisions in ways like humans, making them increasingly vital in ordinary uses. Figure 1 shows significant of drone technology.

Figure 1. Significant of drone technology

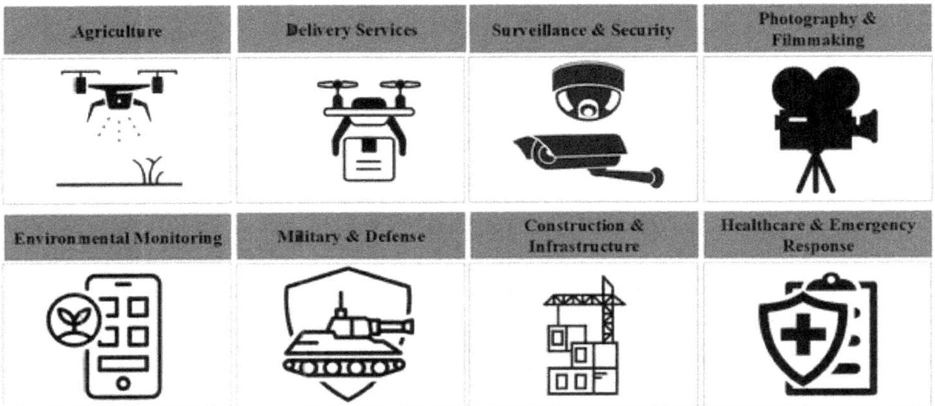

Evolution of Drone Navigation Technologies

Recently, drones used to be remote-control tools. Their motion was entirely reliant upon human input with little ability to act independently. Navigation systems have improved over time to render drones independent and efficient (Arafat et al., 2023). Earlier drones employed GPS to an extensive extent, yet this was not ideal where signals weren't powerful (Floreano & Wood, 2015). Engineers added additional tools in response to this challenge in the shape of LiDAR and cameras to empower drones to "observe" around them and spot impediments in time (Vishweshwaran & Sujatha, 2024; Kiyani et al., 2024). Artificial intelligence was the game-changer. Artificial intelligence through machine learning has been employed to have drones spot patterns, foresee impediments, and even decide independently in the absence of continuous guidance (Heidari et al., 2022). I came across an article about a drone navigating through dense forests by making use only of onboard cameras—it was trained through earlier flights to change the path by whatever "remained in memory" (Schedl et al., 2021; Linqiang et al., 2024). This sort of adaptable navigation is used in search-and-rescue work where drones map through destroyed buildings or rocky landscapes with minimal guidance (Dinelli et al., 2023). Still newer is swarm technology which has added a layer to drones. Drone swarms can now "communicate" with each other to change paths by each other's locations. This is helpful in military uses and in tracking changes in the environment where drones work in tandem to map vast expanses (Mohsan et al., 2023). As computing power expands, future drones can go about navigating with an exactness that is equal to that of human flyers to conduct even more helpful work in regular life.

Importance of Computer Vision in Modern UAVs

Modern drones rely on more than just GPS and basic sensors to navigate—they "see" their surroundings using computer vision (Thomas et al., 2024). This technology allows drones to detect objects, avoid obstacles, and even recognize specific targets. In crowded areas where GPS signals can be unreliable, cameras and AI-powered image processing help drones move safely without constant human control (Jain et al., 2024). A delivery drone, for instance, can identify roads, buildings, and landing spots, ensuring precise drop-offs (Kuru, 2021). One of the most exciting uses of computer vision is in search-and-rescue operations. Drones equipped with thermal cameras can spot missing people in forests or disaster zones, even at night (Tuśnio & Wróblewski, 2021). I once read about a hiker stranded in rough terrain—rescue teams struggling to locate him, but a drone using infrared imaging found him within minutes. This kind of rapid detection would be impossible without advanced visual processing (Khan et al., 2022; Svanstrom et al., 2021). In agriculture, drones scan large fields, detecting crop health issues through colour variations that aren't visible to the human eye (Shaikh et al., 2022). This not only saves farmers time but also improves efficiency by targeting specific areas for treatment. The ability to analyze environments in real time makes computer vision one of the most valuable tools in UAV technology, pushing the limits of what drones can do across different industries.

Chapter Contribution

This chapter is pertinent to the area of fundamentals of drone navigation and the role of computer vision. First, it describes the overview of drone navigation and its significance and the evolution of drone navigation technologies. The chapter then investigates the key components of a drone navigation system and the limitations of traditional navigation systems. Furthermore, the chapter presents why computer vision is crucial for UAV autonomy, and comparison between traditional sensor-based and vision-based navigation. It also discusses edge computing for real-time processing of sensor data and case studies: vision-based autonomous landing, object detection and recognition, optical flow analysis for motion estimation, SLAM (simultaneous localization and mapping, depth estimation using stereo vision and LiDar, lastly, the section provides a glimpse of environmental challenges: lighting, weather, dynamic obstacles, ai-powered real-time perception and decision-making, the role of deep learning in improving navigation accuracy.

Chapter Organization

In the first section, we will cover an insight into the introduction of drone navigation and its significance. In section 2, we go down to the basics of drone navigation. In section 3, we discuss an introduction to computer vision in drone navigation, followed by core computer vision techniques for drone navigation in section 4. Section 5 delves into sensor fusion: integrating vision with other navigation technologies. Section 6 will explain the challenges in vision-based drone navigation. The final section will discuss the future trends in computer vision for drone navigation.

2. BASICS OF DRONE NAVIGATION

Principles of Aerial Navigation

Flying a drone is not just about moving from one point to another—it requires a balance of stability, control, and precision (Okulski & Ławryńczuk, 2022). Aerial navigation follows a set of core principles that ensure smooth flight and accurate positioning (Chai et al., 2024). One of the most important is maintaining altitude. Drones use barometers and GPS to stay at the right height, but in places where these signals are weak, optical sensors help measure distance from the ground (Balestrieri et al., 2021). Another key factor is orientation. Without knowing its exact direction, a drone can drift off course or even crash. Gyroscopes and accelerometers work together to keep drones steady, especially in windy conditions (Kangunde et al., 2021). I have watched a drone struggle in strong winds in a demo—it constantly corrected for tilting automatically in a bid to level out. Such real-time correction is crucial for smooth flight. Obstacle avoiding is another critical ingredient in aerial navigation. The drones have LiDAR, ultrasonic sensors, and cameras for sensing and avoiding obstacles in mid-air (Krishnan & Murugappan, 2023). It is especially crucial in urban environments since buildings, foliage, and power lines offer a dense flight environment (You et al., 2021; Manchuri et al., 2024). Advanced drones have AI for anticipating movement in a bid to avoid collision before it is imminent (Soori et al., 2023). Communication also plays a role. Many drones use radio signals to receive instructions from a ground station, but when operating autonomously, they depend on onboard computing to make instant decisions (Tarekegn et al., 2022). As aerial navigation technology improves, drones are becoming safer and more reliable, allowing them to take on more complex tasks. Figure 2 shows principles of aerial navigation.

Figure 2. Principles of aerial navigation

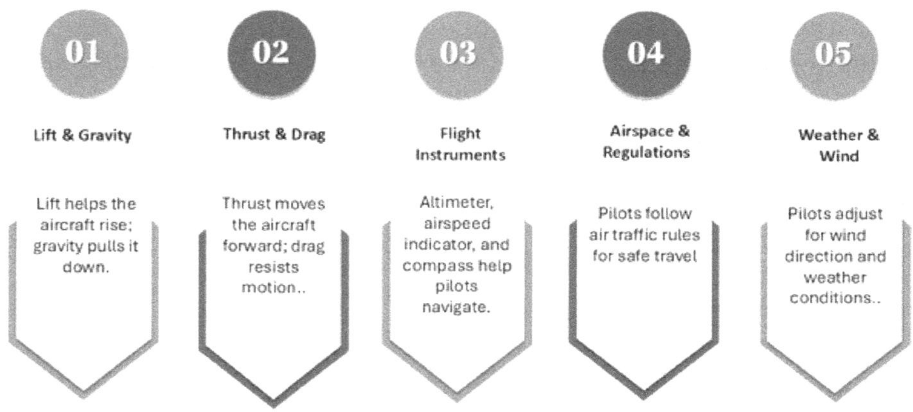

Key Components of a Drone Navigation System

A drone's ability to fly accurately depends on several key components working together (Singh et al., 2022). Without these, it would struggle to maintain direction, altitude, or stability. Among the most important parts of a drone's navigation system are the Inertial Measurement Unit (IMU), GPS and GNSS-based navigation, magnetometers, and barometers (Ali, 2024). The IMU acts as the drone's inner sense of balance. It contains accelerometers and gyroscopes that measure movement, tilt, and rotation. If a drone suddenly shifts due to wind, the IMU detects the change and helps correct its position (Al-Ahmadi, 2024; Ravichandran et al., 2024). During a flight test, I saw a drone stabilize itself mid-air after a sudden gust—it was impressive to watch how quickly it adjusted without any manual input.

GPS and GNSS navigation enable drones to position themselves with high accuracy (Talwandi et al., 2024). GPS depends on U.S. satellites, but GNSS (Global Navigation Satellite System) uses systems such as Europe's Galileo and Russia's GLONASS, with greater accuracy worldwide. Satellite signals in open environments are utilized in drones to maintain pre-programmed routes with minimal drift. Other positioning techniques are required in environments such as in tunnels or forests with heavy cover where signals are low (Koubaa et al., 2019; Riza et al., 2025).

Magnetometers are used as a compass for a drone to detect the Earth's magnetic field and assist with orientation (Javan et al., 2024). Without them, a drone may fly off course, particularly on long-distance flights (Andersen et al., 2024). Pilots usually pre-calibrate the magnetometer to prevent navigation errors. Barometers, on the other hand, measure air pressure to estimate altitude. This is helpful when GPS readings are unreliable or accuracy is required in low altitudes (Scriban et al., 2024). For instance, delivery drones use barometers to give a smooth descent to a desired height.

Collectively, all these elements add to drone stability, accuracy, and responsiveness. With technological advancements, enhancements in all these systems will increasingly improve drone reliability to enable them to perform tasks with greater accuracy in a range of conditions.

Limitations of Traditional Navigation Systems

Traditional drone navigation systems, while effective, have several limitations that can influence performance and reliability. Some systems are highly reliant on GPS, but GPS signals are not always available or reliable. Inside urban environments with high buildings, signals can reflect off surfaces and lead to positioning inaccuracies (Kumar, 2024; Seng et al., 2024). I have seen a drone struggle to land properly due to its GPS continuously adjusting position by several meters. In mission-critical uses like medical delivery, such inaccuracies can lead to serious complications. Another major challenge is environmental interference. GPS and radio signals are weakened in tunnels, forests, or underwater environments, and drone navigation is problematic (Qian et al., 2025; Sheng et al., 2024). In such environments, drones need to rely on alternative technologies such as LiDAR and computer vision, which are not always reliable in low-light or weather conditions (Liu et al., 2024). High winds can pose a challenge as well. Even with stabilization systems, gusts can interfere with balance and lead to drifting or crashes (Sehrawat et al., 2024). Another significant limiting aspect is battery life. Navigation systems require a continuous power source to process position data, adjust flying trajectories, and navigate around obstacles (Suanpang & Jamjuntr, 2024). However, drones have a restricted battery capacity, which restricts flying time—especially when using power-intensive sensors like LiDAR or high-resolution cameras (MahmoudZadeh et al., 2024; Sindiramutty, 2024). This is a serious concern for prolonged missions where constant recharging is inconvenient. Besides, traditional navigation systems lack the capacity to adapt to unforeseen changes. They have preprogrammed routes but fail to effectively cope with unforeseen obstacles. Even as new AI-based systems are being developed to address such issues, traditional systems still struggle with real-time decision-making in dynamic environments.

3. INTRODUCTION TO COMPUTER VISION IN DRONE NAVIGATION

What is Computer Vision?

Computer vision is the science of teaching machines to see and interpret images (Khang et al., 2024). It allows computers to recognize patterns, detect objects, and even understand motion in photos and videos (Javaid et al., 2024). The idea has been around for decades, but improvements in artificial intelligence have made it far more practical. One fascinating use of computer vision is in self-driving cars (Yu, 2024; Sindiramutty, Prabagaran, Akbar, Hussain, & Malik, 2024). These vehicles rely on cameras and sensors to detect road signs, pedestrians, and other cars. The technology must be precise because even a small mistake can lead to accidents. In medical fields, computer vision helps doctors analyze X-rays and MRI scans to spot diseases earlier (Lenka & Tripathy, 2024; Sindiramutty, Prabagaran, Jhanjhi, Ghazanfar, Malik, & Soomro, 2024). It has also changed how businesses operate. Retail stores use it for automated checkouts, and security systems rely on facial recognition (Ocaña et al., 2024; Wolniak et al., 2024; Sindiramutty, Prabagaran, Jhanjhi, Murugesan, Brohi, & Masud, 2024). Training a computer to see is not easy. Unlike humans, who instantly recognize a face in different lighting or angles, a machine must process thousands of examples before learning patterns (Amirgaliyev et al., 2025). The challenge is making the system accurate enough to work in real-world settings. Even small errors can cause major problems, whether in manufacturing or law enforcement.

As this field grows, its impact becomes more visible. From unlocking smartphones with a glance to monitoring traffic in smart cities, the ability to analyze images quickly is reshaping industries. The goal is not just to make machines smarter but to make life safer and more efficient. Figure 3 shows role of computer vision in drone

Figure 3. Role of computer vision in drone

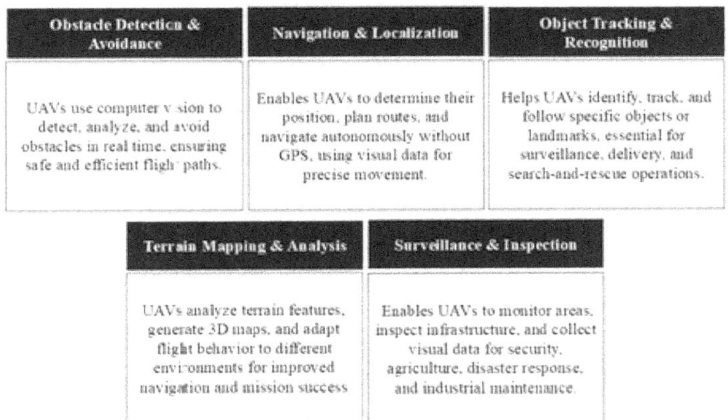

Why Computer Vision is Crucial for UAV Autonomy

Flying a drone manually is one thing, but making it fly on its own—without human control—needs something much smarter. That's where computer vision comes in. Without it, a UAV is just a flying machine that can't truly "see" the world. Imagine a drone trying to avoid obstacles or land safely without cameras and AI to process what's around it. That wouldn't end well. Drones rely on sensors like LiDAR and radar, but cameras provide richer detail (Gong et al., 2024). When a UAV navigates a crowded city or a dense forest, computer vision helps it recognize objects, track movement, and even predict how things will change in real time (Kadam et al., 2024). This is what allows delivery drones to avoid birds or shifting wind conditions mid-flight. Military applications also depend on this technology to identify targets with precision, reducing errors in complex missions (Kadam et al., 2024; Sindiramutty, Prabagaran, Jhanjhi, Murugesan, Brohi, & Wei, 2024), In agriculture, UAVs scan crops and detect diseases early using vision-based analysis, which helps farmers act fast (Agelli et al., 2024b). Search-and-rescue teams also rely on drones with thermal imaging to locate people lost in rough terrain. A personal experience comes to mind—seeing a drone locate a missing hiker in record time, something humans couldn't have done as quickly. The power of computer vision is pushing UAVs beyond simple automation. The moment a drone can understand its surroundings like a human pilot, it opens doors to safer, more efficient flights across industries (Bain & Blaney, 2025).

Comparison Between Traditional Sensor-Based and Vision-Based Navigation

Flying a drone or guiding a robot depends on how well it can understand its surroundings (Thangamani et al., 2024). Traditional sensor-based navigation uses LiDAR, GPS, and IMUs to measure distance and position. These sensors work well in controlled spaces but struggle when signals are weak or obstacles block them (He et al., 2024; Sindiramutty, Prabagaran, Jhanjhi, Murugesan, Malik, & Hussain, 2024). Vision-based navigation, on the other hand, relies on cameras and AI to interpret visual data. It mimics how humans see, making it useful for navigating complex areas without relying on external signals. A personal memory stands out—watching a GPS-guided drone fail under a bridge, while a vision-based drone adjusts itself instantly (Mohyuddin et al., 2024). This difference is why self-driving cars use cameras alongside traditional sensors. Vision-based systems process real-time images, allowing them to recognize road signs or pedestrians (Sarvajcz et al., 2024). In contrast, LiDAR struggles with reflective surfaces, and GPS loses accuracy in urban areas (Shaiba, 2020). Agriculture benefits from both methods. GPS helps tractors follow precise paths, but vision-based AI detects crop health (Vrochidou et al., 2022). The combination of both systems leads to safer and more efficient automation in various industries. Table 1 shows comparison of traditional sensor based navigation and vision based navigation.

Table 1. Comparison of traditional sensor based navigation and vision based navigation

Feature	Traditional Sensor-Based Navigation	Vision-Based Navigation
Main Sensors	GPS, LiDAR, IMU	Cameras, AI Processing
Accuracy	Affected by signal loss	Adapts using visual cues
Performance	Struggles due to reflection	Limited but can be enhanced with AI
Object Recognition	Limited to distance measurement	Identifies objects and movement
Cost	High due to specialized sensors	Lower with consumer-grade cameras

4. CORE COMPUTER VISION TECHNIQUES FOR DRONE NAVIGATION

Object Detection and Recognition (Landmarks, Obstacles, Dynamic Objects)

Drones rely on computer vision to move through their surroundings without crashing into objects or getting lost. One key technique is object detection, which helps drones recognize things like buildings, trees, and vehicles (Cazzato et al., 2020; Waheed et al., 2024). This works by training models to pick out patterns in images. The faster and more accurate this detection is, the better a drone can respond in real-time. I once watched a drone weave through a cluttered alleyway, dodging trash cans and parked bikes (Saif, 2016). The ability to do that comes from deep learning models analyzing visual data at high speed. Landmarks play a big role in navigation, especially when GPS signals are weak. Drones use vision-based localization, matching real-time camera feeds with known maps or stored reference images (Xu et al., 2018). This is how some delivery drones find specific drop-off spots in crowded cities. It's almost like how we recognize familiar streets without checking a map. Another major challenge is detecting moving objects—like birds, other drones, or even people walking through the flight path. Tracking these in real time requires algorithms that predict motion, ensuring the drone doesn't just react but anticipates where obstacles will be (Rezaee et al., 2024; Weiqi et al., 2024). Beyond avoiding crashes, accurate object recognition allows drones to perform complex tasks. For example, agricultural drones can distinguish crops from weeds and target only unwanted plants with pesticides (Qu & Su, 2024). This kind of precision makes drones useful in fields ranging from surveillance to disaster response. Watching a drone adjust its path mid-flight as if it "sees" the world, never fails to impress. The better these vision systems become, the more useful drones will be across industries.

Optical Flow Analysis for Motion Estimation

Drones rely on optical flow analysis to estimate motion, helping them navigate through changing environments. This technique tracks pixel movement across frames, allowing drones to determine speed and direction relative to objects in view (Hansen & De Figueiredo, 2024). Unlike GPS, which struggles in indoor or GPS-denied areas, optical flow provides reliable positioning using just a camera (Al-Jarrah et al., 2024; Wang & Ahmad, 2024). I once saw a small drone avoid obstacles in a dimly lit warehouse, adjusting its path purely through optical flow—a clear example of its effectiveness. One challenge is processing speed. High-resolution images generate vast data, and drones must interpret it in real-time to avoid crashes (S. Kumar et al.,

2024; Xun et al., 2025). The Lucas-Kanade method, often used in an optical flow, assumes small motions and brightness constancy, making it efficient but sometimes inaccurate in fast movements. Some newer approaches combine deep learning with optical flow, refining motion estimation by reducing noise and errors (Guglielmo, 2024). Optical flow isn't just about avoiding walls. In aerial surveying, drones are used to stabilize flight, ensuring clear images for mapping (Peksa & Mamchur, 2024). It also plays a role in search and rescue, guiding drones through smoke-filled environments where GPS fails. With improvements in processing power and AI integration, optical flow will likely remain a key tool in drone navigation, expanding into applications where precision and adaptability matter most.

Figure 4. Core computer vision techniques for drone navigation

Object Detection	Identifies obstacles, landmarks, and other drones. Uses YOLO, Faster R-CNN, etc.
Optical Flow	Tracks motion by analyzing pixel changes. Helps in stabilizing flight and avoiding collisions.
SLAM	Builds real-time maps while tracking drone position. Useful for autonomous navigation in unknown areas
Depth Estimation	Determines distances using stereo vision or LiDAR. Crucial for altitude control and landing
Feature Matching & Tracking	Detects key points (e.g., SIFT, ORB) for precise movement. Helps in recognizing familiar locations

SLAM (Simultaneous Localization and Mapping) for Real-Time Mapping

Drones employ computer vision to map an area, and among the most vital ways is through Simultaneous Localization and Mapping (SLAM) (K. Wang et al., 2024; Ying et al., 2024). This helps a drone map in real-time while knowing where it is. Think about a drone flying through an unfamiliar building—without GPS, it has to operate with cameras and sensors to detect walls, obstructions, and pathways to avoid collisions. This functionality is priceless in search-and-rescue where drones map through rubble to locate victims. SLAM is an aggregate process including various methods including feature detection, estimation of depth, and closure in loops. Feature detection is employed to detect unique aspects in an area, e.g., edges around items or unique patterns, to inform the drone where it is (Kashi et al., 2023). Range estimation through stereo cams or LiDAR helps the drone detect range, necessary in navigating through crowded areas (Cerutti, 2024). Loop closure is employed to recover errors by recognizing locations visited earlier to have better accuracy over time (X. Wang et al., 2024).

The challenge in speed in SLAM is processing speed. A drone has to process in time to avoid delays that can lead to collisions. Developments in machine learning have improved SLAM performance to be faster and better (Li et al., 2024). Edge computing has been developed to empower drones to process locally in place of servers elsewhere to reduce latency to achieve real-time mapping in practice (Alanhdi & Toka, 2024). SLAM is changing drones to map unfamiliar areas independently. Whether in search-and-rescue or warehousing automation, this tech is making autonomous flight efficient and powerful.

Depth Estimation Using Stereo Vision and LiDAR

Flying over unfamiliar environments is done through accurate estimation of depths. Stereo vision and LiDAR are essential technologies to accomplish this. Stereo vision utilizes two cameras to determine depths by comparing slightly offset views (Sun et al., 2024). When lighting is dynamic or surfaces have minimal to no texture, this is an issue because finding corresponding points becomes increasingly complicated (Guo et al., 2024; Alex et al., 2022). LiDAR projects beams to determine distances directly. This is perfect in darkness because it is not vulnerable to changing lighting. However, reflective or glassy material can confuse it. Some systems incorporate both to get increased reliability. I have seen drones fly over forests where tree branches that otherwise would have been difficult to detect with only one technique were safely dodged around. Stereo vision provided shape context while LiDAR confirmed distances to prevent errors (Sten et al., 2025; Alferidah & Jhanjhi, 2020). Utilizing both is helpful but adds processing power requirements, something that is problematic in lightweight drones (Aldossri et al., 2024). Advances in AI-assisted estimation have developed these systems to be employed in faster decisions by drones. As development continues to advance, drones can increasingly be employed in search-and-rescue or precision agriculture where accurate depth is required.

5. SENSOR FUSION: INTEGRATING VISION WITH OTHER NAVIGATION TECHNOLOGIES

Combining Vision With IMU, GPS, and LiDAR

Flying through intricate environments is not done through only one source. Cameras provide rich vision data but lose effectiveness in darkness or unfavourable weather (Thakur & Mishra, 2024). This is where sensor fusion comes in—combining vision with GPS and inertial Measurement Units (IMU) to achieve better reliability and accuracy. IMUs keep track of motion through gyroscopes and accelerometers to determine where an object is in motion (Molaei et al., 2024). GPS provides worldwide positioning but is hindered in tunnels or urban canyons where signals can be blocked (Jiang et al., 2025). LiDAR uses laser pulses to determine distances to construct accurate 3D maps, which is particularly beneficial to autonomous vehicles and drones (Leong & Ahmad, 2024). Navigation systems can offset each sensor's weakness by combining these technologies.

Consider an autonomous car moving through a tunnel. GPS drops out, but the IMU keeps tracking motion. As the car exits, GPS corrects any drift in the IMU's estimate. Meanwhile, LiDAR refines the vehicle's surroundings, detecting obstacles with high precision (Gupta et al., 2024; Alkinani et al., 2021). A camera, processing road signs and lane markings, adds another layer of verification. This fusion of data ensures a smoother and safer ride. The combination of these sensors is already shaping industries. Drones rely on it for mapping disaster zones. Robots use it in warehouses to navigate tight spaces. Self-driving technology is advancing rapidly because of these integrations. The challenge now is improving how these sensors communicate, reducing errors, and making systems more efficient. As technology advances, the possibilities continue to expand. Figure 5 shows drone's other technology.

Figure 5. Drone's other technology

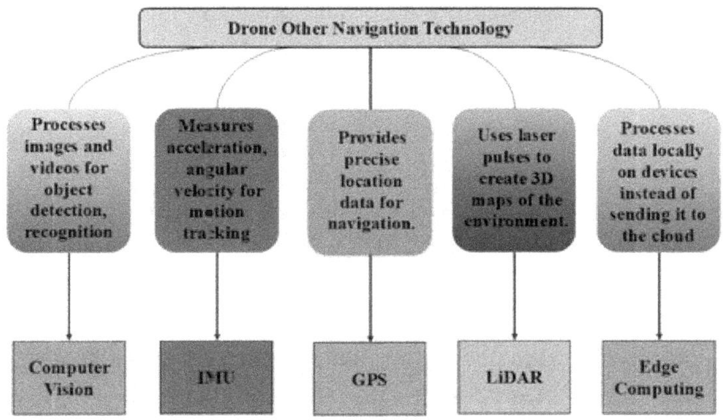

Edge Computing for Real-Time Processing of Sensor Data

Walking through a crowded city with self-driving cars, drones, and delivery robots, it's easy to see how machines need sharp awareness. Cameras catch movement, but they can struggle in poor lighting. LiDAR builds 3D maps, yet it can be thrown off by rain. GPS works well outdoors but loses accuracy in tunnels. No single sensor is perfect, so combining them improves navigation (Cai et al., 2024).

Edge computing steps in by handling sensor data where it's collected instead of waiting on distant servers. A drone scanning a disaster zone can process LiDAR, GPS, and camera inputs instantly, avoiding delays that could risk missing obstacles (Hussain et al., 2024; Babbar et al., 2021). Self-driving cars do the same, reacting in milliseconds to a child running into the street. Centralized cloud systems wouldn't keep up. Edge devices filter unnecessary data, cutting power use while improving speed (Bhoi et al., 2024; Brohi et al., 2020). A real-world example is Amazon's warehouse robots. They use depth cameras and inertial sensors, making real-time decisions without needing constant cloud communication (Campini, 2024). The same approach helps in agriculture, where autonomous tractors adjust paths on uneven ground. Military applications also rely on it—battlefield drones interpret sensor input without exposing data to cyber threats (Alquwayzani & Albuali, 2024; Chesti et al., 2020). While challenges remain—like processing power limits and sensor synchronization—advancements in AI are making edge-based fusion smarter. The future isn't just about collecting data but acting on it instantly, keeping machines as sharp as human reflexes.

Case Study: Vision-Based Autonomous Landing

A case study on vision-based autonomous landing highlights the benefits of merging sensors. Cameras provide real-time terrain mapping, while LiDAR enhances depth perception (Zhang et al., 2023). Radar helps detect obstacles, ensuring safer descent. IMUs compensate for sudden movements, stabilizing calculations . A test on an unmanned aerial vehicle (UAV) showed improved landing precision when vision data was merged with LiDAR and IMU readings (Sabatini et al., 2012; Dogra et al., 2021). The aircraft adjusted dynamically, reducing error margins significantly. Unexpected factors often complicate landings. Glare, fog, or shadows can distort vision-based readings. Sensor fusion counteracts these issues by allowing different sensors to compensate for weaknesses. In one trial, an AI model trained on multi-sensor data adapted to changing conditions, achieving safer touchdowns (Tomasicchio et al., 2024). This suggests a future where autonomous aircraft land reliably in varying environments. Real-world applications extend beyond UAVs. Space missions rely on similar principles for lunar and planetary landings. A Mars Lander study revealed that fusing optical data with radar improved descent stability (Trawny et al., 2024). With ongoing advancements, this approach will likely influence aviation and drone logistics, making autonomous landings safer and more practical.

6. CHALLENGES IN VISION-BASED DRONE NAVIGATION

Computational Limitations and Real-Time Processing Constraints

Vision-based navigation drones face serious challenges in processing vast amounts of visual information within milliseconds (Le et al., 2024). High-resolution cameras create massive data streams, which onboard processing cannot handle efficiently (Chung et al., 2024). Unlike ground-based robots, drones have limited battery lifetimes, and energy-intensive computations are not feasible (Majumdar et al., 2024; Fatima-Tuz-Zahra et al., 2020). Accordingly, compromises between speed and accuracy are inevitable. I experimented with a drone with a highly developed vision system, but with moving objects, there was a visible lag in processing images and hence making it unreliable. Real-time object detection and avoidance require optimized algorithms, but state-of-the-art deep learning models bring in latencies (Hosain et al., 2024).

Another issue is environmental conditions that continuously change. A drone flying in an urban environment is faced with shadows, reflective surfaces, and varying light levels, all affecting visual perception (Xi et al., 2025). Classical image processing algorithms struggle to cope with such uncertain factors. Some researchers turn to neuromorphic computing and edge AI to minimize latency, but such technologies are in their infancy (Z. Zhang et al., 2023; Hemmati et al., 2024; Gopi et al., 2021). It is not a question of enhancing processing speed but reliability in real-world applications. If a drone misidentifies an object due to processing latency, the outcome can range from small navigation errors to mission failure. Research for efficient hardware and light models goes on, but providing real-time performance with energy constraints is a persisting challenge. Figure 6 illustrates vision-based drone navigation challenges.

Figure 6. Challenges in vision-based drone navigation

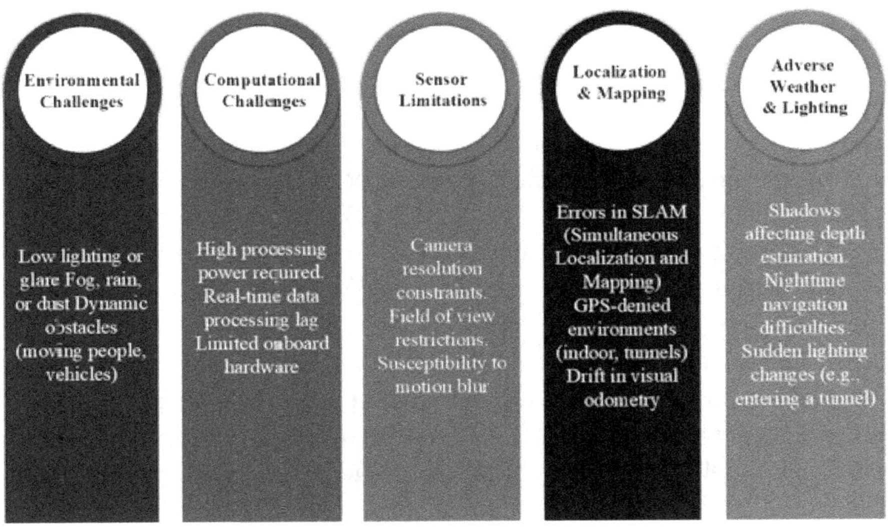

Environmental Challenges: Lighting, Weather, Dynamic Obstacles

Flying a drone with only vision-based navigation sounds like something out of a sci-fi movie, but in reality, it's tough. Light, weather, and moving obstacles make it tricky for drones to see and react fast enough. Imagine trying to drive a car with fogged-up glasses, that's what poor lighting does to drones. Shadows, glare, and low light mess with their cameras making it hard to track objects (Shokri et al., 2024). Rain, snow, and fog add another layer of trouble. Raindrops on the camera lens blur images, and fog confuses depth perception (Pahk et al., 2024). Strong winds push drones off course, forcing their vision systems to work overtime to correct errors (Fahlstrom & Gleason, 2013; Gouda et al., 2022).

Then there's the issue of moving objects. A drone flying through a city has to dodge birds, other drones, and even people. A split-second delay in processing means a collision. Some systems use AI to predict motion, but fast-moving obstacles still pose a challenge. A drone might see a cyclist one second and lose sight of them the next due to a parked bus blocking the view. These gaps in vision make navigation unreliable, especially in tight spaces. Researchers are exploring ways to improve these systems, like combining multiple cameras or using infrared sensors for low-light conditions. But for now, vision-based navigation alone struggles with the unpredictability of real-world environments.

Data Processing Bottlenecks and the Need for Edge Computing

Processing large volumes of visual data in drone navigation creates delays, making real-time decision-making difficult. A drone relying on cameras must handle constant image streams, detect objects, and adjust its path—all within fractions of a second (Zhu et al., 2021; Singhal et al., 2020). Traditional cloud-based methods struggle because transmitting data introduces lag. A personal experience with a small quadcopter showed how delays in processing made it unresponsive to sudden obstacles. It's frustrating when a drone "sees" an object but reacts too late. The problem worsens with high-resolution cameras that generate massive data loads (McEwen et al., 2023; Humayun et al., 2022). Edge computing helps by processing data closer to the drone itself (McEnroe et al., 2022). Instead of sending everything to a distant server, compact onboard processors analyze images instantly. This reduces lag and improves efficiency. Military and industrial drones already use edge computing to enhance autonomous flight in complex spaces (Fatima et al., 2022; Jhanjhi et al., 2021). Yet, hardware limitations remain a challenge. Small drones struggle to carry high-performance chips due to weight constraints, and energy efficiency becomes a major concern (R. Singh & Kumar, 2025). Battery life drains quickly when advanced processors work continuously. Looking ahead, lightweight AI models and specialized processors may help. Some researchers are developing chips designed specifically for drones, offering powerful computing without excessive power consumption. These innovations could make real-time navigation more reliable, allowing drones to operate smoothly in unpredictable settings. The shift toward edge computing is not just a technical upgrade—it's a necessity for making drones smarter and safer.

7. FUTURE TRENDS IN COMPUTER VISION FOR DRONE NAVIGATION

AI-Powered Real-Time Perception and Decision-Making

A drone flying through a dense forest dodging branches in split seconds is now not sci-fi. Computer vision through AI is driving drone navigation to unprecedented levels, making autonomous flight efficient and secure. Deep learning algorithms in present drones decode video streams to detect objects and forecast movements in an instant (Purcell et al., 2022; Kumar et al., 2021). This instant processing allows drones to react in real-time, necessary in disaster response, search-and-rescue, and urban deliveries (Rejeb et al., 2021). One giant leap is through neural networks modelling human vision. No longer pre-scripted to take pre-designated routes, drones gain experience through experience, adapting to unfamiliar environments. One paper published in recent times indicated that reinforcement learning can get drones to master complicated moves with no interference from the user (Vaezi et al., 2023; Lim et al., 2019). Within urban landscapes where GPS is unreliable, vision-based SLAM (Simultaneous Localization and Mapping) allows drones to map landscapes and fly independently (Couturier & Akhloufi, 2024; Nayyar et al., 2021). Research in intelligent drones has been accelerated by growing demand in edge computing where AI algorithms run locally in the drone and not in remote servers (Chang et al., 2021). This eliminates delay and offers instant responses, making drones fueled by AI suitable for time-critical deployments. As these systems mature further, drones not only fly through landscapes but predict changes in surroundings leading to uses that seem to be in imagination.

Figure 7. Future trends in computer vision for drone navigation

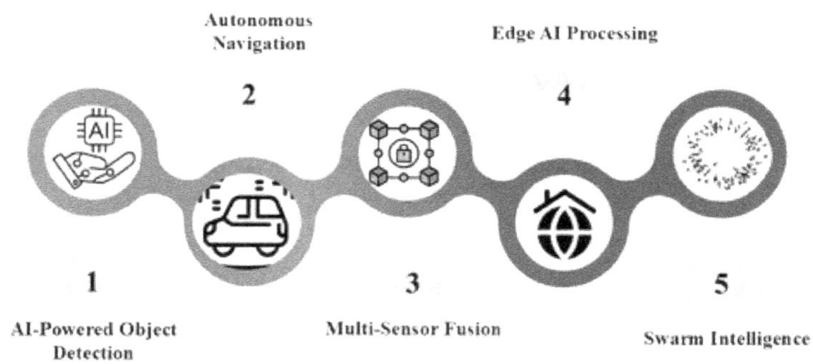

The Role of Deep Learning in Improving Navigation Accuracy

Computer vision is making drones smarter by allowing better navigation. Deep learning plays an integral part in this by allowing drones to understand their surroundings. Utilizing convolutional neural networks (CNNs), drones can spot things, avoid collisions, and react quickly in flight (Arshad et al., 2022; Shah et al., 2022). When drones fly over dense forests or urban landscapes, traditional navigation is not an option. Deep learning-powered algorithms trained over extensive sets of images better allow drones to spot and react to hindrances in real-time (Rovira-Sugranes et al., 2022).

One emerging technique is self-supervised learning. Unlike relying upon tagged sets of data, drones learn patterns by experience. This reduces extensive sets of human-tagged sets to accelerate training. Sensor fusion is an integral trend. Alone in darkness, cameras perform poorly, yet by combining data from LiDAR, infrared, and radar, drones perform in every situation (Bayomi & Fernandez, 2023; Sharma et al., 2021). Performance upgrades aside, deep learning is allowing drones to adapt to changing conditions. Reinforcement learning allows drones to perfect navigating through practice and testing by trial and error, just as humans perfect walking (Fagundes-Junior et al., 2024). Such algorithms get better with each subsequent flight, making navigating increasingly accurate. Organizations and researchers have been testing these ideas. Delivery drones, search-and-rescue operations, and agricultural monitoring benefit. Deep learning-powered vision navigation is proving invaluable in disaster-stricken areas where GPS is unreliable. As development continues forward, drones increasingly resort to vision-based AI to be autonomous and reliable.

8. CONCLUSION

Summary of Key Takeaways

Drone systems have evolved from simple remote-controlled toys to autonomous systems with capabilities to navigate in complex environments with minimal control from humans. GPS and inertial sensors can use standard techniques to provide good navigation, but in environments with weak signals or obstacles changing randomly, they do not work. Computer vision transcends such limitations as it enables drones to interpret their environments, recognize objects, avoid obstacles and build real-time maps. Techniques like optical flow, SLAM (Simultaneous Localization and Mapping), and perception with deep learning have become commonplace in high-end drone operations. Vision systems can enhance accuracy and maneuverability with additional sensors like LiDAR (Light Detection and Ranging), radar, and GPS, and can navigate urban canyons, forests, and below-ground environments more effectively. But processing large volumes of visual information in real time is still a challenge. Computational constraints, power consumption, and varying light conditions still impair performance. Edge computing innovations, AI-based perception, and small neural networks are making drones more efficient in information processing, faster in decision-making, and more effective in navigation. Vision-powered drones in search-and-rescue, agricultural management, and logistics have now increasingly been relying on various key jobs. From tracking down missing people through thermal vision to optimizing farm management through automated farm scanning, computer vision-powered drones have been proving to be invaluable. As increased amounts of AI and real-time perception get integrated into drones, autonomous capabilities continue to get stretched to newer boundaries.

Final Thoughts on the Impact of Computer Vision in Drone Navigation

The ability to "see" and respond to the surroundings has entirely altered drones' applications in various spheres. Utilizing computer vision-based perception, drones no longer have to fly pre-scripted paths but can react to reality in real-time. This is particularly helpful in areas where GPS is unreliable, e.g., urban canyons, disaster-stricken areas, or indoors. As processing power expands and algorithms continue to get better, drones' reliability and accuracy will continue to get better. Future development can result in drones mirroring humans in navigating through pathways by making decisions in an instant. Computer vision, sensor fusion, and machine learning have formed an enabling ground for drones to not only be reliable and safe but to perform complicated jobs with greater accuracy. Though limitations continue to restrain development, work in AI and processing in real-time is increasingly making drone navigation an imperative future technology.

REFERENCES

Agelli, M., Corona, N., Maggio, F., & Moi, P. V. (2024a). Unmanned ground vehicles for continuous crop monitoring in agriculture: Assessing the readiness of current ICT technology. *Machines (Basel)*, *12*(11), 750. DOI: 10.3390/machines12110750

Agelli, M., Corona, N., Maggio, F., & Moi, P. V. (2024b). Unmanned ground vehicles for continuous crop monitoring in agriculture: Assessing the readiness of current ICT technology. *Machines (Basel)*, *12*(11), 750. DOI: 10.3390/machines12110750

Akbari, Y., Almaadeed, N., Al-Maadeed, S., & Elharrouss, O. (2021). Applications, databases and open computer vision research from drone videos and images: A survey. *Artificial Intelligence Review*, *54*(5), 3887–3938. DOI: 10.1007/s10462-020-09943-1

Al-Ahmadi, A. (2024). Drone attitude and position prediction via stacked hybrid deep learning model for massive MIMO applications. *IEEE Access : Practical Innovations, Open Solutions*, *1*, 1. Advance online publication. DOI: 10.1109/ACCESS.2024.3516126

Al-Jarrah, O. Y., Shatnawi, A. S., Shurman, M. M., Ramadan, O. A., & Muhaidat, S. (2024). Exploring deep learning-based visual localization techniques for UAVs in GPS-Denied environments. *IEEE Access : Practical Innovations, Open Solutions*, *12*, 113049–113071. DOI: 10.1109/ACCESS.2024.3440064

Alanhdi, A., & Toka, L. (2024). A survey on integrating edge computing with AI and blockchain in maritime domain, aerial systems, IoT, and industry 4.0. *IEEE Access : Practical Innovations, Open Solutions*, *12*, 28684–28709. DOI: 10.1109/ACCESS.2024.3367118

Aldossri, R., Aljughaiman, A., & Albuali, A. (2024). Advancing Drone Operations through Lightweight Blockchain and Fog Computing Integration: A Systematic Review. *Drones (Basel)*, *8*(4), 153. DOI: 10.3390/drones8040153

Alex, S. A., Jhanjhi, N., Humayun, M., Ibrahim, A. O., & Abulfaraj, A. W. (2022). Deep LSTM Model for Diabetes Prediction with Class Balancing by SMOTE. *Electronics (Basel)*, *11*(17), 2737. DOI: 10.3390/electronics11172737

Alferidah, D. K., & Jhanjhi, N. (2020). Cybersecurity Impact over Bigdata and IoT Growth. *2020 International Conference on Computational Intelligence (ICCI)*. DOI: 10.1109/ICCI51257.2020.9247722

Ali, S. (2024). *Sensor fusion-based indoor navigation in GNSS-denied environments - Espace ETS*. https://espace.etsmtl.ca/id/eprint/3442/

Alkinani, M. H., Almazroi, A. A., Jhanjhi, N., & Khan, N. A. (2021). 5G and IoT Based Reporting and Accident Detection (RAD) System to Deliver First Aid Box Using Unmanned Aerial Vehicle. *Sensors (Basel)*, *21*(20), 6905. DOI: 10.3390/s21206905 PMID: 34696118

Alquwayzani, A. A., & Albuali, A. A. (2024). A Systematic Literature review of zero trust architecture for military UAV security systems. *IEEE Access : Practical Innovations, Open Solutions*, *1*, 176033–176056. Advance online publication. DOI: 10.1109/ACCESS.2024.3503587

Amirgaliyev, B., Mussabek, M., Rakhimzhanova, T., & Zhumadillayeva, A. (2025). A Review of Machine Learning and Deep Learning Methods for Person Detection, Tracking and Identification, and Face Recognition with Applications. *Sensors (Basel)*, *25*(5), 1410. DOI: 10.3390/s25051410

Andersen, T. S., Andersen, N. A., Ravn, O., & Fumagalli, M. (2024). A long distance mono optical localization system for unmanned aerial vehicles. *2022 17th International Conference on Control, Automation, Robotics and Vision (ICARCV)*, 947–953. DOI: 10.1109/ICARCV63323.2024.10821626

Arafat, M. Y., Alam, M. M., & Moh, S. (2023). Vision-Based Navigation Techniques for Unmanned Aerial Vehicles: Review and challenges. *Drones (Basel)*, *7*(2), 89. DOI: 10.3390/drones7020089

Arshad, M. A., Khan, S. H., Qamar, S., Khan, M. W., Murtza, I., Gwak, J., & Khan, A. (2022). Drone navigation using region and Edge Exploitation-Based Deep CNN. *IEEE Access : Practical Innovations, Open Solutions*, *10*, 95441–95450. DOI: 10.1109/ACCESS.2022.3204876

Babbar, H., Rani, S., Masud, M., Verma, S., Anand, D., & Jhanjhi, N. (2021). Load balancing algorithm for migrating switches in software-defined vehicular networks. *Computers, Materials & Continua/Computers. Computers, Materials & Continua*, *67*(1), 1301–1316. DOI: 10.32604/cmc.2021.014627

Bain, G., & Blaney, M. (2025). *Managing safety in the drone industry: A Practical Guide*.

Balestrieri, E., Daponte, P., De Vito, L., Picariello, F., & Tudosa, I. (2021). Sensors and Measurements for UAV safety: An overview. *Sensors (Basel)*, *21*(24), 8253. DOI: 10.3390/s21248253 PMID: 34960347

Bayomi, N., & Fernandez, J. E. (2023). Eyes in the Sky: Drones Applications in the Built Environment under Climate Change Challenges. *Drones (Basel)*, *7*(10), 637. DOI: 10.3390/drones7100637

Bhoi, S. K., Chakraborty, S., Verbrugge, B., Helsen, S., Robyns, S., Baghdadi, M. E., & Hegazy, O. (2024). Intelligent data-driven condition monitoring of power electronics systems using smart edge–cloud framework. *Internet of Things : Engineering Cyber Physical Human Systems, 26*, 101158. DOI: 10.1016/j.iot.2024.101158

Brohi, S. N., Jhanjhi, N., Brohi, N. N., & Brohi, M. N. (2020). Key Applications of State-of-the-Art Technologies to Mitigate and Eliminate COVID-19.pdf. *TECHRxiv*. DOI: 10.36227/techrxiv.12115596.v1

Cai, Y., Gao, F., Huang, J., & Qiao, G. (2024). GPS-aided Stereo Inertial Navigation Localization Algorithm for outdoor scenarios. *IEEE Sensors Journal, 1*. Advance online publication. DOI: 10.1109/JSEN.2024.3519880

Campini, M. (2024). *GPS-Driven Autonomous Navigation for Precision Agriculture with AWS - Webthesis*. https://webthesis.biblio.polito.it/31580/

Cazzato, D., Cimarelli, C., Sanchez-Lopez, J. L., Voos, H., & Leo, M. (2020). A Survey of Computer Vision Methods for 2D Object Detection from Unmanned Aerial Vehicles. *Journal of Imaging, 6*(8), 78. DOI: 10.3390/jimaging6080078 PMID: 34460693

Cerutti, A. (2024). *Enhancing Drone Detection and Tracking Using LiDAR Sensor with Adaptive Motion Strategies - Webthesis*. https://webthesis.biblio.polito.it/33086/

Chai, R., Guo, Y., Zuo, Z., Chen, K., Shin, H., & Tsourdos, A. (2024). Cooperative motion planning and control for aerial-ground autonomous systems: Methods and applications. *Progress in Aerospace Sciences, 146*, 101005. DOI: 10.1016/j.paerosci.2024.101005

Chang, Z., Liu, S., Xiong, X., Cai, Z., & Tu, G. (2021). A survey of recent advances in Edge-Computing-Powered artificial intelligence of Things. *IEEE Internet of Things Journal, 8*(18), 13849–13875. DOI: 10.1109/JIOT.2021.3088875

Chesti, I. A., Humayun, M., Sama, N. U., & Jhanjhi, N. (2020). Evolution, Mitigation, and Prevention of Ransomware. *2020 2nd International Conference on Computer and Information Sciences (ICCIS)*. DOI: 10.1109/ICCIS49240.2020.9257708

Chung, H. J., Kang, B., & Yang, Y. S. (2024). N-DriverMotion: Driver motion learning and prediction using an event-based camera and directly trained spiking neural networks on Loihi 2. *IEEE Open Journal of Vehicular Technology*, •••, 1–14. DOI: 10.1109/OJVT.2024.3504481

Couturier, A., & Akhloufi, M. A. (2024). A review on deep learning for UAV Absolute Visual Localization. *Drones (Basel), 8*(11), 622. DOI: 10.3390/drones8110622

Dass, S. M., Mahalingam, S. R., Hang, P. L., Kelvin, K., Wingsky, J., Mohamed, A. M. M. A., Rebecca, A. J., Hatmal, N. I. Y. B. A., Ali, A. R., Tjahyosedjati, E., Wei, T., & Sindiramutty, S. R. (2025). The Role of Cryptography and AI in Enhancing Blockchain Security. *preprints.org*. DOI: 10.20944/preprints202501.0386.v1

Dinelli, C., Racette, J., Escarcega, M., Lotero, S., Gordon, J., Montoya, J., Dunaway, C., Androulakis, V., Khaniani, H., Shao, S., Roghanchi, P., & Hassanalian, M. (2023). Configurations and Applications of Multi-Agent Hybrid Drone/Unmanned Ground Vehicle for Underground Environments: A review. *Drones (Basel)*, *7*(2), 136. DOI: 10.3390/drones7020136

Dogra, V., Singh, A., Verma, S., Kavita, N., Jhanjhi, N. Z., & Talib, M. N. (2021). Analyzing DistilBERT for Sentiment Classification of Banking Financial News. In *Lecture notes in networks and systems* (pp. 501–510). DOI: 10.1007/978-981-16-3153-5_53

Fagundes-Junior, L. A., De Carvalho, K. B., Ferreira, R. S., & Brandão, A. S. (2024). Machine Learning for Unmanned Aerial Vehicles Navigation: An Overview. *SN Computer Science*, *5*(2), 256. Advance online publication. DOI: 10.1007/s42979-023-02592-5

Fahlstrom, P. G., & Gleason, T. J. (2013). Introduction to UAV systems. *Choice (Chicago, Ill.)*, *50*(11), 50–6201. DOI: 10.5860/CHOICE.50-6201

Fatima, N., Saxena, P., & Gupta, M. (2022). Integration of multi access edge computing with unmanned aerial vehicles: Current techniques, open issues and research directions. *Physical Communication*, *52*, 101641. DOI: 10.1016/j.phycom.2022.101641

Fatima-Tuz-Zahra, N., Jhanjhi, N., Brohi, S. N., Malik, N. A., & Humayun, M. (2020). Proposing a Hybrid RPL Protocol for Rank and Wormhole Attack Mitigation using Machine Learning. *2020 2nd International Conference on Computer and Information Sciences (ICCIS)*. DOI: 10.1109/ICCIS49240.2020.9257607

Foudeh, H. A., Luk, P. C., & Whidborne, J. F. (2021). An Advanced Unmanned Aerial Vehicle (UAV) Approach via Learning-Based Control for Overhead Power Line Monitoring: A Comprehensive Review. *IEEE Access : Practical Innovations, Open Solutions*, *9*, 130410–130433. DOI: 10.1109/ACCESS.2021.3110159

Gong, W., He, C., Su, K., & Li, Q. (2024). DIDLM:A Comprehensive Multi-Sensor Dataset with Infrared Cameras, Depth Cameras, LiDAR, and 4D Millimeter-Wave Radar in Challenging Scenarios for 3D Mapping. *arXiv (Cornell University)*. https://doi.org//arxiv.2404.09622DOI: 10.48550

Gopi, R., Sathiyamoorthi, V., Selvakumar, S., Manikandan, R., Chatterjee, P., Jhanjhi, N. Z., & Luhach, A. K. (2021). Enhanced method of ANN based model for detection of DDoS attacks on multimedia internet of things. *Multimedia Tools and Applications*, *81*(19), 26739–26757. DOI: 10.1007/s11042-021-10640-6

Gouda, W., Almurafeh, M., Humayun, M., & Jhanjhi, N. Z. (2022). Detection of COVID-19 based on chest x-rays using deep learning. *Health Care*, *10*(2), 343. DOI: 10.3390/healthcare10020343 PMID: 35206957

Grenier, A., Lohan, E. S., Ometov, A., & Nurmi, J. (2023). Towards Smarter Positioning through Analyzing Raw GNSS and Multi-Sensor Data from Android Devices: A Dataset and an Open-Source Application. *Electronics (Basel)*, *12*(23), 4781. DOI: 10.3390/electronics12234781

Guglielmo, D. C. (2024, September 30). *Refining forest mapping drones for swift movement and reliable precision via deep learning.* https://aaltodoc.aalto.fi/items/9d5f9e0d-ef2f-4df2-9be0-e4be137349d8

Guo, Q., Liu, Z., Yang, Z., Jiang, Y., Sun, Y., Xu, J., Zhao, W., Wang, W., Wang, W., Ren, Q., & Shu, C. (2024). Development, challenges and future trends on the fabrication of micro-textured surfaces using milling technology. *Journal of Manufacturing Processes*, *126*, 285–331. DOI: 10.1016/j.jmapro.2024.07.112

Gupta, A., Jain, S., Choudhary, P., & Parida, M. (2024). Dynamic object detection using sparse LiDAR data for autonomous machine driving and road safety applications. *Expert Systems with Applications*, *255*, 124636. DOI: 10.1016/j.eswa.2024.124636

Hansen, J. G., & De Figueiredo, R. P. (2024). Active object detection and tracking using gimbal mechanisms for autonomous drone applications. *Drones (Basel)*, *8*(2), 55. DOI: 10.3390/drones8020055

He, B., Ji, X., Li, G., & Cheng, B. (2024). Key Technologies and Applications of UAVs in Underground Space: A review. *IEEE Transactions on Cognitive Communications and Networking*, *10*(3), 1026–1049. DOI: 10.1109/TCCN.2024.3358545

Heidari, A., Navimipour, N. J., Unal, M., & Zhang, G. (2022). Machine Learning Applications in Internet-of-Drones: Systematic review, recent deployments, and open Issues. *ACM Computing Surveys*, *55*(12), 1–45. DOI: 10.1145/3571728

Hemmati, A., Raoufi, P., & Rahmani, A. M. (2024). Edge artificial intelligence for big data: A systematic review. *Neural Computing & Applications*, *36*(19), 11461–11494. DOI: 10.1007/s00521-024-09723-w

Hosain, M. T., Zaman, A., Abir, M. R., Akter, S., Mursalin, S., & Khan, S. S. (2024). Synchronizing Object Detection: Applications, advancements and existing challenges. *IEEE Access : Practical Innovations, Open Solutions*, *12*, 54129–54167. DOI: 10.1109/ACCESS.2024.3388889

Humayun, M., Sujatha, R., Almuayqil, S. N., & Jhanjhi, N. Z. (2022). A Transfer Learning Approach with a Convolutional Neural Network for the Classification of Lung Carcinoma. *Health Care*, *10*(6), 1058. DOI: 10.3390/healthcare10061058 PMID: 35742109

Hussain, A., Li, S., Hussain, T., Lin, X., Ali, F., & AlZubi, A. A. (2024). Computing Challenges of UAV Networks: A Comprehensive survey. *Computers, Materials & Continua/Computers. Computers, Materials & Continua*, *81*(2), 1999–2051. DOI: 10.32604/cmc.2024.056183

Hussain, K., Rahmatyar, A. R., Riskhan, B., Sheikh, M. U., & Sindiramutty, S. R. (2024). Threats and Vulnerabilities of Wireless Networks in the Internet of Things (IoT). *IEEE 1st Karachi Section Humanitarian Technology Conference (KHI-HTC)*, *2*, 1–8. DOI: 10.1109/KHI-HTC60760.2024.10482197

Jain, N., Gambhir, A., & Pandey, M. (2024). Unmanned Aerial Networks—UAVs and AI. In *Recent Trends in Artificial Intelligence Towards a Smart World* (pp. 321–351). DOI: 10.1007/978-981-97-6790-8_12

Javaid, M., Haleem, A., Singh, R. P., & Ahmed, M. (2024). Computer Vision to Enhance Healthcare Domain: An Overview of features, implementation, and opportunities. *Intelligent Pharmacy*, *2*(6), 792–803. DOI: 10.1016/j.ipha.2024.05.007

Javan, F. D., Samadzadegan, F., Toosi, A., & Van Der Meijde, M. (2024). Unmanned Aerial Geophysical Remote Sensing: A Survey. *Preprints.org*. DOI: 10.20944/preprints202411.0058.v1

Jhanjhi, N., Humayun, M., & Almuayqil, S. N. (2021). Cyber security and privacy issues in industrial internet of things. *Computer Systems Science and Engineering*, *37*(3), 361–380. DOI: 10.32604/csse.2021.015206

Jiang, S., Xu, Q., Wang, W., Peng, P., & Li, J. (2025). Vehicle positioning systems in tunnel environments: A review. *Complex & Intelligent Systems*, *11*(2), 164. Advance online publication. DOI: 10.1007/s40747-024-01744-1

Jun, A. Y. M., Jinu, B. A., Seng, L. K., Maharaiq, M. H. F. B. Z., Khongsuwan, W., Junn, B. T. K., Hao, A. W., & Sindiramutty, S. R. (2024). Exploring the Impact of Crypto-Ransomware on Critical Industries: Case Studies and Solutions. *Preprints.org*. DOI: 10.20944/preprints202409.1325.v1

Kadam, P., Fang, G., & Zou, J. J. (2024). Object Tracking Using Computer Vision: A review. *Computers, 13*(6), 136. DOI: 10.3390/computers13060136

Kangunde, V., Jamisola, R. S.Jr, & Theophilus, E. K. (2021). A review on drones controlled in real-time. *International Journal of Dynamics and Control, 9*(4), 1832–1846. DOI: 10.1007/s40435-020-00737-5 PMID: 33425650

Kashi, R. N., Prashanth, A., Kashi, S. R., & Prabhakara, G. (2023). A survey and analysis of drone detection systems using a systems approach superposed on scenarios. *Systems Engineering, 27*(3), 598–636. DOI: 10.1002/sys.21735

Khan, M. A., Menouar, H., Eldeeb, A., Abu-Dayya, A., & Salim, F. D. (2022). On the Detection of Unauthorized Drones—Techniques and Future Perspectives: A Review. *IEEE Sensors Journal, 22*(12), 11439–11455. DOI: 10.1109/JSEN.2022.3171293

Khang, A., Abdullayev, V., Litvinova, E., Chumachenko, S., Alyar, A. V., & Anh, P. T. N. (2024). Application of computer Vision (CV) in the healthcare ecosystem. In *CRC Press eBooks* (pp. 1–16). DOI: 10.1201/9781003429609-1

Kiyani, F. F., Hamid, B., Humayun, M., Sindiramutty, S. R. L., & Chowdhury, S. (2024). Discovery of Influential Publications Using Research Article's Usage Context. *2024 International Conference on Emerging Trends in Networks and Computer Communications (ETNCC)*, 1–7. DOI: 10.1109/ETNCC63262.2024.10767576

Krishna, K. R. (2017). *Push button agriculture: robotics, drones, Satellite-Guided soil and crop management.*

Krishnan, S., & Murugappan, M. (2023). *Internet of drones: Applications, Opportunities, and Challenges*. CRC Press. DOI: 10.1201/9781003252085

Kumar, A. (2024). Geo-Tagged 3D geometric modeling of urban structures by mitigating reflected GPS signals using a laser range sensor. In *Lecture notes in networks and systems* (pp. 396–414). DOI: 10.1007/978-3-031-62281-6_29

Kumar, M. S., Vimal, S., Jhanjhi, N., Dhanabalan, S. S., & Alhumyani, H. A. (2021). Blockchain based peer to peer communication in autonomous drone operation. *Energy Reports, 7*, 7925–7939. DOI: 10.1016/j.egyr.2021.08.073

Kumar, S., Poyyamozhi, M., Murugesan, B., Rajamanickam, N., Alroobaea, R., & Nureldeen, W. (2024). Investigation of unsafe construction site conditions using deep learning algorithms using unmanned aerial vehicles. *Sensors (Basel), 24*(20), 6737. DOI: 10.3390/s24206737 PMID: 39460217

Kuru, K. (2021). Planning the future of smart cities with swarms of fully autonomous unmanned aerial vehicles using a novel framework. *IEEE Access : Practical Innovations, Open Solutions, 9*, 6571–6595. DOI: 10.1109/ACCESS.2020.3049094

Le, N., Thai, H., Yoon, C., & Huh, J. (2024). Recent development of Drone Technology Software Engineering: A Systematic survey. *IEEE Access : Practical Innovations, Open Solutions, 12*, 128729–128751. DOI: 10.1109/ACCESS.2024.3454546

Lenka, A. K., & Tripathy, H. K. (2024). 5 Computer vision for medical diagnosis and surgery. In *De Gruyter eBooks* (pp. 101–124). DOI: 10.1515/9783110750942-005

Leong, P. Y., & Ahmad, N. S. (2024). LiDar-based Obstacle Avoidance with Autonomous Vehicles: A Comprehensive Review. *IEEE Access : Practical Innovations, Open Solutions, 1*, 164248–164261. Advance online publication. DOI: 10.1109/ACCESS.2024.3493238

Li, Y., Shen, H., Fu, Y., & Wang, K. (2024). A method of dense point cloud SLAM based on improved YOLOV8 and fused with ORB-SLAM3 to cope with dynamic environments. *Expert Systems with Applications, 255*, 124918. DOI: 10.1016/j.eswa.2024.124918

Lim, M., Abdullah, A., Jhanjhi, N., Khan, M. K., & Supramaniam, M. (2019). Link Prediction in Time-Evolving Criminal Network with deep Reinforcement learning technique. *IEEE Access : Practical Innovations, Open Solutions, 7*, 184797–184807. DOI: 10.1109/ACCESS.2019.2958873

Linqiang, Y., Sindiramutty, S. R. L., Ashraf, H., Muzammal, S. M., Balakrishnan, S. P., Gupta, S., & Kavita, N. (2024). Intelligent Household Waste Classification System Based on Machine Learning. *2024 International Conference on Emerging Trends in Networks and Computer Communications (ETNCC)*, 760–768. DOI: 10.1109/ETNCC63262.2024.10767563

Liu, Z., An, P., Yang, Y., Qiu, S., Liu, Q., & Xu, X. (2024). Vision-Based Drone Detection in Complex Environments: A survey. *Drones (Basel), 8*(11), 643. DOI: 10.3390/drones8110643

MahmoudZadeh, S., Yazdani, A., Kalantari, Y., Ciftler, B., Aidarus, F., & Al Kadri, M. O. (2024). Holistic Review of UAV-Centric Situational Awareness: Applications, limitations, and Algorithmic challenges. *Robotics (Basel, Switzerland), 13*(8), 117. DOI: 10.3390/robotics13080117

Majumdar, P., Mitra, S., Bhattacharya, D., & Bhushan, B. (2024). Enhancing sustainable 5G powered agriculture 4.0 : Summary of low power connectivity, internet of UAV things, AI solutions and research trends. *Multimedia Tools and Applications*. Advance online publication. DOI: 10.1007/s11042-024-19728-1

Manchuri, A., Kakera, A., Saleh, A., & Raja, S. (2024). pplication of Supervised Machine Learning Models in Biodiesel Production Research - A Short Review. *Borneo Journal of Sciences and Technology*. DOI: 10.35370/bjost.2024.6.1-10

McEnroe, P., Wang, S., & Liyanage, M. (2022). A survey on the convergence of edge computing and AI for UAVs: Opportunities and challenges. *IEEE Internet of Things Journal*, *9*(17), 15435–15459. DOI: 10.1109/JIOT.2022.3176400

McEwen, A., Byrne, S., Hansen, C., Daubar, I., Sutton, S., Dundas, C., Bardabelias, N., Baugh, N., Bergstrom, J., Beyer, R., Block, K., Bray, V., Bridges, J., Chojnacki, M., Conway, S., Delamere, W., Ebben, T., Espinosa, A., Fennema, A., & Wilson, S. (2023). The high-resolution imaging science experiment (HiRISE) in the MRO extended science phases (2009–2023). *Icarus*, *419*, 115795. DOI: 10.1016/j.icarus.2023.115795

Mohsan, S. H., Othman, N. Q. H., Li, Y., Alsharif, M. H., & Khan, M. A. (2023). Unmanned aerial vehicles (UAVs): Practical aspects, applications, open challenges, security issues, and future trends. *Intelligent Service Robotics*. Advance online publication. DOI: 10.1007/s11370-022-00452-4 PMID: 36687780

Mohyuddin, G., Khan, M. A., Haseeb, A., Mahpara, S., Waseem, M., & Saleh, A. M. (2024). Evaluation of Machine Learning Approaches for Precision Farming in Smart Agriculture System: A Comprehensive Review. *IEEE Access : Practical Innovations, Open Solutions*, *12*, 60155–60184. DOI: 10.1109/ACCESS.2024.3390581

Molaei, A., Kolu, A., Lahtinen, K., & Geimer, M. (2024). Automatic recognition of excavator working cycles using supervised learning and motion data obtained from inertial measurement units (IMUs). *Construction Robotics*, *8*(2), 14. Advance online publication. DOI: 10.1007/s41693-024-00130-0

Nayyar, A., Gadhavi, L., & Zaman, N. (2021). Machine learning in healthcare: review, opportunities and challenges. In *Elsevier eBooks* (pp. 23–45). DOI: 10.1016/B978-0-12-821229-5.00011-2

Nyaaba, A. A., & Ayamga, M. (2021). Intricacies of medical drones in healthcare delivery: Implications for Africa. *Technology in Society*, *66*, 101624. DOI: 10.1016/j.techsoc.2021.101624

Ocaña, R. T., Peñaranda, S. M., & Falcon, V. M. P. (2024). Smart Shopping Carts: Reducing Checkout Queues with RFID and AWS Facial Recognition. *2024 IEEE XXXI International Conference on Electronics, Electrical Engineering and Computing (INTERCON)*, 1–7. DOI: 10.1109/INTERCON63140.2024.10833475

Okulski, M., & Ławryńczuk, M. (2022). How Much Energy Do We Need to Fly with Greater Agility? Energy Consumption and Performance of an Attitude Stabilization Controller in a Quadcopter Drone: A Modified MPC vs. PID. *Energies*, *15*(4), 1380. DOI: 10.3390/en15041380

Pahk, J., Park, S., Shim, J., Son, S., Lee, J., An, J., Lim, Y., & Choi, G. (2024). Lane segmentation data augmentation for heavy rain sensor blockage using realistically translated raindrop images and CARLA Simulator. *IEEE Robotics and Automation Letters*, *9*(6), 5488–5495. DOI: 10.1109/LRA.2024.3390587

Peksa, J., & Mamchur, D. (2024). A review on the state of the art in copter drones and flight control systems. *Sensors (Basel)*, *24*(11), 3349. DOI: 10.3390/s24113349 PMID: 38894139

Purcell, C. R., Walsh, A. J., Colefax, A. P., & Butcher, P. (2022). Assessing the ability of deep learning techniques to perform real-time identification of shark species in live streaming video from drones. *Frontiers in Marine Science*, *9*, 981897. Advance online publication. DOI: 10.3389/fmars.2022.981897

Qian, B., Guo, X., Boateng, G. O., An, R., Liu, H., Lai, Z., & Chen, C. (2025). Magnetoquasistatic Positioning: Recent advances, applications, potential solutions, and future prospects. *IEEE Internet of Things Journal*, *1*(6), 6250–6274. Advance online publication. DOI: 10.1109/JIOT.2024.3521084

Qu, H., & Su, W. (2024). Deep Learning-Based Weed–Crop Recognition for Smart Agricultural Equipment: A review. *Agronomy (Basel)*, *14*(2), 363. DOI: 10.3390/agronomy14020363

Ravichandran, N., Tewaraja, T., Rajasegaran, V., Kumar, S. S., Gunasekar, S. K. L., & Sindiramutty, S. R. (2024). Comprehensive Review Analysis and Countermeasures for Cybersecurity Threats: DDoS, Ransomware, and Trojan Horse Attacks. *Preprints.org*. DOI: 10.20944/preprints202409.1369.v1

Rejeb, A., Rejeb, K., Simske, S., & Treiblmaier, H. (2021). Humanitarian Drones: A review and research agenda. *Internet of Things : Engineering Cyber Physical Human Systems*, *16*, 100434. DOI: 10.1016/j.iot.2021.100434

Rezaee, M. R., Hamid, N. W. A., Hussin, M., & Zukarnain, Z. A. (2024). Comprehensive Review of Drones Collision Avoidance Schemes: Challenges and open issues. *IEEE Transactions on Intelligent Transportation Systems*, *25*(7), 6397–6426. DOI: 10.1109/TITS.2024.3375893

Riza, A. Z. B. M., Jennsen, L., Anggani, P., Rafeen, A. I., Ruth, P. N. J., Sookun, D., Sookun, V., Yusri, N. Z. B. M., Sern, L. J., Luximon, L., Omer, M. L., & Sindiramutty, S. R. (2025). Leveraging Machine Learning and AI to Combat Modern Cyber Threats. *preprints.org*. DOI: 10.20944/preprints202501.0360.v1

Rovira-Sugranes, A., Razi, A., Afghah, F., & Chakareski, J. (2022). A review of AI-enabled routing protocols for UAV networks: Trends, challenges, and future outlook. *Ad Hoc Networks*, *130*, 102790. DOI: 10.1016/j.adhoc.2022.102790

Sabatini, R., Bartel, C., Shaid, T., & Kaharkar, A. (2012). Low-cost Vision Sensors and Integrated Systems for Unmanned Aerial Vehicle Navigation. *Journal of Systems and Software*, *2*(11), 323–349. https://scientific-journals.org/journalofsystemsandsoftware/archive/vol2no11/vol2no11_5.pdf

Saif, A. A. (2016). *The Drone Eats with Me: A Gaza Diary*. Beacon Press.

Sarvajcz, K., Ari, L., & Menyhart, J. (2024). AI on the Road: NVIDIA Jetson Nano-Powered Computer Vision-Based System for Real-Time Pedestrian and Priority Sign Detection. *Applied Sciences (Basel, Switzerland)*, *14*(4), 1440. DOI: 10.3390/app14041440

Schedl, D. C., Kurmi, I., & Bimber, O. (2021). An autonomous drone for search and rescue in forests using airborne optical sectioning. *Science Robotics*, *6*(55), eabg1188. Advance online publication. DOI: 10.1126/scirobotics.abg1188 PMID: 34162744

Scriban, A., Nabeneza, S., Cornelis, D., Delay, E., Vayssières, J., Cesaro, J., & Salgado, P. (2024). GPS-Based hidden Markov models to document pastoral mobility in the Sahel. *Sensors (Basel)*, *24*(21), 6964. DOI: 10.3390/s24216964 PMID: 39517860

Sehrawat, N., Vashisht, S., Singh, A., Dhiman, G., Viriyasitavat, W., & Alghamdi, N. S. (2024). A power prediction approach for a solar-powered aerial vehicle enhanced by stacked machine learning technique. *Computers & Electrical Engineering*, *115*, 109128. DOI: 10.1016/j.compeleceng.2024.109128

Seng, Y. J., Cen, T. Y., Raslan, M. H. B. M., Subramaniam, M. R., Xin, L. Y., Kin, S. J., Long, M. S., & Sindiramutty, S. R. (2024). In-Depth Analysis and Countermeasures for Ransomware Attacks: Case Studies and Recommendations. *Preprint.org*. DOI: 10.20944/preprints202408.2261.v1

Shah, I. A., Jhanjhi, N. Z., & Laraib, A. (2022). Cybersecurity and blockchain usage in contemporary business. In *Advances in information security, privacy, and ethics book series* (pp. 49–64). DOI: 10.4018/978-1-6684-5284-4.ch003

Shaiba, O. A. (2020). *Update urban basemap by using the LiDAR mobile mapping system : the case of Abu Dhabi municipal system*. DOI: 10.5821/dissertation-2117-330389

Shaikh, T. A., Mir, W. A., Rasool, T., & Sofi, S. (2022). Machine Learning for smart agriculture and Precision Farming: Towards making the fields talk. *Archives of Computational Methods in Engineering*, *29*(7), 4557–4597. DOI: 10.1007/s11831-022-09761-4

Sharma, R., Singh, A., Kavita, N., Jhanjhi, N. Z., Masud, M., Jaha, E. S., & Verma, S. (2021). Plant disease diagnosis and image classification using deep learning. *Computers, Materials & Continua/Computers. Materials & Continua (Print)*, *71*(2), 2125–2140. DOI: 10.32604/cmc.2022.020017

Sheng, L. T., Azman, A. B., Khan, N. A., Sindiramutty, S. R., Singh, S., & Verma, S. (2024). Reinforcement learning-based traffic signal control. *2024 International Conference on Emerging Trends in Networks and Computer Communications (ETNCC)*, 792–798. DOI: 10.1109/ETNCC63262.2024.10767501

Shokri, D., Larouche, C., & Homayouni, S. (2024). Proposing an Efficient Deep Learning Algorithm Based on Segment Anything Model for Detection and Tracking of Vehicles through Uncalibrated Urban Traffic Surveillance Cameras. *Electronics (Basel)*, *13*(14), 2883. DOI: 10.3390/electronics13142883

Sindiramutty, S. R. (2024). Autonomous Threat Hunting: a future paradigm for AI-Driven Threat intelligence. *arXiv (Cornell University)*. https://doi.org//arxiv.2401.00286DOI: 10.48550

Sindiramutty, S. R., Prabagaran, K. R. V., Akbar, R., Hussain, M., & Malik, N. A. (2024). Overview of Generative AI techniques for cybersecurity. In *Advances in information security, privacy, and ethics book series* (pp. 1–52). DOI: 10.4018/979-8-3693-5415-5.ch001

Sindiramutty, S. R., Prabagaran, K. R. V., Jhanjhi, N. Z., Ghazanfar, M. A., Malik, N. A., & Soomro, T. R. (2024). Security Considerations in Generative AI for web Applications. In *Advances in information security, privacy, and ethics book series* (pp. 281–332). DOI: 10.4018/979-8-3693-5415-5.ch009

Sindiramutty, S. R., Prabagaran, K. R. V., Jhanjhi, N. Z., Murugesan, R. K., Brohi, S. N., & Masud, M. (2024). Generative AI in network security and intrusion detection. In *Advances in information security, privacy, and ethics book series* (pp. 77–124). DOI: 10.4018/979-8-3693-5415-5.ch003

Sindiramutty, S. R., Prabagaran, K. R. V., Jhanjhi, N. Z., Murugesan, R. K., Brohi, S. N., & Wei, G. W. (2024). Generative AI for threat intelligence and information sharing. In *Advances in digital crime, forensics, and cyber terrorism book series* (pp. 191–234). DOI: 10.4018/979-8-3693-8944-7.ch006

Sindiramutty, S. R., Prabagaran, K. R. V., Jhanjhi, N. Z., Murugesan, R. K., Malik, N. A., & Hussain, M. (2024). Ethics and transparency in secure web model generation. In *Advances in information security, privacy, and ethics book series* (pp. 411–464). DOI: 10.4018/979-8-3693-5415-5.ch012

Singh, R., & Kumar, S. (2025). A Comprehensive Insights into Drones: History, Classification, Architecture, Navigation, Applications, Challenges, and Future Trends. *arXiv (Cornell University)*. https://doi.org//arxiv.2501.10066DOI: 10.48550

Singh, S., Zuber, M., Hamidon, M. N., Mazlan, N., Basri, A. A., & Ahmad, K. A. (2022). Classification of actuation mechanism designs with structural block diagrams for flapping-wing drones: A comprehensive review. *Progress in Aerospace Sciences*, *132*, 100833. DOI: 10.1016/j.paerosci.2022.100833

Singhal, V., Jain, S. S., Anand, D., Singh, A., Verma, S., Kavita, N., Rodrigues, J. J. P. C., Jhanjhi, N. Z., Ghosh, U., Jo, O., & Iwendi, C. (2020). Artificial Intelligence Enabled Road Vehicle-Train Collision Risk Assessment Framework for Unmanned railway level crossings. *IEEE Access : Practical Innovations, Open Solutions*, *8*, 113790–113806. DOI: 10.1109/ACCESS.2020.3002416

Soori, M., Arezoo, B., & Dastres, R. (2023). Artificial intelligence, machine learning and deep learning in advanced robotics, a review. *Cognitive Robotics*, *3*, 54–70. DOI: 10.1016/j.cogr.2023.04.001

Sten, G., Feng, L., & Möller, B. (2025). Enhancing Off-Road Topography Estimation by Fusing LIDAR and Stereo Camera Data with Interpolated Ground Plane. *Sensors (Basel)*, *25*(2), 509. DOI: 10.3390/s25020509 PMID: 39860879

Suanpang, P., & Jamjuntr, P. (2024). Optimizing Autonomous UAV Navigation with D* Algorithm for Sustainable Development. *Sustainability (Basel)*, *16*(17), 7867. DOI: 10.3390/su16177867

Sun, H., Xue, J., Zhang, Y., Li, H., Liu, R., Song, Y., & Liu, S. (2024). Novel method of rapid and accurate tree trunk location in pear orchard combining stereo vision and semantic segmentation. *Measurement*, *242*, 116127. DOI: 10.1016/j.measurement.2024.116127

Svanstrom, F., Englund, C., & Alonso-Fernandez, F. (2021). Real-Time drone detection and tracking with visible, thermal and acoustic sensors. *2022 26th International Conference on Pattern Recognition (ICPR)*, 7265–7272. DOI: 10.1109/ICPR48806.2021.9413241

Talwandi, N. S., Thakur, P., & Khare, S. (2024). An Automatic Navigation System for New Technical Advanced Drones for Different Alpplications. *2024 IEEE International Conference on Computing, Power and Communication Technologies (IC2PCT)*, 736–741. DOI: 10.1109/IC2PCT60090.2024.10486524

Tarekegn, G. B., Juang, R., Lin, H., Munaye, Y. Y., Wang, L., & Bitew, M. A. (2022). Deep-Reinforcement-Learning-Based drone base station deployment for wireless communication services. *IEEE Internet of Things Journal*, *9*(21), 21899–21915. DOI: 10.1109/JIOT.2022.3182633

Thakur, A., & Mishra, S. K. (2024). An in-depth evaluation of deep learning-enabled adaptive approaches for detecting obstacles using sensor-fused data in autonomous vehicles. *Engineering Applications of Artificial Intelligence*, *133*, 108550. DOI: 10.1016/j.engappai.2024.108550

Thangamani, R., Suguna, R. K., & Kamalam, G. K. (2024). Drones and Autonomous Robotics Incorporating Computational Intelligence. *Wiley*, 243–296. DOI: 10.1002/9781394175437.ch8

Thomas, G. S., Muthukaruppasamy, S., Kumar, S. S., Karthikeyan, B. J., & Saravanan, K. (2024). Navigating the Nexus: Unravelling Challenges, Ethics, and Applications of Embodied AI in Drone Technology Through the Lens of Computer Vision. In *Building Embodied AI Systems: The Agents, the Architecture Principles, Challenges, and Application Domains* (pp. 61–78). DOI: 10.1007/978-3-031-68256-8_3

Tomasicchio, G., Andolfi, L., Brancati, M., Di Donna, A. M., Giannattasio, S., Del Prete, R., Ostrogovich, L., Renga, A., Grassi, M., Ceresoli, M., Silvestrini, S., & Lavagna, M. (2024). Multi-Sensor fusion and resilient PVT techniques for safe lunar landing missions. *Proceedings of the Satellite Division's International Technical Meeting (Online)/Proceedings of the Satellite Division's International Technical Meeting (CD-ROM)*, 3125–3143. DOI: 10.33012/2024.19849

Trawny, N., Johnson, A. E., Bailey, E. S., Massone, G., Reid, M., Setterfield, T. P., Cheng, Y., Sellar, G., & Soto, J. (2024). The Enhanced Lander Vision System for Mars sample return lander entry descent and landing. *AIAA SCITECH 2022 Forum*. DOI: 10.2514/6.2024-0314

Tuśnio, N., & Wróblewski, W. (2021). The Efficiency of Drones Usage for Safety and Rescue Operations in an Open Area: A Case from Poland. *Sustainability (Basel)*, *14*(1), 327. DOI: 10.3390/su14010327

Vaezi, M., Lin, X., Zhang, H., Saad, W., & Poor, H. V. (2023). Deep Reinforcement Learning for Interference Management in UAV-Based 3D Networks: Potentials and challenges. *IEEE Communications Magazine*, *62*(2), 134–140. DOI: 10.1109/MCOM.001.2200973

Vishweshwaran, M., & Sujatha, E. R. (2024). *A review on Applications of drones in geotechnical engineering*. Indian Geotechnical Journal., DOI: 10.1007/s40098-024-01071-9

Vrochidou, E., Oustadakis, D., Kefalas, A., & Papakostas, G. A. (2022). Computer vision in Self-Steering tractors. *Machines (Basel)*, *10*(2), 129. DOI: 10.3390/machines10020129

Waheed, A., Seegolam, B., Jowaheer, M. F., Sze, C. L. X., Hua, E. T. F., & Sindiramutty, S. R. (2024). Zero-Day Exploits in Cybersecurity: Case Studies and Countermeasure. *Preprints.org*. DOI: 10.20944/preprints202407.2338.v1

Wang, K., Kooistra, L., Pan, R., Wang, W., & Valente, J. (2024). UAV-based simultaneous localization and mapping in outdoor environments: A systematic scoping review. *Journal of Field Robotics*, *41*(5), 1617–1642. DOI: 10.1002/rob.22325

Wang, S., & Ahmad, N. S. (2024). A comprehensive review on sensor fusion techniques for localization of a dynamic target in GPS-Denied environments. *IEEE Access : Practical Innovations, Open Solutions*, *1*. Advance online publication. DOI: 10.1109/ACCESS.2024.3519874

Wang, X., Zheng, S., Lin, X., Zhang, Q., & Liu, X. (2024). Robust loop closure detection and relocalization with semantic-line graph matching constraints in indoor environments. *International Journal of Applied Earth Observation and Geoinformation*, *129*, 103844. DOI: 10.1016/j.jag.2024.103844

Weiqi, X., Hooi, S. T. C., Sindiramutty, S. R. L., Asirvatham, D. L., Kumar, D., & Verma, S. (2024). Surface Anomaly Detection Using Machine Learning Technique. *2024 International Conference on Emerging Trends in Networks and Computer Communications (ETNCC)*, 1–7. DOI: 10.1109/ETNCC63262.2024.10767562

Wolniak, R., Stecuła, K., & Aydın, B. (2024). Digital Transformation of Grocery In-Store Shopping-Scanners, Artificial intelligence, Augmented Reality and beyond: A review. *Foods*, *13*(18), 2948. DOI: 10.3390/foods13182948 PMID: 39335877

Xi, H., Ru, L., Tian, J., Lu, B., Hu, S., Wang, W., & Luan, X. (2025). URADV: A novel framework for generating Ultra-Robust adversarial patches against UAV object detection. *Mathematics*, *13*(4), 591. DOI: 10.3390/math13040591

Xu, Y., Pan, L., Du, C., Li, J., Jing, N., & Wu, J. (2018). Vision-based UAVs Aerial Image Localization. *ACM*, 9–18. DOI: 10.1145/3281548.3281556

Xun, A. T., En, L. Z., Shen, L. T., Xin, A. N., Soon, W. H., Jun, W. Z., Ramachandra, H., Xinghao, G., Khant, N. M., Weitao, F., & Sindiramutty, S. R. (2025). Building Trust in Cloud Computing:Strategies for Resilient Security. *Preprints.org*. DOI: 10.20944/preprints202501.0716.v1

Ying, X., Murugesan, R. K., Sindiramutty, S. R., Wei, G. W., Balakrishnan, S., Kumar, D., & Verma, S. (2024). Scene Text Recognition using Deep Learning Techniques. *2024 International Conference on Emerging Trends in Networks and Computer Communications (ETNCC)*, 1–9. DOI: 10.1109/ETNCC63262.2024.10767484

You, H., Li, S., Xu, Y., He, Z., & Wang, D. (2021). Tree Extraction from Airborne Laser Scanning Data in Urban Areas. *Remote Sensing (Basel)*, *13*(17), 3428. DOI: 10.3390/rs13173428

Yu, P. (2024). Self-driving car implementation guided by computer vision for detection. *Applied and Computational Engineering*, *74*(1), 1–6. DOI: 10.54254/2755-2721/74/20240420

Zhang, L., Yu, X., Adu-Gyamfi, Y., & Sun, C. (2023). Spatio-Temporal fusion of LiDAR and camera data for omnidirectional depth perception. *Transportation Research Record: Journal of the Transportation Research Board*, *2678*(4), 308–322. DOI: 10.1177/03611981231184187

Zhang, Z., Liu, X., Zhou, H., Xu, S., & Lee, C. (2023). Advances in Machine-Learning Enhanced Nanosensors: From cloud artificial intelligence toward future edge computing at chip level. *Small Structures*, *5*(4), 2300325. Advance online publication. DOI: 10.1002/sstr.202300325

Zhu, P., Wen, L., Du, D., Bian, X., Fan, H., Hu, Q., & Ling, H. (2021). Detection and Tracking Meet Drones challenge. *IEEE Transactions on Pattern Analysis and Machine Intelligence*, *44*(11), 7380–7399. DOI: 10.1109/TPAMI.2021.3119563 PMID: 34648430

Chapter 7

Future Implications of the Drone Industry: Emerging Trends and Potential Challenges

Muhammad Tahir
School of Software Technology, Dalian University of Technology, China

Muhammad Aamir
Huanggang Normal University, China

Sajjad Ali Rajper
https://orcid.org/0009-0009-8150-4959
Iqra University, Karachi, Pakistan

Rajesh Kumar
Department of Computer Science, University of Milano Bicocca, Italy

ABSTRACT

The primary objective of this chapter is to explore the future implications of drone technology and focus on emerging trends such as autonomous, significant growth of the drone industry, and transformation sectors such as healthcare, disaster management, logistics, and agriculture. Further, the focus should be on the evolving drone policies, urban air traffic control, signal interference, hacking, and data breaches. In view of this, the Real-time data processing, increased use cases, and improved operational efficiency are all anticipated benefits of these developments. Additionally, the need for UAVs to have greater mobility, autonomy, and range led to the development of innovative battery-swapping, docking station, and precision landing systems. As the drone industry can transform a variety of industries, such as

DOI: 10.4018/979-8-3693-8497-8.ch007

logistics and public safety, actively dealing with the challenges and navigating these trends. It will help to create a future that is more robust, connected and efficient.

INTRODUCTION

As the drone industry can transform a variety of industries, such as logistics and public safety, actively dealing with the challenges and navigating these trends. Improved mission safety, lower operating costs, significant payload capacity gains, ease of reuse, and improvements in battery endurance, propulsion systems, stabilisation strategies, navigation, and onboard sensor technologies are just a few of the many benefits that unmanned aerial systems (UAS) offer over manned vehicles. These systems give access to physically difficult areas and allow long-range operations beyond human observers' detection range(Yadav et al., 2024, Sulaiman et al., 2024). The aerial vehicle itself is explicitly referred to by the abbreviation UAV. On the other hand, the abbreviation UAS refers to the complete system, which consists of the aircraft, a GCS, and a communication subsystem (data link) for information transmission and reception. The usage of UAVs for a variety of operational duties in urban, suburban, and rural locations has been gradually growing in recent years(-Sulaiman et al., 2024, Ariante and Papa, 2024). Figureure 1 Overview of the future implications of the drone industry.

Figure 1. Overview of future implications of ID

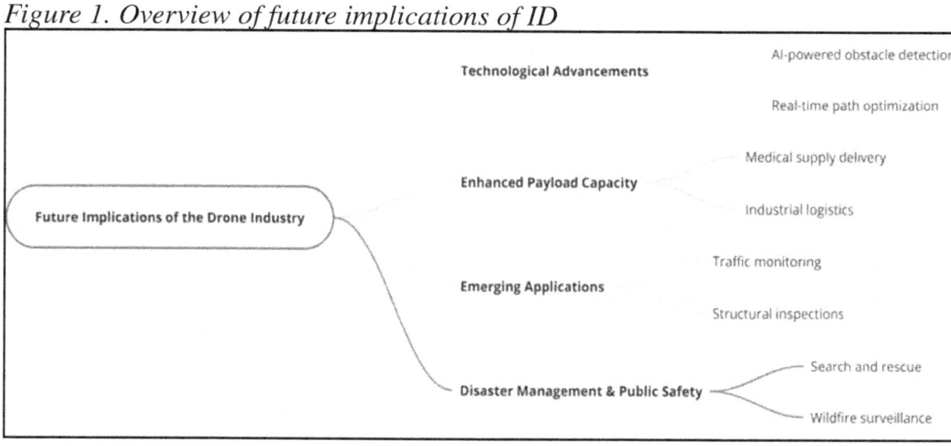

Precision farming, emergency medical supply delivery to remote areas, surveillance, disaster management, SAR operations, coastguard and sea-lane monitoring, fire detection, Drones 2025, 9, x for peer review response services, environmental conservation, pollution monitoring, and land surveying are just a few of the applications. New guidelines have been created to guarantee the accessibility and safe operation of UAVs in urban settings, hence facilitating their safe incorporation(-Goyal et al., 2018, Goyal et al., 2021). The idea of UAM, a unique air transportation paradigm that permits the use of drones for passenger and cargo transportation within urban areas, was born out of this development. Civilian UAVs are regarded as recoverable drones. If properly regulated, unmanned aerial vehicles (UAVs) typically provide a safer operating environment than manned aerial vehicles because of their unmanned capability. For more hazardous missions and activities, a UAV may be a viable alternative without endangering the lives of human operators. One significant operational disadvantage of UAVs is their very short flying times when compared to conventional aircraft(Cohen et al., 2021, Shah, 2024c). Furthermore, because UAVs are often smaller than manned aerial vehicles, they are more susceptible to weather-related issues, and in inclement weather, they are still difficult to manoeuvre. UAVs are already being used in many different applications, starting with aerial recordings of the planet. These include mapping wildfires, doing agricultural surveys, inspecting buildings and bridges, monitoring pipelines, surveying powerlines, and many more. In the near future, the UAV sector will be essential to the development of technology worldwide.

As UAVs have cameras on board, they are frequently utilised for surveillance. They are also capable of delivering light goods and carrying equipment other than cameras. UAVs have been widely deployed by the military for both humanitarian and war purposes. The military uses over 70% of all UAVs, followed by the consumer sector (17%) and the commercial sector (13%). The commercial business sector has the fastest growth potential for UAVs, despite the military sector currently being the largest market(Hasan, 2019, Menichino et al., 2022, Fellner, 2021). At the moment, it appears that the infrastructure sector has the most potential for growth, followed by the transportation and agricultural sectors. The tremendous growth potential and commercial interest of UAVs, along with the quick development of technology and laws, have led to a sharp rise in research on UAV-related applications and their capabilities.

LITERATURE REVIEW

Unmanned Aerial Vehicles (UAVs) have become a game-changing technology in traffic monitoring and management systems because of their quick deployment times, low operating costs, and great mobility. Static infrastructure, including cameras, sensors, and reliable communication networks, are essential components of traditional ground-based systems for traffic control and road condition assessments.

Synonymous with the War on Terror, frequently referred to as the conflict's emblematic platform. The asymmetric imagery of the War on Terror is changing, even though the discourse surrounding it has traditionally concentrated on the massive platforms that the US Air Force and Central Intelligence Agency (CIA) use extensively(Capitán et al., 2019, Shah et al., 2024b, Schaefer et al., 2022). Commercially available off-the-shelf (COTS) drones are becoming a more common sight on the battlefield, when non-state actors use them to visualise such areas and direct and cause damage.

After all, drones are becoming more common, tested, and developed for use in domestic airspace. Here, drones are used and expected to be used for a variety of tasks, including infrastructure monitoring and inspection, security, emergency support, photo and videography, and the transportation of commercial and medical supplies. An aerial vehicle without a human operator on board is called a UAV. Often called a "drone," a UAV is a machine that can fly on its own or be controlled remotely(Castrillo et al., 2022, Shah et al., 2025, Abro et al., 2022). Kamikaze drones, which are made to attack targets by colliding with them, are an example of expendable UAVs. The use of UAVs in commercial businesses has increased recently, and they are now being used for a wide range of diverse jobs across many industries. UAVs are being used by farmers to collect crop data in order to increase crop yields. Because UAVs can provide comprehensive maps of water and forest vegetation, they are becoming increasingly important in environmental conservation initiatives(Khan et al., 2020, Ullah et al., 2020, Shahzadi et al., 2021). In order to increase worker safety, the mining industry is also using them for indoor mapping and inspection. UAVs are being utilised more and more in the construction sector to map and survey building sites. Figureure 3 Overview of the taxonomy of literature review

Figure 2. Overview of the taxonomy of literature review

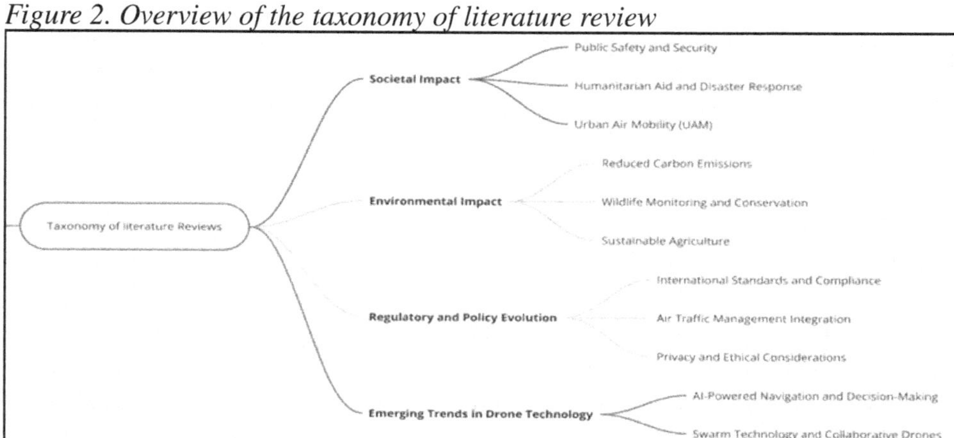

Furthermore, even in the absence of conventional network access, UAVs can create ad hoc networks to send crucial data to other AVs and control centres by utilising sophisticated communication protocols. In situations where conventional, centralised data processing and transmission techniques would be unfeasible, these strategies allow the UAVs to function efficiently and independently. Through the integration of various technologies, (Yang et al., 2021, Jhanjhi et al., 2025) the system can analyse data efficiently, respond with low latency, and rely less on high-bandwidth communication routes, making it suitable for both urban and remote settings.

OVERVIEW OF THE DRONE INDUSTRY

Drones have evolved from military equipment to commonplace commercial tools that top businesses use to save lives, money, and time in a variety of industries. Small businesses and entrepreneurs can now use drones for a growing range of applications thanks to the democratisation of access brought about by falling costs for essential parts like sensors, cameras, and navigation systems(Boiteau et al., 2024, Ariante et al., 2021b, Ariante et al., 2021a). Drones are proving to be particularly well-suited for precise, repetitive, or dangerous activities that are typically completed by expensive and risky manual means, such as filming blockbuster films, monitoring vital infrastructure, or evaluating the condition of crops. Drones are being used in the agriculture sector for a variety of tasks, including accurate pesticide spraying and crop health analysis. Farmers can get extremely detailed maps that show variables like soil variance, irrigation requirements, and pest damage across fields thanks to specialised drone sensors(Lerro et al., 2020, Ariante et al., 2020). Drones enable precision agriculture, which optimises inputs for increased production and profitability by segmenting land into highly customised management zones. Compared to tractor operations, automated drone spraying can help reduce agricultural pesticide usage by up to 20 times. Adoption is accelerating due to cost reductions and higher yields(Shah, 2024a). The global market for agricultural drones is now worth over $1 billion a year, according to some studies.

Manual, risky inspections are frequently necessary for infrastructure that is hazardous and difficult to access. Drones use distant sensing to solve this problem. Energy companies can use drones to quickly examine pipelines, power lines, wind turbines, and offshore oil rigs without interfering with service. Additionally, public agencies use high-resolution drone cameras to survey water, rail, bridges, and roadways(Peng et al., 2020, Liu et al., 2021). The global market for drones used for infrastructure inspection is expected to grow to a value of over $7 billion annually. Drone technology has also been adopted by private security companies and law enforcement for improved response coordination and surveillance activities.

Companies like UPS, Amazon, and Alphabet's Wing are working hard to create drone delivery systems to cut expenses and expedite order fulfilment. In a matter of minutes as opposed to hours, lightweight drones with vertical takeoff and landing capabilities may deliver a variety of tiny products straight to homes and businesses. According to one estimate, drone deliveries may save global supply networks a total of $50 billion a year. Commercial drone delivery trials have been authorised in several nations, including Australia, Iceland, and Singapore. As enabling laws and technology advance, full-scale rollouts are anticipated in the next three to five years(Jiang et al., 2021, Gyagenda et al., 2022).

By offering exciting new camera viewpoints and mobility at a tenth of the price of more traditional equipment like cranes and helicopters, drones have completely changed the film and entertainment industries. Drone cinematography is used in well-known films and television series like Game of Thrones to create breathtaking images. Drones are now frequently used by professional photographers and videographers for content creation, events, and real estate listings. Beyond film and television, new professional drone racing leagues made possible by innovations like FPV goggles are becoming a spectator sport and an industry with millions of viewers worldwide(Besada et al., 2021, Cioffi et al.). Drones are being used by the insurance business in place of manual in-person inspections to quickly analyse claims following disasters. Sophisticated drones are used by oil and gas companies to identify pipeline problems early and maximise environmental protection. Utility firms use drones to inspect gas, electricity, and district heating infrastructure, which speeds up maintenance and improves dependability. Drones are used by construction businesses for project supervision, site planning, and surveying. The examples show how drones can be used in a variety of industries as costs come down and capabilities advance via consistent innovation. Beyond upending long-standing industries, the widespread use of drones raises both opportunities and issues for society at large. By increasing automation, expanding commercial drone operations seeks to support environmental sustainability, productivity, and economic progress(Yasin et al., 2020, Yu and Zhang, 2015, Shah et al., 2024a). However, as consumer access increases, particularly in metropolitan areas, governments must address citizen concerns about privacy, safety, and nuisance complaints. Although research on public opinion shows that opinions on drones are mixed, sceptics may eventually be won over by sensible rules and proper use that produce real local advantages.

Overview of the Drone Industry and Its Emerging Trends

UAVs are starting to emerge and are slowly being integrated into the current transportation infrastructures. In the near future, UAVs have the potential to drastically change and become an integral part of the transportation sector. As a result of a decrease in road capacity and an increase in the number of cars on the road, the current ground transportation infrastructure has become more congested every year.

The global market for UAV-assisted logistics and transportation is expected to grow from 5.3 billion dollars to 11 billion dollars by 2026(Wang et al., 2021, Aldao et al., 2022). UAVs have been used for backyard delivery, e-commerce, and medical logistic delivery. The burgeoning e-commerce industry and rising consumer demands for quicker delivery times are the main causes of the UAV usage's explosive expansion. Other transportation-related fields, such as traffic monitoring, urban planning and management, and disaster response or relief, are also incorporating UAVs. UAVs will be essential in creating a fully automated transportation system in the future, as more and more cars are becoming automated. They will help with road surveys, automated field support teams, traffic police, and rescue teams. A UAV generates vehicle lift using aerodynamic forces(Ponte et al., 2020, Shah, 2024c, Ji et al., 2021). Based on their vertical lift technique, UAVs can be divided into two main groups: fixed-wing and rotary-wing. In essence, rotary-wing UAVs, sometimes referred to as multi-rotors, are rotorcraft that use two or more rotors to create lift. The number of propellers that provide the UAV with vertical thrust can subsequently be used to divide them into subdivisions. Figureure 4 Overview future trajectories.

Figure 3. Overview future trajectories

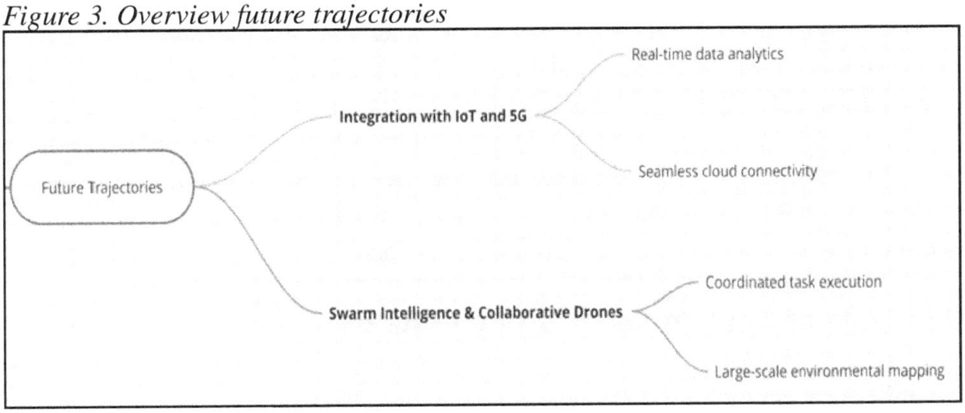

Aerodynamics-Based Categorisation Drone efficiency, range, and adaptability are all impacted by aerodynamic designs. Because of their long endurance and vast range, the aerodynamic High-Altitude Platforms (HAP) are perfect for atmospheric research, climate monitoring, and strategic reconnaissance. They operate at altitudes exceeding 20 km. Medium Altitude Platforms (MAP) are mostly utilised for border patrol, surveillance, and other mid-range activities that require continuous observation. They operate between 5 and 20 km.

Weight has a direct impact on payload capacity, range, and endurance, making it a crucial consideration in drone classification. The kind of weight-based classification. With a weight of less than 200 g, nano drones are incredibly light and made for close-quarters missions; they are frequently employed in reconnaissance and surveillance(Shah et al., 2024b, Ariante et al., 2019). Due to their portability and low weight (less than 5 kg), micro drones are ideal for outdoor photography and simple surveillance duties. Light drones, which weigh between 5 and 50 kg, can carry small loads and heavy sensors, making them useful for precision farming and environmental monitoring. For industrial inspections and mid-range logistics, medium drones—which weigh between 50 and 200 kg—offer a compromise between portability and capability. On the higher end of the spectrum, super heavy drones—which weigh more than 2000 kg—are designed to carry massive cargo and military-grade payloads, assisting with complex missions and strategic operations. Open spaces. Additionally, drones can be grouped according to how they take off and land, which dictates their operational settings and flexibility in deployment(Shah, 2024b).

Other important grading criteria are weather resistance and durability. Drones that can survive rain, wind, and extremely high or low temperatures include the DJI Matrice 300 RTK. The BlueROV2 and other underwater drones are made for submarine operations. Based on their intended usage, drones are categorised, demonstrating their adaptability to a wide range of sectors. The main functions of military drones, such as the MQ-9 Reaper, are target acquisition and observation. Commercial drones, such as the DJI Inspire 2, are frequently used for delivery and aerial photography. Precision farming and inspections are frequent uses for industrial drones like the SenseFly Albris. While medical drones like Zipline are utilised for emergency medical supplies delivery, environmental drones like the Zephyr S are made for animal conservation and climate monitoring. Drones for recreation, such as the DJI Mini 3, are widely used for hobby flying and aerial photography.

Overview of Challenges in Drone Industry

Utilising technologies that make use of computer resources not available in traditional or cloud data centres, edge computing shifts the load closer to the site of data generation. It makes it possible to respond to data analysis. Edge computing leverages and manages the processing power available at remote locations, including production plants, shops, storage facilities, hotels, and distribution hubs. An edge device is a piece of equipment with a specific purpose and integrated processing power. An automobile, an ATM, an intelligent camera, or an assembly machine on a production floor are examples of edge devices that can be utilised for engaging tasks. Economic considerations often motivate the use of edge devices, which typically possess limited computing resources(Ran et al., 2022, Shang et al., 2023). Although edge devices have the potential to be more powerful, they are now the exception rather than the rule. A multipurpose computer used for information technology that is located in a remote operations facility, such as a hotel, retail store, or factory. An edge gateway is typically an edge cluster or server that also hosts services that perform network functions like protocol translation, network termination, tunnelling, firewall protection, or wireless connectivity, in addition to hosting shared services and business application workloads. Although some edge devices might serve as restricted gateways or host network services, edge gateways are usually different from edge devices. Figureure 1 Overview of challenges in advance in DI.

Figure 4. Overview of challenges in advance in DI

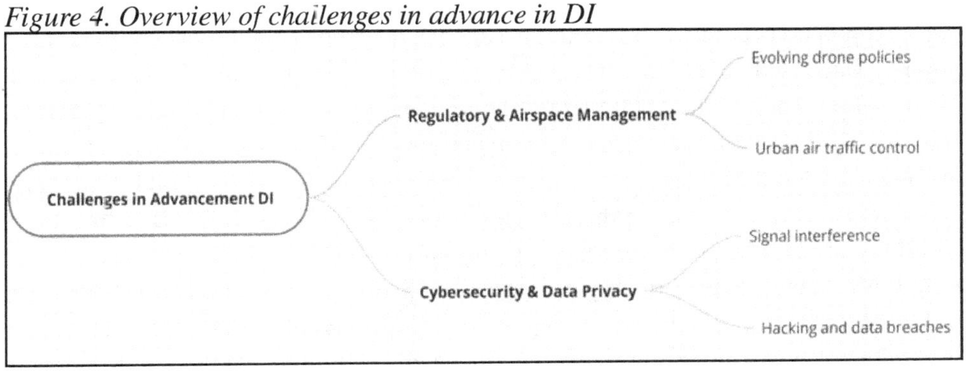

IoT sensors are fixed-function gadgets that collect and transmit data to an edge or cloud without internal processing, memory, or storage. Containers cannot be installed on them as a result. Identifying and preventing collisions with manned aircraft is one of the most crucial duties in UAV operation, particularly in shared airspace. Both piloted and unpiloted aircraft must be able to locate other aircraft that are safe to fly. Recent research around aeroplane discovery for UAVs has focused on sophisticated systems that use a range of detectors and optic cameras to detect and track aircraft in real-time. However, using standard radar systems to represent sensitive aircraft, such as general aviation aircraft, might be risky. The link between drone assiduity and pall waiters has led to a new era of technological innovation and practical efficacy(Ariante et al., 2022). The employment of drones in a variety of tasks, including husbandry, guidance, environmental monitoring, structure inspection, and logistics, is being revolutionised by this group. Pall computing provides dependable data operating systems and storage capacities to handle the enormous amounts of data generated by drones. This enables faultless messaging, advanced dissection, and improved real-time processing. Strong computers known as "edge servers" are positioned near a network's edge, where data processing needs to take place. An edge server's physical location is frequently near the systems or applications that generate the data that the edge servers use or store. A more conventional server setup, on the other hand, usually keeps data in the cloud or a data centre. Important computers known as "byte waiters" are positioned at the network's byte, where data processing must take place.

Risk Assessment and Disaster Management

During tasks that may be dangerous or impossible for humans, UAVs are crucial support tools in catastrophe situations, whether they be man-made or natural, such as fires, tsunamis, or terrorist attacks. With sensors like radars, cameras, and LiDAR, as well as sophisticated onboard avionics systems, UAVs help rescue teams by detecting damage, starting recovery efforts quickly, and delivering vital supplies like first aid kits or supporting manned helicopters. Their capacity to cover large regions in real time greatly improves operational effectiveness while maintaining employee safety and security. Furthermore, UAV-generated early warnings are essential for saving lives and protecting species in threatened locations(Siewert et al., 2020). Since they use their capacity to glide to reduce the amount of fuel or energy they use, fixed-wing UAVs are incredibly energy efficient. The primary constraint of an unmanned aerial vehicle is the need for a runway or catapult for takeoff and landing. They can only take off horizontally, just like an aeroplane. Although fixed-wing UAVs are often larger and more stable, with superior flying capabilities, cargo capacity, and endurance, their bulk makes them less nimble than multirotor(Sier et al., 2023).

As technology develops, the usage of UAVs in several industry sectors is expanding quickly. However, before the widespread use of UAVs in ITS, privacy and security issues such as vulnerability to theft or hacking, invasion of privacy, and collision responsibility must be considered. UAVs frequently fly in populous areas and cover wide areas. Because of this, the system is vulnerable to hacking, and an unprotected UAV is harmful. Investigations were conducted into UAV security flaws(Riordan et al., 2021, Yan et al., 2019). The majority of the five methods a cyber-terrorist could hack and interfere with UAV use are quite likely to happen. Signal jamming is the initial technique. To interfere with the UAV's ability to receive communications, the perpetrator creates interference signals at a similar frequency band.

CONCLUSION AND FUTURE WORK

the substantial expansion of the drone market, which is revolutionising industries like agriculture, logistics, healthcare, and disaster relief. Therefore, it is expected that these improvements would lead to real-time data processing, more use cases, and increased operational efficiency. The requirement for UAVs to have increased mobility, autonomy, and range also prompted the creation of cutting-edge precise landing, docking station, and battery-swapping technologies. highway traffic control, wireless communication, remote sensing, precision farming, fire detection and prevention, power-line monitoring, and disaster monitoring. Studies have switched to imagining UAVs' position in the future of passenger and cargo transportation because of their growing integration into transportation networks. The integration of ATM and UTM systems, which permits smooth coordination between manned and unmanned vehicles operating in shared airspace, is one of the main new problems.

As the drone industry can transform a variety of industries, such as logistics and public safety, actively dealing with the challenges and navigating these trends. It will help to create a future that is more robust, connected and efficient.

REFERENCES

Abro, G., Zulkifli, S., Masood, R., Asirvadam, V. & Laouti, A. (2022). Comprehensive Review of UAV Detection, Security, and Communication Advancements to Prevent Threats. Drones 2022, 6, 284. *Advances in UAV Detection, Classification and Tracking*, 63.

Aldao, E., González-De Santos, L. M., & González-Jorge, H. (2022). LiDAR based detect and avoid system for UAV navigation in UAM corridors. *Drones (Basel)*, *6*(8), 185. DOI: 10.3390/drones6080185

Ariante, G., & Papa, U. (2024). *Embedded Platforms for UAS Landing Path and Obstacle Detection II: Navigation Support Systems for Urban Air Mobility Operations*. Springer Nature. DOI: 10.1007/978-3-031-59754-1

Ariante, G., Papa, U., Ponte, S., & Del Core, G. (2019, June). UAS for positioning and field mapping using LIDAR and IMU sensors data: Kalman filtering and integration. In 2019 IEEE 5th International Workshop on Metrology for AeroSpace (MetroAeroSpace) (pp. 522-527). IEEE.

Ariante, G., Papa, U., Ponte, S., & Del Core, G. (2020, June). Velocity and attitude estimation of a small unmanned aircraft with micro Pitot tube and Inertial Measurement Unit (IMU). In 2020 IEEE 7th International Workshop on Metrology for AeroSpace (MetroAeroSpace) (pp. 314-318). IEEE.

Ariante, G., Ponte, S., Papa, U., & Del Core, G. (2021). 2021a. Estimation of Airspeed, Angle of Attack, and Sideslip for Small Unmanned Aerial Vehicles (UAVs) Using a Micro-Pitot Tube. [s Note: MDPI stays neutral with regard to jurisdictional claims in published....]. *Electronics (Basel)*, *10*(19), 2325. DOI: 10.3390/electronics10192325

Ariante, G., Ponte, S., Papa, U., & Del Core, G. (2021, June). Safe landing area determination (slad) for unmanned aircraft systems by using rotary lidar. In 2021 IEEE 8th international workshop on metrology for AeroSpace (MetroAeroSpace) (pp. 110-115). IEEE.

Ariante, G., Ponte, S., Papa, U., Greco, A., & Del Core, G. (2022). Ground control system for UAS safe landing area determination (SLAD) in urban air mobility operations. *Sensors (Basel)*, *22*(9), 3226. DOI: 10.3390/s22093226 PMID: 35590916

Besada, J. A., Campaña, I., Carramiñana, D., Bergesio, L., & De Miguel, G. (2021). Review and simulation of counter-UAS sensors for unmanned traffic management. *Sensors (Basel)*, *22*(1), 189. DOI: 10.3390/s22010189 PMID: 35009730

Boiteau, S., Vanegas, F., & Gonzalez, F. (2024). Framework for autonomous uav navigation and target detection in global-navigation-satellite-system-denied and visually degraded environments. *Remote Sensing (Basel)*, *16*(3), 471. DOI: 10.3390/rs16030471

Capitán, C., Capitán, J., Castano, A. R., & Ollero, A. (2019, June). Risk assessment based on SORA methodology for a UAS media production application. In *2019 International Conference on Unmanned Aircraft Systems (ICUAS)* (pp. 451-459). IEEE.

Castrillo, V. U., Manco, A., Pascarella, D., & Gigante, G. (2022). A review of counter-UAS technologies for cooperative defensive teams of drones. [CIOFFI, J. R., CERQUEIRA, C. S. & DOS SANTOS LIMA, J. S. A brief review of concepts and technological solutions to perfom DAA functions in the context of SIMUA Project.]. *Drones (Basel)*, *6*(3), 65. DOI: 10.3390/drones6030065

Cohen, A. P., Shaheen, S. A., & Farrar, E. M. (2021). Urban air mobility: History, ecosystem, market potential, and challenges. *IEEE Transactions on Intelligent Transportation Systems*, *22*(9), 6074–6087. DOI: 10.1109/TITS.2021.3082767

Fellner, R. (2021). *Unmanned Aircraft Systems Risk Assessment Based on SORA for First Responders and Disaster Management. Janik, P., Zawistowski, M., Fellner, R., & Zawistowski, G.* Unmanned Aircraft Systems Risk Assessment Based on SORA for First Responders and Disaster Management.

Goyal, R., Reiche, C., Fernando, C., & Cohen, A. (2021). Advanced air mobility: Demand analysis and market potential of the airport shuttle and air taxi markets. *Sustainability (Basel)*, *13*(13), 7421. DOI: 10.3390/su13137421

Goyal, R., Reiche, C., Fernando, C., Serrao, J., Kimmel, S., Cohen, A. & Shaheen, S. (2018). Urban air mobility (UAM) market study.

Gyagenda, N., Hatilima, J. V., Roth, H., & Zhmud, V. (2022). A review of GNSS-independent UAV navigation techniques. *Robotics and Autonomous Systems*, *152*, 104069. DOI: 10.1016/j.robot.2022.104069

Hasan, S. (2019). Urban air mobility (UAM) market study.

Jhanjhi, N., Shah, I. A., & Brohi, S. N. (2025). *Hardware Vulnerabilities: Taxonomy and Business Security Models. Generative AI for Web Engineering Models.* IGI Global.

Ji, Y., Li, S., Peng, C., Xu, H., Cao, R., & Zhang, M. (2021). Obstacle detection and recognition in farmland based on fusion point cloud data. *Computers and Electronics in Agriculture*, *189*, 106409. DOI: 10.1016/j.compag.2021.106409

Jiang, S., Jiang, W., & Wang, L. (2021). Unmanned Aerial Vehicle-Based Photogrammetric 3D Mapping: A survey of techniques, applications, and challenges. *IEEE Geoscience and Remote Sensing Magazine*, *10*(2), 135–171. DOI: 10.1109/MGRS.2021.3122248

Khan, S. K., Farasat, M., Naseem, U., & Ali, F. (2020). Performance evaluation of next-generation wireless (5G) UAV relay. *Wireless Personal Communications*, *113*(2), 945–960. DOI: 10.1007/s11277-020-07261-x

Lerro, A., Battipede, M., Gili, P., & Brandl, A. (2020). Aerodynamic angle estimation: Comparison between numerical results and operative environment data. *CEAS Aeronautical Journal*, *11*(1), 249–262. DOI: 10.1007/s13272-019-00417-x

Liu, X., Li, X., Shi, Q., Xu, C., & Tang, Y. (2021). UAV attitude estimation based on MARG and optical flow sensors using gated recurrent unit. *International Journal of Distributed Sensor Networks*, *17*(4), 15501477211009814. DOI: 10.1177/15501477211009814

Menichino, A., Di Vito, V., Dziugiel, B., Liberacki, A., Hesselink, H., & Giannuzzi, M. (2022, April). Urban air mobility perspectives over mid-term time horizon: main enabling technologies readiness review. In 2022 Integrated Communication, Navigation and Surveillance Conference (ICNS) (pp. 1-13). IEEE.

Peng, J., Zhang, P., Zheng, L., & Tan, J. (2020). UAV positioning based on multi-sensor fusion. *IEEE Access : Practical Innovations, Open Solutions*, *8*, 34455–34467. DOI: 10.1109/ACCESS.2020.2974285

Ponte, S., Ariante, G., Papa, U., & Del Core, G. (2020). An embedded platform for positioning and obstacle detection for small unmanned aerial vehicles. *Electronics (Basel)*, *9*(7), 1175. DOI: 10.3390/electronics9071175

Ran, W., Lan, Y., Dai, X., Gu, J., Liu, B., Geng, L., Han, X. & Yi, L. (2022). Obstacle detection system for autonomous vineyard robots based on passthrough filter. *International Journal of Precision Agricultural Aviation,* 5.

Riordan, J., Manduhu, M., Black, J., Dow, A., Dooly, G., & Matalonga, S. (2021, June). LiDAR simulation for performance evaluation of UAS detect and avoid. In *2021 International Conference on Unmanned Aircraft Systems (ICUAS)* (pp. 1355-1363). IEEE.

Schaefer, D., Hately, A., Ferrara, G., Lambers, D., Schweiger, K. & Rushton, A. P. (2022). The european concept of operations for U-space and urban air mobility. *SESAR Innovation Days 2022*.

Shah, I., Jhanjhi, N., & Ray, S. (2024a). IoT Devices in Drones: Security Issues and Future Challenges. Cybersecurity Issues and Challenges in the Drone Industry. *IGI Global.DOI, 10*, 979–8.

Shah, I. A. (2024a). Autonomous Shipping: Security Issues and Challenges. *Cybersecurity in the Transportation Industry*, 187-210.

Shah, I. A. (2024b). IoT-Based Smart Transportation Industry: Security Challenges. *Cybersecurity in the Transportation Industry*, 211-239.

Shah, I. A. (2024c). Privacy and security challenges in unmanned aerial vehicles (UAVs). *Cybersecurity in the Transportation Industry*, 93-115.

Shah, I. A., Jhanjhi, N., & Brohi, S. N. (2024b). *IoT Smart Healthcare Security Challenges and Solutions. Advances in Computational Intelligence for the Healthcare Industry 4.0*. IGI Global Scientific Publishing.

Shah, I. A., Jhanjhi, N., & Brohi, S. N. (2025). *Machine Learning Models for Detecting Software Vulnerabilities. Generative AI for Web Engineering Models*. IGI Global.

Shahzadi, R., Ali, M., Khan, H. Z., & Naeem, M. (2021). UAV assisted 5G and beyond wireless networks: A survey. *Journal of Network and Computer Applications, 189*, 103114. DOI: 10.1016/j.jnca.2021.103114

Shang, Y., Wang, H., Qin, W., Wang, Q., Liu, H., Yin, Y., Song, Z., & Meng, Z. (2023). Design and test of obstacle detection and harvester pre-collision system based on 2D lidar. *Agronomy (Basel), 13*(2), 388. DOI: 10.3390/agronomy13020388

Sier, H., Yu, X., Catalano, I., Queralta, J. P., Zou, Z., & Westerlund, T. (2023, June). Uav tracking with lidar as a camera sensor in gnss-denied environments. In *2023 International Conference on Localization and GNSS (ICL-GNSS)* (pp. 1-7). IEEE.

Siewert, S. B., Andalibi, M., Bruder, S., Buchholz, J. M., Chamberlain, D., Lindsey, A. L., & Stockhouse, D. (2020). Comparison of RADAR, passive optical with acoustic, and fused multi-modal active and passive sensing for UAS traffic management compliance and urban air mobility safety. In *AIAA Scitech 2020 Forum* (p. 1456).

Sulaiman, M. M., Matloub, F. K., Ajjam, S. K. A., & Alwan, H. H. (2024). Innovative: International Multi-disciplinary Journal of Applied Technology. *Journal of Applied Technology, 2*, 2024.

Ullah, Z., Al-Turjman, F., & Mostarda, L. (2020). Cognition in UAV-aided 5G and beyond communications: A survey. *IEEE Transactions on Cognitive Communications and Networking, 6*(3), 872–891. DOI: 10.1109/TCCN.2020.2968311

Wang, J., Liu, Y., & Song, H. (2021). Counter-unmanned aircraft system (s)(C-UAS): State of the art, challenges, and future trends. *IEEE Aerospace and Electronic Systems Magazine*, *36*(3), 4–29. DOI: 10.1109/MAES.2020.3015537

Yadav, R., Dahal, K. R., Adhikari, B. B., Ghimire, P., & Belbase, N. P. (2024). Effect of Nutrient Sources on Productivity of Spring Maize (Zea Mays L.) at Khairahani, Chitwan, Nepal. *Multi-Disciplinary Explorations: The Kasthamandap College Journal*, *2*(2), 149–160. DOI: 10.3126/mdekcj.v2i2.74190

Yan, C., Fu, L., Zhang, J., & Wang, J. (2019). A comprehensive survey on UAV communication channel modeling. *IEEE Access : Practical Innovations, Open Solutions*, *7*, 107769–107792. DOI: 10.1109/ACCESS.2019.2933173

Yang, Y., Khalife, J., Morales, J. J., & Kassas, Z. M. (2021). UAV waypoint opportunistic navigation in GNSS-denied environments. *IEEE Transactions on Aerospace and Electronic Systems*, *58*(1), 663–678. DOI: 10.1109/TAES.2021.3103140

Yasin, J. N., Mohamed, S. A., Haghbayan, M.-H., Heikkonen, J., Tenhunen, H., & Plosila, J. (2020). Unmanned aerial vehicles (uavs): Collision avoidance systems and approaches. *IEEE Access : Practical Innovations, Open Solutions*, *8*, 105139–105155. DOI: 10.1109/ACCESS.2020.3000064

Yu, X., & Zhang, Y. (2015). Sense and avoid technologies with applications to unmanned aircraft systems: Review and prospects. *Progress in Aerospace Sciences*, *74*, 152–166. DOI: 10.1016/j.paerosci.2015.01.001

Chapter 8
Integrating Computer Vision and Edge Computing in the Drone Industry

Muhammad Tahir
Dalian University of Technology, China

Sajjad Ali Rajper
https://orcid.org/0009-0009-8150-4959
Iqra University, Karachi, Pakistan

Qurban Ali
https://orcid.org/0009-0003-0087-9270
University of Milano-Bicocca, Italy

ABSTRACT

This chapter's primary objective is to focus on computer vision and edge computing (EC) advancements, challenges and future trends, further highlighting their role in reshaping industry applications in the drone industry. UAVs have garnered significant interest from both academics and industry as a crucial facilitator of IoT services. In this post, we thoroughly examine the function of edge AI for UAVs. With a focus on the function of edge AI, we examined several important UAV technological issues and applications. We also investigated the ideas of UAVs, Computer Vision and edge computing, artificial intelligence, and edge AI. Additionally, the difficulties in implementing AI on UAV edges were examined, and future directions and lessons learnt were deliberated. The device edge consists of edge devices; the local edge, which consists of the application and network layer; and the cloud edge.

DOI: 10.4018/979-8-3693-8497-8.ch008

INTRODUCTION

An aircraft without a pilot that is controlled from the ground or by an onboard computer is In recent years, drone technology has emerged as a game-changing innovation with important commercial implications. Drones, also known as unmanned aerial vehicles (UAVs), have developed from their original military application to become indispensable tools in a wide range of commercially viable fields(Aibibu et al., 2023; Bhat, 2024). This specialised excrescence is being driven by rapid advancements in tackle, software, and legitimate fabrics, which are also enhancing the capabilities and availability of drones for commercial operations. Figureure 1 Overview Computer Vision & computing, drones.

Figure 1. Overview of computer vision & computing drones

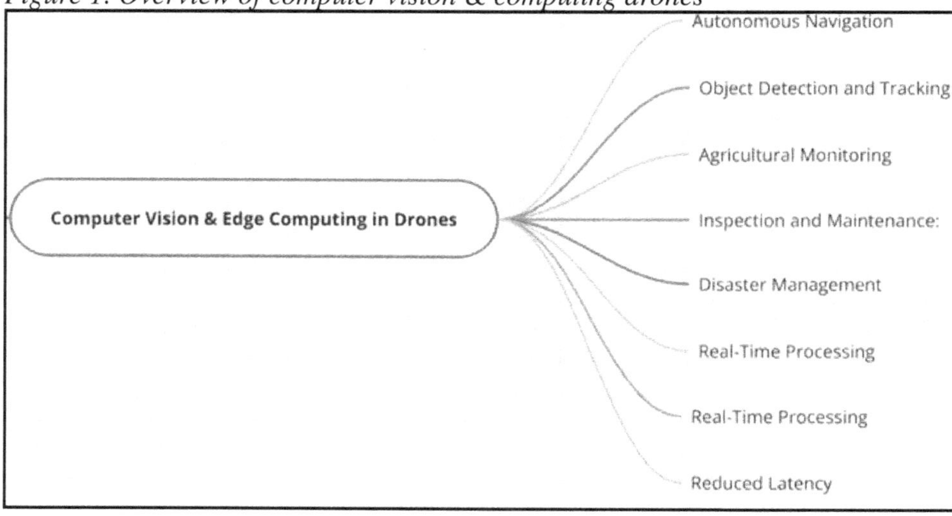

A multidisciplinary strategy incorporating knowledge from domains like robotics, computer vision, wireless communications, and transportation engineering is needed to address these issues (Bhat & Kavasseri, 2024; Coifman et al., 2004). The shortcomings of conventional traffic monitoring systems may be addressed by UAVs, especially in dynamic and resource-constrained contexts. UAVs can help traffic management and autonomous vehicle operations by offering real-time, actionable data using advanced sensing technology, AI-driven analytics, and energy-efficient operation. However, overcoming major operational, legal, and technical obstacles is necessary to fully utilise UAVs in these fields. Technologies like edge computing, cloud computing, IoT, and AI have emerged because of Industry 4.0's rapid evolution of industrial processes, enabling smart factories and increasing productivity.

A new paradigm that emphasises revolutions starting with mechanical innovation in Industry 1.0, mass production in Industry 2.0, and automation in Industry 3.0 is emerging as industries transition to Industry 5.0. Intelligent, networked, and smooth human-machine collaboration is a top priority for Industry 5.0(Savkin & Huang, 2021; Wang et al., 2021; Yang et al., 2019). This new era places more emphasis on a value-centric form of manufacturing, where cognitive and response talents are essential, rather than only on process-driven production. Figureure 2 Overview of challenges of CV & EC in the drone industry.

Figure 2. Overview of challenges of CV & EC in the drone industry

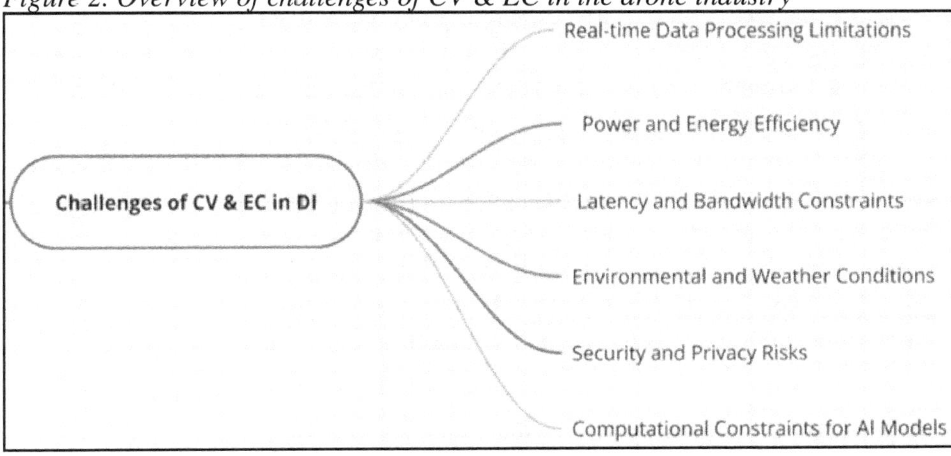

In contrast to previous industrial and adaptive systems, which function in real-time to satisfy intricate and ever-changing needs. In settings where real-time insights are crucial, edge computing, which is at the heart of this innovation, empowers systems by enabling local data processing with low latency, hence supporting quick decision-making and high reliability. For instance, edge computing in manufacturing may forecast maintenance requirements by analysing real-time sensor data, guaranteeing uninterrupted operation, and minimising downtime(Hossain et al., 2022; Wang et al., 2023). Edge computing is therefore becoming more and more important in attaining the agility, effectiveness, and resilience needed for Industry 5.0, enabling improved human-machine interaction and intelligent, responsive industrial processes. examines the architecture, various applications, and upcoming difficulties of edge computing in this data-intensive, decentralised industrial environment. The exponential growth of IoT devices and smart technologies, which require low-latency applications, has made edge computing a paradigm shift. Context-sensitive as cities and businesses adopt edge-enabled IoT solutions, edge computing puts processing power close to data sources—sensors, actuators, and end devices—to enhance response times and quality of service across a range of critical applications(Dosovitskiy et al., 2020; Wang et al., 2022). Because cloud computing usually does not enable applications that need fast data processing and dynamic adaption, edge computing is necessary for anything from intelligent mobility and real-time surveillance to healthcare monitoring and environmental tracking.

Although cloud infrastructures are rich in resources, they are unable to satisfy the latency, mobility, and privacy requirements of the contemporary IoT ecosystem. Rather, edge computing provides decentralised processing power at the edge of the network, facilitating data optimisation and real-time decision-making in complex IoT frameworks and smart cities. This approach lessens dependency on distant cloud servers and provides devices with adaptability to unanticipated changes in the environment or malfunctions(Shah, 2024a, 2024b). Without servers, edge nodes improve safety and coordination by assisting autonomous cars in making snap decisions. Systems for edge computing need to be self-organising and self-adaptive in order to function well in dynamic, resource-constrained environments. Self-edge computing systems are required to maintain efficiency, dependability, and functioning in unpredictable situations (Hao et al., 2024; Muzammul et al., 2024; Shi et al., 2023). Through thorough analysis, this study classifies edge-enabled IoT applications, assesses how well they integrate with self- systems design concepts, and identifies research opportunities and challenges for adaptive edge computing systems.

LITERATURE REVIEW

As the Internet of Things (IoT) manufacturing technology and wireless networks have grown, so have IoT applications. These applications are now common in many areas, including agriculture, industry, transportation, environmental monitoring, and smart city applications of today. For instance, IoT devices like wireless sensors are frequently placed on the ground and can sense and detect their surroundings to help with environmental monitoring, traffic control, and industrial inspection. However, because IoT devices usually have limited computation and storage capacity, they must send data via wireless networks to data centres or nearby base stations for processing. IoT devices typically use multi-hop relays to transfer sensed data and rely on static nodes to deliver services in traditional wireless networks(An et al., 2023; Niu & Yan, 2023; Yang et al., 2024). Because of their portability, affordability, and quick deployment, unmanned aerial vehicles (UAVs) can be extremely important on the Internet of Things. UAVs can support wireless networks by becoming airborne base stations that can fly to remote locations to deliver services. Additionally, UAVs can create line-of-sight (LoS) links by hovering over IoT devices, greatly enhancing the communication quality of IoT devices. As a result, UAVs are a prospective IoT technology that can support wireless networks. In general, UAV technology is used extensively in two IoT scenarios: UAV-assisted data collection (DC) and UAV-assisted mobile edge computing (MEC). Figureure 3 Overview of pros & cons of computer vision of edge computers in the drone industry.

Figure 3. Overview of pros & cons of computer vision and edge computer in the drone industry

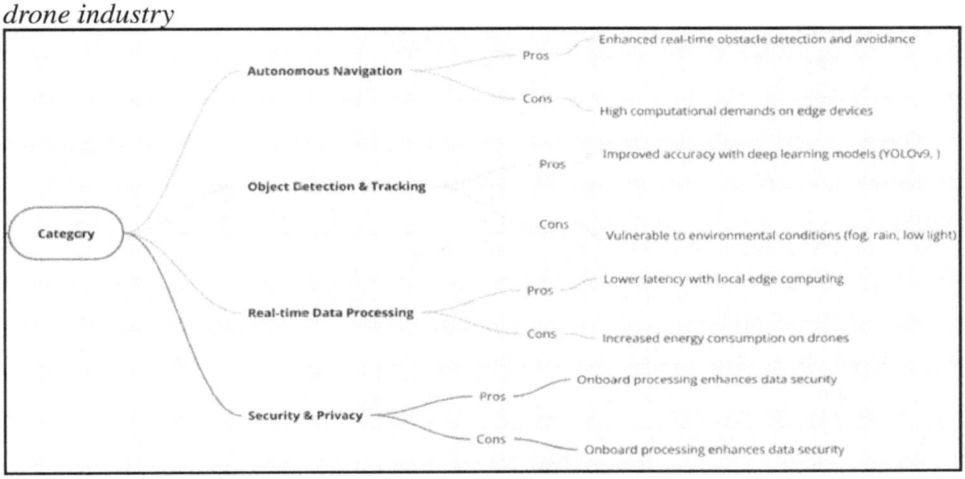

Furthermore, there are situations like post-disaster aid, traffic control, and agricultural management when both computation-intensive activities and DC requirements exist, necessitating the simultaneous use of numerous UAVs for MEC and DC(Xiao et al., 2024; Yang et al., 2022). UAVs, for instance, gather data, which is subsequently offloaded to a local data centre or used for edge computing. However, because of the UAVs' energy limitations and mission priorities.

Real-time and continuous computing support is typically needed for certain MEC tasks with strict deadlines. A decrease in the real-time performance of the MEC activities and an increase in energy consumption could result from an MEC-UAV doing DC concurrently. Moreover, UAVs from several operators may perform distinct jobs in the same location, such as environmental monitoring and edge computing within manufacturing (Shah, 2024a; Shah, Jhanjhi, & Ashraf, 2024). One of the most efficient ways to preserve data security and privacy is to isolate MEC and DC data onto separate UAVs. There are numerous significant obstacles to overcome when building the hybrid MEC-DC system(Elfwing et al., 2018; Liu et al., 2023). First, the lack of spectrum resources is becoming a bigger issue in both present and future networks, making spectrum-sharing strategies especially crucial. However, doing so can result in overuse of energy, possibly going beyond energy limits. As a result, it is challenging to strike a balance between the objectives of MEC and DC because they are conflicting and interdependent within the same situation. In contrast to earlier research that just focused on optimising individual MEC or DC systems, this paper suggests a collaborative optimisation strategy to enhance MEC and DC system performance concurrently.

OVERVIEW OF THE DRONE INDUSTRY AND COMPUTER VISION

The onboard computational power is the main constraint on UAV real-time embedded computer vision processing. claimed that the computational demands of cutting-edge UAV computer vision algorithms frequently clash with the constraints of hardware resources (Jhanjhi et al., 2025; Muzammul et al., 2024). Both aboard the UAV and at edge servers, edge AI plays a part in computer vision processing. All the fundamental benefits of edge computing and additional benefits of edge AI apply when the AI processing may be done on-device or at edge servers rather than under the standard cloud architecture. The benefit of low latency will be emphasised because it is especially pertinent to computer vision applications.

Edge computing is especially useful for applications with strict latency restrictions. For instance, the minimum end-to-end delay for cloud access is more than 80 ms, making cloud computing unacceptable when considering real-time packet delivery between self-driving cars that requires an end-to-end delay of less than 10 ms (Li et al., 2023; Shah, Jhanjhi, & Ray, 2024). This requirement can be satisfied using edge computing. With firms like Intel, IBM, Google, and Microsoft presenting pilot projects to illustrate edge computing's benefits in paving the final mile of AI, edge AI has attracted industry attention. Numerous AI applications, including live video analytics, smart homes, precision agriculture, and industrial IoT, have advanced because of these initiatives. Definitions of edge AI can vary widely, despite interest from both the academic and business communities. Some definitions, for instance, limit edge AI to the idea of local AI algorithm execution on endpoints. Edge AI is defined as "the capability to enable edges to execute AI algorithms".

Overview of UAV Systems Challenges

The past few years have seen significant advancements in the commercialisation of drone technology. Unmanned aerial vehicles (UAVs), sometimes referred to as drones, have transformed many fields of study by providing novel outcomes that enhance data efficacy, security, and delicacy. Two rapidly evolving fields that have a significant impact on the corporate sector are drone technology and artificial intelligence. Examining the latest developments in bite computing and their relationship to drone technology, as well as how they impact different business disciplines, is the aim of this literature study(Lee et al., 2020; Tsouros et al., 2019). Bite computing involves processing data close to where it is generated rather than relying on a centralised, cloud-based system. This system reduces quiescence, improves real-time data processing, and promotes overall system efficacy. In response to the increasing amount of data generated by IoT, byte computing has had to be developed in order to process data more effectively and efficiently. Figureure 4 Overview of technical challenges of UAV systems.

Figure 4. Overview of technical challenges of UAV systems

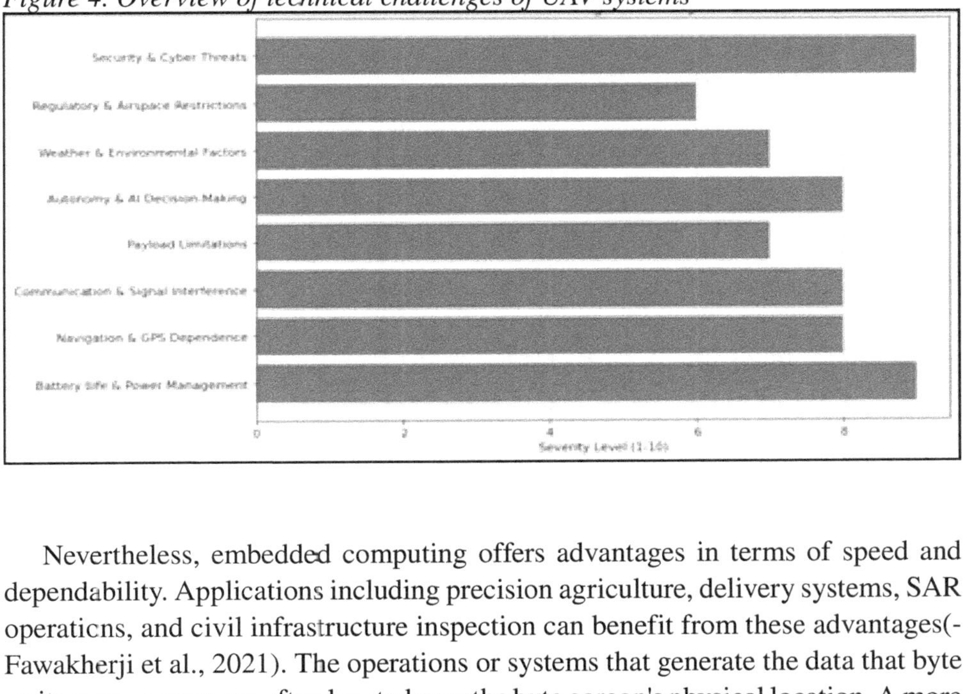

Nevertheless, embedded computing offers advantages in terms of speed and dependability. Applications including precision agriculture, delivery systems, SAR operations, and civil infrastructure inspection can benefit from these advantages(-Fawakherji et al., 2021). The operations or systems that generate the data that byte waiters use or save are often located near the byte garçon's physical location. A more conventional garçon design, on the other hand, continuously stores data in the pall or a data locus. A key component of edge computing, a computer paradigm that improves response times and conserves bandwidth by relocating processing and data storage closer to the point of demand, is edge servers. Identifying and preventing collisions with manned aircraft is one of the most crucial duties in UAV operation, particularly in shared airspace. To operate safely, both manned and unmanned aircraft need to be able to recognise other aircraft. Recent research in aircraft discovery for UAVs has focused on sophisticated systems that use a range of detectors and optical cameras to detect and track aircraft in real-time. Like ordinary aeronautic aircraft, it can be risky to use standard radar systems to characterise delicate aircraft. The link between drone assiduity and pall waiters has led to a new era of technological innovation and practical efficacy(Rasti et al., 2022).

The employment of drones in a variety of tasks, including husbandry, guidance, environmental monitoring, structure inspection, and logistics, is being revolutionised by this group. Pall computing provides dependable data operating systems and storage capacities to handle the enormous amounts of data generated by drones. This enables faultless messaging, advanced dissection, and improved real-time processing. Strong computers known as "edge servers" are positioned near a network's edge, where data processing needs to take place. An edge server's physical location is frequently near the systems or applications that generate the data that the edge servers use or store.

Overview UAV Applications

The use of edge AI can also be very beneficial for UAV-based IoT applications. Like the technical challenges of UAVs, many of the advantages of edge AI for UAV applications are like those of edge computing. Figureure 5 UAV applications across the industry.

Figure 5. UAV applications across the industry

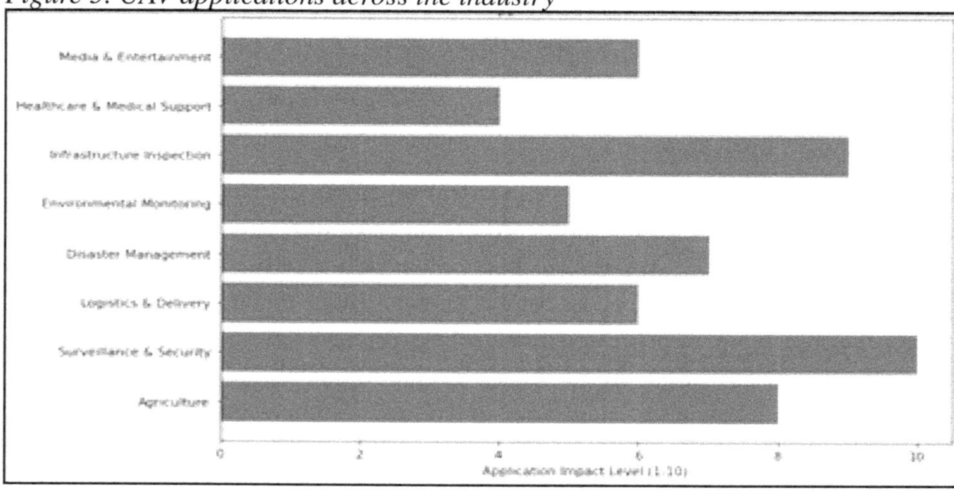

In contrast, edge servers used with edge AI are situated close to or at the users, reducing the likelihood of a network outage(Beheshti & Johnsson, 2020). Additionally, edge AI lowers energy consumption by facilitating preprocessing at the edge, which minimises the amount of data that must be transmitted to the distant cloud and, consequently, the energy consumption related to data transport. For applications like precision agriculture, aerial wireless BSs, SAR operations, and civil infrastructure inspection, this lower energy consumption can enable extended coverage, searches, and inspections before requiring recharging. Although UAVs in drone light shows typically interact directly with a GCS, edge AI supports these performances in a more limited way than it does for other applications. Having said that, edge AI's higher latency compared to cloud-based AI could make drone light displays safer and more dependable.

Overview Security and Privacy in UAVs

Compared to other privacy-invading devices, UAVs provide a unique range of nimble access approaches, which attracts thieves. However, they are also susceptible to attacks that target wireless links, cyber elements, physical components, and interfaces between cyber and physical elements. Furthermore, UAVs produce a lot of data at the network edge, and there is a significant danger involved in transferring original data sets directly between several edge nodes. Ensuring that privacy and security measures are implemented in accordance with system and user needs. ensuring that anomalous behaviour is recognised in accordance with operator and user standards(Kiran et al., 2021; Yu et al., 2018). Designing standard deployment interfaces or languages for various UAV systems could be one way to ensure that security and privacy methods are implemented in accordance with user expectations and system requirements for UAV edge AI.

CONCLUSION

UAVs have garnered significant interest from both academics and industry as a crucial facilitator of IoT services. In this post, we thoroughly examine the function of edge AI for UAVs. With a focus on the function of edge AI, we examined several important UAV technological issues and applications. We also investigated the ideas of UAVs, edge computing, artificial intelligence, and edge AI. Additionally, the difficulties in implementing AI on UAV edges were examined, and future directions and lessons learnt were deliberated. This Chapter study will likely serve as a helpful resource for scholars interested in the convergence of edge AI and UAVs, given their growing popularity. Autonomous navigation can range from fully autonomous vehicles that can navigate from point A to point B without requiring human intervention to vehicles that are operated remotely by a human but have basic onboard algorithms that take over and keep the vehicle from colliding.

REFERENCES

Aibibu, T., Lan, J., Zeng, Y., Lu, W., & Gu, N. (2023). An efficient rep-style gaussian–wasserstein network: Improved uav infrared small object detection for urban road surveillance and safety. *Remote Sensing (Basel)*, *16*(1), 25. DOI: 10.3390/rs16010025

An, J., Putro, M. D., Priadana, A., & Jo, K.-H. (2023). Improved yolov5 network with cbam for object detection vision drone. 2023 IEEE International Conference on Industrial Technology (ICIT), Beheshti, N., & Johnsson, L. (2020). Squeeze u-net: A memory and energy efficient image segmentation network. Proceedings of the IEEE/CVF conference on computer vision and pattern recognition workshops, Bhat, S. (2024). Leveraging 5g network capabilities for smart grid communication. *Journal of Electrical Systems*, *20*(2), 2272–2283.

Bhat, S., & Kavasseri, A. (2024). Multi-source data integration for navigation in gps-denied autonomous driving environments. *International Journal of Electrical and Electronics Research*, *12*(3), 863–869. DOI: 10.37391/ijeer.120317

Coifman, B., McCord, M., Mishalani, R. G., & Redmill, K. (2004). Surface transportation surveillance from unmanned aerial vehicles. *Proc. of the 83rd Annual Meeting of the Transportation Research Board*

Dosovitskiy, A., Beyer, L., Kolesnikov, A., Weissenborn, D., Zhai, X., Unterthiner, T., Dehghani, M., Minderer, M., Heigold, G., & Gelly, S. (2020). An image is worth 16x16 words. arXiv preprint arXiv:2010.11929, 7.

Elfwing, S., Uchibe, E., & Doya, K. (2018). Sigmoid-weighted linear units for neural network function approximation in reinforcement learning. *Neural Networks*, *107*, 3–11. DOI: 10.1016/j.neunet.2017.12.012 PMID: 29395652

Fawakherji, M., Potena, C., Pretto, A., Bloisi, D. D., & Nardi, D. (2021). Multi-spectral image synthesis for crop/weed segmentation in precision farming. *Robotics and Autonomous Systems*, *146*, 103861. DOI: 10.1016/j.robot.2021.103861

Hao, C., Zhang, H., Song, W., Liu, F., & Wu, E. (2024). Slinet: Slicing-aided learning for small object detection. *IEEE Signal Processing Letters*, *31*, 790–794. DOI: 10.1109/LSP.2024.3373261

Hossain, M. A., Hossain, A. R., & Ansari, N. (2022). Numerology-capable UAV-MEC for future generation massive IoT networks. *IEEE Internet of Things Journal*, *9*(23), 23860–23868. DOI: 10.1109/JIOT.2022.3189945

Jhanjhi, N., Shah, I. A., & Nawaz, S. (2025). Generative AI for Visualization. In *Generative AI for Web Engineering Models* (pp. 63–82). IGI Global.

<jrn>Niu, K., & Yan, Y. (2023). A small-object-detection model based on improved yolov8 for uav aerial images. 2023 2nd International Conference on Artificial Intelligence and Intelligent Information Processing (AIIIP)

Kiran, S., Rajper, S., Shaikh, R. A., Shah, I. A., & Danwar, S. H. (2021). Categorization of CVE Based on Vulnerability Software By Using Machine Learning Techniques. *International Journal (Toronto, Ont.)*, *10*(3).

Lee, J., Nazki, H., Baek, J., Hong, Y., & Lee, M. (2020). Artificial intelligence approach for tomato detection and mass estimation in precision agriculture. *Sustainability (Basel)*, *12*(21), 9138. DOI: 10.3390/su12219138

Li, Y., Miao, N., Ma, L., Shuang, F., & Huang, X. (2023). Transformer for object detection: Review and benchmark. *Engineering Applications of Artificial Intelligence*, *126*, 107021. DOI: 10.1016/j.engappai.2023.107021

Liu, S., Zha, J., Sun, J., Li, Z., & Wang, G. (2023). EdgeYOLO: An edge-real-time object detector. 2023 42nd Chinese Control Conference (CCC).

Muzammul, M., Algarni, A., Ghadi, Y. Y., & Assam, M. (2024). Enhancing UAV aerial image analysis: Integrating advanced SAHI techniques with real-time detection models on the VisDrone dataset. *IEEE Access : Practical Innovations, Open Solutions*, *12*, 21621–21633.

Rasti, S., Bleakley, C. J., Holden, N., Whetton, R., Langton, D., & O'Hare, G. (2022). A survey of high resolution image processing techniques for cereal crop growth monitoring. *Information Processing in Agriculture*, *9*(2), 300–315.

Savkin, A. V., & Huang, H. (2021). Navigation of a UAV network for optimal surveillance of a group of ground targets moving along a road. *IEEE Transactions on Intelligent Transportation Systems*, *23*(7), 9281–9285. DOI: 10.1109/TITS.2021.3077880

Shah, I. A. (2024a). Autonomous Shipping: Security Issues and Challenges. *Cybersecurity in the Transportation Industry*, 187-210.

Shah, I. A. (2024b). Privacy and security challenges in unmanned aerial vehicles (UAVs). *Cybersecurity in the Transportation Industry*, 93-115.

Shah, I. A., Jhanjhi, N., & Ashraf, H. (2024). Logistics With the Internet of Things: Challenges, Perspectives, and Applications. In *Navigating Cyber Threats and Cybersecurity in the Logistics Industry* (pp. 172-195). IGI Global.

Shah, I. A., Jhanjhi, N. Z., & Ray, S. K. (2024). IoT Devices in Drones: Security Issues and Future Challenges. In *Cybersecurity Issues and Challenges in the Drone Industry* (pp. 217-235). IGI Global.

Shi, T., Ding, Y., & Zhu, W. (2023). YOLOv5s_2E: Improved YOLOv5s for aerial small target detection. *IEEE Access : Practical Innovations, Open Solutions*, *11*, 80479–80490. DOI: 10.1109/ACCESS.2023.3300372

Tsouros, D. C., Bibi, S., & Sarigiannidis, P. G. (2019). A review on UAV-based applications for precision agriculture. *Information (Basel)*, *10*(11), 349. DOI: 10.3390/info10110349

Wang, J., Liu, W., Zhang, W., & Liu, B. (2022). LV-YOLOv5: A light-weight object detector of Vit on Drone-captured Scenarios. 2022 16th IEEE International Conference on Signal Processing (ICSP)

Wang, T., Huang, P., & Dong, G. (2021). Cooperative persistent surveillance on a road network by multi-UGVs with detection ability. *IEEE Transactions on Industrial Electronics*, *69*(11), 11468–11478. DOI: 10.1109/TIE.2021.3121729

Wang, Y., He, Y., Yu, F. R., Song, B., & Leung, V. C. (2023). Efficient resource allocation for building the metaverse with uavs: A quantum collective reinforcement learning approach. *IEEE Wireless Communications*, *30*(5), 152–159. DOI: 10.1109/MWC.009.2300029

Xiao, Y., Xu, T., Yu, X., Fang, Y., & Li, J. (2024). A Lightweight Fusion Strategy with Enhanced Inter-layer Feature Correlation for Small Object Detection. *IEEE Transactions on Geoscience and Remote Sensing*, *62*, 1–11. DOI: 10.1109/TGRS.2024.3457155

Yang, C., Li, X., Zhang, H., Meng, Y., Zhang, R., & Yuan, R. (2024). Uav small target detection in complex scenes based on improved yolov8s. 2024 39th Youth Academic Annual Conference of Chinese Association of Automation (YAC)

Yang, J., Zhang, J., Ye, F., & Cheng, X. (2019). A uav based multi-object detection scheme to enhance road condition monitoring and control for future smart transportation. Artificial Intelligence for Communications and Networks: First EAI International Conference, AICON 2019, Harbin, China, May 25–26, 2019, Proceedings, Part II DOI: 10.1007/978-3-030-22971-9_23

Yang, Y., Gao, X., Wang, Y., & Song, S. (2022). VAMYOLOX: An accurate and efficient object detection algorithm based on visual attention mechanism for UAV optical sensors. *IEEE Sensors Journal*, *23*(11), 11139–11155. DOI: 10.1109/JSEN.2022.3219199

Yu, J., Yang, L., Xu, N., Yang, J., & Huang, T. (2018). Slimmable neural networks. *arXiv preprint arXiv:1812.08928.*

Chapter 9
Integration of Edge AI With Cloud Computing and Drone Technology: An Analysis

Anuj Rawat
Sharda University, India

Vishal Jain
https://orcid.org/0000-0003-1126-7424
Infrastructure University Kuala Lumpur, Malaysia & Sharda University, Greater Noida, India

Ng Khai Mun
https://orcid.org/0000-0002-8263-9275
Infrastructure University Kuala Lumpur, Malaysia

Danish Ather
https://orcid.org/0000-0003-1596-5553
Amity University, Tashkent, Uzbekistan

ABSTRACT

Artificial Intelligence is rapidly becoming an integral part of our daily lives. There is a growing interest in artificial intelligence and related technologies. The chapter focuses on the integration of edge AI with cloud computing in the drone technology system, highlighting how this combination enhances overall efficiency. It provides a comparative analysis of edge AI and cloud AI, evaluating their impact on performance, setup costs, and scalability. This study offers a thorough analysis of the most relevant academic works concerning the uses of UAV technology in terms of

DOI: 10.4018/979-8-3693-8497-8.ch009

enabled or aided Edge AI systems. It describes how UAV-enabled Mobile Edge Computing architecture might be useful for new applications and how it can help overcome a variety of constraints like latency, task offloading, energy consumption, and security. The reader also learns about the advantages and the critical challenges to be tackled when using AI

1. INTRODUCTION

Artificial intelligence (AI) has revolutionized many industries by speeding up advancements in automation, data processing, and decision-making. The two primary types of AI systems are edge AI devices and cloud AI systems. They serve distinct goals and have different functionalities.

Large datasets and complex models are managed by these systems, which are frequently installed at research facilities and data centres. Without question, one of the sectors in the world that is expanding the fastest right now is the drone industry.Unmanned aerial vehicles (UAVs) are an autonomous aircraft class that does not require a human operator to fly it. Another popular name for these aircraft is drones. Drones are currently widely used in a wide range of commercial and military scenarios, including transportation, online shopping, cinematography, surveillance, crop protection, and surveying construction projects. Technology is changing everything, upending entire sectors and regions. Digital technologies are being used by businesses and government agencies to improve operational efficiency and save costs. Drones, also known as unmanned aerial vehicles (UAVs), are one such technology whose use and applications have grown over time to support a wider range of industries(Edulakanti & Ganguly, 2023).The synergy between drones and AI has led to notable progressin the autonomy of drones, which have become capable of completing complex missions withoutdirect human supervision(Caballero-Martin et al., 2024).

2. APPLICATIONS FOR DRONES

Using AI with Drones is not a big challenge nowadays.The integration of Artificial Intelligence (AI) tools and techniques has provided a significant advance in drone technology. Besides military applications, drones are being increasingly used for logistics and cargo transportation, agriculture, construction, security and surveillance, exploration, and mobile wireless communication. The synergy between drones and AI has led to notable progress in the autonomy of drones, which have become capable of completing complex missions without direct human supervision(Westheider et al., 2023).

Some applications of automaton invention are listed below:

1. **Airportal Imagery:** The computer vision component follows the typical methodology for a deep-learning-based SW: dataset creation, election and training of the most accurate object detection model, and testing. The result is the creation of a new dataset with a wider variety of traffic signs and an object detection-based system using Faster R-CNN to enable the detection and geo-location of traffic signs from drone-captured images. Additionally, columnists are considering using robots to capture video and data during live broadcasts(Lopatin &Poikonen, 2023; Naranjo et al., 2023).
2. **Precision Agriculture:** Farmers and ranchers are always looking for cheap and effective ways to regularly test their crops. The drones manufactured these days are becoming smarter by integrating open-source technology. Automatons equipped with infrared sensors that can be adjusted to detect crop health allow ranchers to better harvest conditions on a local level by applying bug sprays or adding manure. It enhances the board more and results in higher harvest yields(Chin et al., 2023).
3. **Searching For Rescue:** UAVs have night vision thanks to the proximity of warm sensors, which makes them a valuable tool for surveillance.Particularly in harsh environments or demanding landscapes, automatons are capable of locating the location of missing persons and tragically fallen soldiers. In addition to locating tragic casualties, an automaton can deliver supplies to remote locations in countries devastated by conflict or natural disasters. For example, before salvage organizations can relocate stranded unfortunate casualties, an automaton can deliver a walkie-talkie, GPS finder, medication, and supplies for food, clothing, and water(Erdelj et al., 2017).
4. **Transportation:** There are several research studies were done in which if a company uses drones in their delivery chain how it benefits them. Findings show that last-mile delivery by drones can be successfully used to deliver small and light packages, reducing environmental and social impacts, and ensuring profits for the transport provider(Borghetti et al., 2022).Categorization of Drones based onthe number of propellers.

Table 1. Types Of drones

Drone Type	Number Of Propellers
Tricopter	3
Quadcopter	4
Hexacopter	6
Octocopter	8

3. CLOUD COMPUTING

Due to the large amount of data generated and the computational need to be treated and evaluated, cloud simulation is presented as a novel aspect, an essential tool in the development and research of intelligent systems.Artificial intelligence (AI) is given enormous power by cloud computing, which is also regarded as one of the key drivers of the creation of innovative smart applications. The potential of artificial intelligence (AI) to transform the way data is stored and processed across geographical boundaries has expanded the field's application and effect marketplace. With so many cloud models available, consumers and developers of AI began to produce an ecosystem that makes millions of people's lives better. Nowadays, voice assistants such as Siri, our lives are infused with AI and cloud computing through Google Home and Amazon's Alexa. each and every day. Practitioners of AI utilizing the cloud approach of Infrastructure as a Service (IaaS) does not need to wait for an infrastructure team to set everything up before using sophisticated infrastructure resources like CPU, GPU, RAM, disk, network, and operating system.

The use of artificial intelligence in certain industries has a considerable impact on how successful cloud services techniques are, resulting in a consequence, the two innovations' cumulative effect boosts the prosperity of such businesses. The use of smart - device and computer vision models improve the effectiveness of public clouds. It has also demonstrated implementing AI into public cloud strategies may benefit enterprises in a variety of ways(Mohamed et al., 2023).

3.1 Applications of Cloud AI

1. **Scalable Data Storage and Management:** Cloud AI offers scalable storage for structured as well as unstructured data for any kind of organization. This is a highly imperative technology feature for business-most business organizations require quick access to data through the application of AI algorithms on it for effective indexing and retrieval that subsequently enhances decision-making.

2. **Fraud Detection and Risk Management:** Cloud AI allows strengthening the mechanism of fraud detection at banks. Algorithms based on AI monitor the pattern of behavior of customers in real-time and pin-point deviations which are suspicious-that is, represent possible fraudulent activities. This application will provide security with the minimum possible losses of financial nature and result in a sound framework for risk management.
3. **Supply Chain Optimization:** Organizations can make use of cloud AI to study and optimize the supply chain processes. These applications enable it to predict demand, handle inventory, and manage logistics. It enhances operational efficiency, reduces cost, and assists in time deliveries of products to the consumers.
4. **Distributed Collaborative Research and Development**: Cloud AI generates a collaborative infrastructure to exchange and analyze data. It can analyze massive geographically remote databases around the world, bestowing tremendous impacts on health, environmental sciences, and technologies. This application speeds up discovery and fosters inter-organization and inter-institutional collaboration.(Mohammed et al., 2023).

Figure 1. Mobile edge computing architecture (Petrović &Žižović, 2019; Verma & Verma, n.d.)

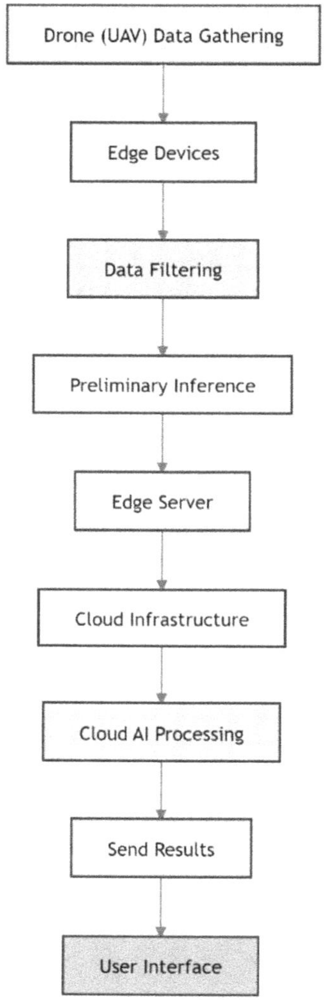

3.2 Edge AI

With the growing volume of data being generated by connected devices, Edge AI found its place as a key means of processing real-time information right where the data was generated. The new approach makes it possible for the capabilities of AI to be directly embedded in the edge devices such as drones, sensors, or IoT, thereby very significantly enhancing efficiency with which data is handled when needed for real-time applications in decision-making.

With the ability to analyze data locally, that means Edge AI can minimize dependence on centralized cloud resources and reduce latency for quicker responsiveness in a whole host of applications. Such an architecture shift in data processing has already revolutionized many industries by allowing real-time analytics and action sans the delay of sending data to the cloud for processing. Edge AI brings this distributed framework of computing, which allows computation at the edge, where devices can run independently. The paradigms have to increase the functionality and operational capabilities but have the drawback of data privacy since sensitive information is processed locally rather than moving on networks.

Edge AI has increasingly become a valuable resource in agriculture, healthcare, and logistics among others. Drones empowered with Edge AI can do autonomous inspection and crop health monitoring, whereas wearable health devices can rapidly process biometric data on-the-fly in a way that it provides information to the users and the healthcare providers regarding any vital health changes in real-time. Furthermore, Edge AI manages to encourage a greener data management solution by reducing the number of data being transferred to the cloud. These reduced bandwidth needs translate into not only saving operational costs but also to a further decrease in the environmental impact data transfer eventually brings. As institutions increasingly embrace Edge AI, the synergy between cloud and edge computing continues to evolve-the systematic composition of both helps build a strong ecosystem for robust real-time data processing while still serving to retain the benefits of centralized processing where such is warranted.

With the collaboration of edge AI and traditional cloud systems, it has now become feasible to design wiser and more efficient systems that are shaping the direction of technology(R. Singh & Gill, 2023).

3.3 Applications of Edge AI

Edge AI is particularly effective in various sectors where real-time processing is essential. For instance:

- **Smart Drones:** Drones equipped with Edge AI can analyze environmental data, detect obstacles, and make navigation decisions autonomously, enhancing their operational efficiency in sectors such as agriculture and search and rescue.
- **Industrial IoT:** In manufacturing, Edge AI enables predictive maintenance by monitoring equipment health in real-time and facilitating immediate corrective actions, thus reducing downtime(Kommineni & Baseer, 2024).

- **Healthcare:** Wearable devices powered by Edge AI can continuously monitor patients' health and alert caregivers to any anomalies, ensuring timely medical intervention.

Figure 2. Mobile-edge computing architecture (Shen & Brower, 2021; Verma & Verma, n.d.)

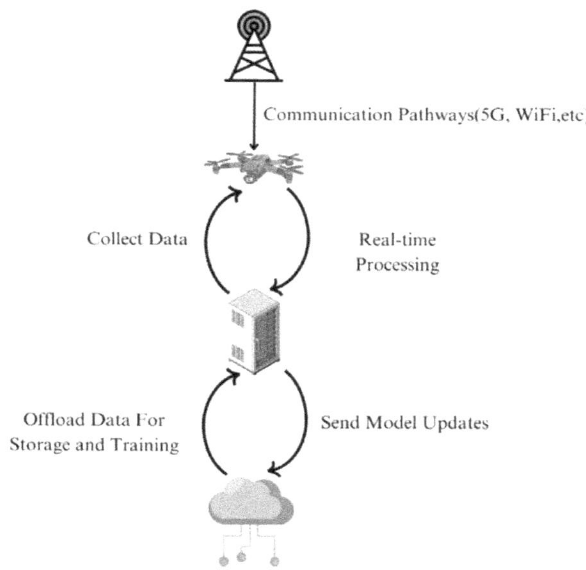

4. LITERATURE SURVEY

Research on integrating Edge AI and Cloud AI into UAV systems is a new generation of research. In the last few years, there has been a lot of attention given toward improving drones' capabilities using AI, especially post-deployment of edge computing for real-time processing and cloud computing for big data analytics. This paper reviews the most recent academic studies that have been published between 2020 and 2024.

4.1 Edge AI in UAV Systems

Edge AI handles local processing of data, either on or near the UAV, thus enabling true real-time decision-making. As might be understood, latency and also networking considerations can be extremely sensitive in drone technology itself. Some relevant work in this area are:

- The studies focused on the need for Edge AI in UAVs clearly explain how they work towards latency reduction in enhancing real-time performance, especially in applications such as disaster management and surveillance. These studies explain how drones can function if there is a lack of network connectivity or that network connectivity is poor using AI-based edge computing and further provides rapid responses in cases of real-time scenarios(Caballero-Martin et al., 2024; Yazid et al., 2021).
- Research discussing the design of a novel architecture for Edge AI deployment in UAV-based surveillance systems presents how offloading AI computations to edge devices installed on drones will lead to precision in object detection and tracking while reducing energy consumption associated with constant communications with cloud servers. This study underscores energy-saving algorithms to extend the lifetime of the UAV(Iyengar & Portilla, 2020; Yazid et al., 2021).

4.2 Cloud AI in UAV Systems

Unlike Edge AI, Cloud AI offers superior computational strength in that resource-intensive tasks may be pushed to central cloud servers. This has an advantage in large-scale data analysis or processing, especially when there are several drones or a large amount of data involved. Important contributions in this area include:

- Research that develops a ubiquitous, cloud-based AI architecture for swarm UAVs, demonstrating the application of Cloud AI for the coordination of multiple UAVs in collaborative missions, like environmental monitoring and agriculture applications. The authors discuss how machine learning models running in the cloud can process data gathered by drones in real-time to offer better scalability than purely edge-based solutions.
- Hybrid Edge-Cloud AI: The concept of dividing computational tasks between edge nodes and the cloud in a hybrid Edge-Cloud AI model has been inferences showing that this method not only reduces latency but also handles resource-intensive tasks, such as deep learning-based data analysis, ef-

ficiently. Further, these studies address cost-saving mechanisms in Cloud AI through dynamic workload management.

4.3 Hybrid Edge-Cloud AI Systems

This hybrid approach that combines Edge and Cloud AI advantages of low-latency processing with strong analytics in the cloud has gained great attention in the emerging area of UAV system performance optimization.

- Research proposing an integrated Edge-Cloud AI system for UAV-based mobile edge computing (MEC) highlights how task offloading between edge devices and the cloud can significantly reduce power consumption and increase efficiency in applications such as aerial surveillance and precision agriculture. This study also provides detailed performance metrics comparing edge-only, cloud-only, and hybrid systems(Hao et al., 2023; Verma & Verma, n.d.).
- Research on the use of federated learning in hybrid Edge-Cloud systems shows how UAVs can train a model collectively without sharing raw data with the cloud, improving both efficiency and privacy. This research exemplified federated learning use cases in drone fleets applied to anomaly detection and classification of data.

Figure 3. Benefits of hybrid approach

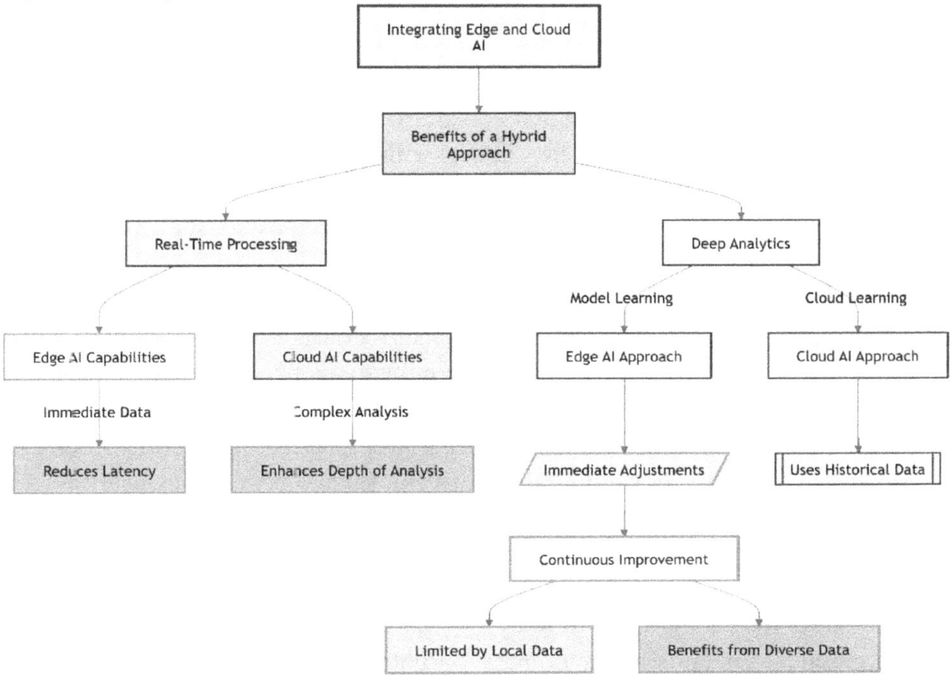

4.4 Limitations in Current Research

Despite significant progress in integrating AI with UAV systems, several challenges remain. Key limitations identified across recent studies include:

- **Energy Consumption**: Most studies recognize energy consumption as a major challenge, particularly for edge-based AI. Prolonged UAV operation necessitates energy-efficient processing techniques.
- **Task Offloading**: Numerous works highlight the challenges in determining which tasks should be offloaded to the cloud and which should be processed on the edge, especially in real-time applications.
- **Security**: The increased utilization of AI in drones raises security concerns, particularly regarding data integrity, privacy, and potential cyberattacks on both edge devices and cloud infrastructure.

5. TECHNICAL OVERVIEW OF EDGE AI AND CLOUD AI

5.1 Edge AI

The Edge AI is the setup in which actual artificial intelligence algorithms are implemented directly on devices like UAVs. This is an approach that enables decisions to be made directly without the need for a network connection but poses challenges of computational, energy expenditure, and security issues:

- Low latency is the major advantage of Edge AI. Because calculations happen locally, delays during network transmission are not substantial, which allows it to meet real-time demands of applications like surveillance, obstacle detection, and navigation. In smart-city intersection studies, low-latency object detection is expected to be achieved through edge-cloud computing. For example, evidence in keeping track of vehicles and pedestrians using bird's-eye cameras denotes the challenge the researchers face in real-time processing, with a performance objective of beating 33.3 ms to detect vehicles. This shows that there's some low latency that the researchers have worked on, but the target for delivering extremely low latencies, like 1-5 ms, is still difficult in practice, given its heavy reliance on both hardware capabilities and optimized algorithms (Yang et al., 2020).
- Energy Efficiency: While Edge AI makes it quicker to make decisions, it commonly suffers from high power consumption since the data locally observed has to be processed incessantly on the edge. This may challenge the already short battery life of UAVs to a great extent. Experiments involving UAVs using object recognition models such as MobileNet and YOLO demonstrated that, in certain cases, edge processing may consume much more energy than offloading to the cloud. The extra energy was maybe due to the continuous working invocations over onboard hardware, i.e., GPUs or TPUs. Nevertheless, the exact energy overhead will depend on efficiency properties of the hardware, model complexity, and optimization of the entire system.
- Hardware Constraints and Scalability: Onboard hardware places limitations in the form of computational capabilities for Edge AI. There have been experiments on using different types of hardware accelerators, TPUs and FPGAs, for embedding into UAVs. Findings indicate that while FPGAs provide higher performance-per-watt compared to GPUs, they come with increased development complexity. UAVs equipped with TPUs for deep learning inference exhibited a performance boost over CPU-based systems, though this was accompanied by higher power consumption.

- Data Privacy and Security: Processing data locally with Edge AI reduces the volume of sensitive information transmitted over the network, thereby diminishing exposure to potential cybersecurity threats. Edge AI reduces network exposure, but its security benefits rely on several elements. These include how well data is encrypted, who can access the system, and how devices are handled. However, the physical vulnerability of UAVs poses a significant security challenge; capturing a UAV could grant an attacker access to onboard data and algorithms.

5.2 Cloud AI

Cloud AI employs centralized cloud servers for AI model training and inference, allowing UAVs to offload complex computations. While cloud systems provide virtually unlimited resources and scalability, they are subject to higher latencies due to network transmission.

- Scalability and Computational Power:Cloud AI offers extensive scalability and computational power, making it ideal for tasks that involve large datasets or require significant processing, such as environmental monitoring, disaster response, and complex AI model training. Studies have shown that cloud systems can scale to handle data from up to 1,000 UAVs operating simultaneously, processing terabytes of data without noticeable performance degradation. Additionally, UAVs can execute complex deep learning models like ResNet50 and GPT-based models on the cloud, which are computationally unfeasible to run on the edge devices. Complex tasks that require the coordination of multiple UAVs with simultaneous processing, for example, multi-UAV coordination and data global fusion, are very suitable for cloud AI.
- Latency and Network Dependency: Network dependency still stands as a major threat to Cloud AI. Latency levels in a cloud-based UAV system are bound to be higher because that delay in traveling data from the UAV to the cloud servers will hinder its performance in real-time activities such as collision avoidance and real-time video analysis. Much research is being conducted on hybrid approaches that distribute computations between edges and clouds, to ensure completion of the task. Hybrid architectures thus rely on lightweight models on UAVs, in real-time processing, and offload resource-intensive computing tasks to the cloud. This strategy enhances responsiveness and may reduce end-to-end latency by large margins as well.
- Energy Efficiency: In UAV systems with Cloud AI architectures, various processing jobs are processed in cloud servers to consequently reduce the computational overhead on the system on the UAV per se. Over the long haul, this

could yield evident gains in energy efficiency, especially in operations that span longer times. Some studies have shown that cloud solutions can keep UAV energy usage lower than edge solutions, mainly because such UAVs usually need lighter processors and are not forced to run resource-hogging AI models locally. However, the fact must be remembered: cloud offloading indeed reduces energy consumption on UAVs, but once again, energy consumption is shifted toward data transmission and operations of the cloud server, potentially offsetting some gains based on application requirements and network conditions.

- Security: Cloud AI systems centralize data, making it easy to implement robust security measures such as central encryption and intrusion detection systems. Research also concludes that cloud systems are less vulnerable to physical attacks because the data cannot be found in the UAVs. However, centralized data storage creates a single point of failure, making the system susceptible to distributed denial of service (DDoS) attacks or large-scale data breaches.

5.3 Comparison of Edge AI and Cloud AI

A comparative analysis of Edge AI and Cloud AI reveals trade-offs between latency, computational power, processing of data, storage, and security. The table below summarizes these key differences based on computing systems of both technologies:

Table 2. Comparison of edge computing and cloud computing (Verma & Verma, n.d.)

Features	Edge Computing	Cloud Computing
Latency	As the processing is at the edge the latency is highly reduced.	Latency is very high as the cloud is very far from the source of data
Capacity of computation	Comparatively Low	Higher computation capacity
Data Processing	Real-time data processing is very efficient in edge	Sometimes it takes longer time to process the data in real-time
Data Handling	It can store less amount of data.	It is best for storing humongous amounts of data set.
Communication of Frontend and Backend	devices can get cached content directly from edge gateway.	devices need to be connected to internet throughout entire duration.
Suited Application	Projects which require very less latency are suited.	Projects which require a large amount of storage are suited.
Security risks	More vulnerable to physical attacks	Centralized security, but a potential single point of failure

In Comparison on Latency and Energy Efficiency: Edge AI and Cloud AI appear to have great performance altogether in respect of latency and energy efficiency. Processing data at the edge allows for the experience of low latency on time-critical tasks like detection or avoidance of collision. Cloud-based systems may do so by offloading any computational tasks on a centralized server, which can further save the power consumption of UAVs. However, it comes with a price for added latency due to network transmission delay, which is the energy-saving advantage transferring into a disadvantage almost all the time.

Computational Power and Scalability: Cloud AI has greater computational power than Edge AI on average. Cloud servers are made to harness and compute bigger databases and manage larger populations of UAV networks. While Edge AI can expand, it's constrained by the resources of UAV hardware alone and can't scale to higher operations.

Centralized security provided by Cloud AI is more robust, but it also makes such systems more prone to wide-scale cyberattacks, especially DDoS attacks. While network attacks are very little a concern for Edge AI, local data storage has brought a problem with it-literally: a thief might steal the entire device.

6. INTEGRATION OF EDGE AI AND CLOUD AI IN UAV SYSTEMS

This has emerged to be the principal direction for enhancing the effectiveness of capabilities in UAV systems by integrating Edge AI with Cloud AI technologies. The subsequent sections will demonstrate how these technologies complement each other in UAV operations-whether in making real-time decisions, processing big data, or fleet coordination.

6.1 Introduction to UAV Systems

Unmanned Aerial Vehicles, or UAVs, are increasingly applied in challenging applications across many domains, ranging from the other end of agricultural precision farming to disaster management. By and large, a generic UAV contains onboard processors, communication modules, sensors like cameras or LiDAR, and actuators controlling movement. The computation architecture, on which the level of autonomy in UAVs is highly dependent-be it fully onboard, offloaded to a remote cloud, or a hybrid of both-can vary extensively.

- UAV Autonomy and AI Integration: To achieve the highest UAV autonomy possible, integration with AI is required. By applying AI models, such as route optimization, identification of targets, and more, UAVs can conduct complex autonomous operations. Whereas some systems implement strictly Edge AI in decision-making, other systems feature Cloud AI for more intensive operations; therefore, UAVs may be deployed in any challenging environment, including disaster areas and massive agricultural tracks.

- Key UAV Applications: UAVs have various applications that extend to:

1. **Surveillance:** Real-time security monitoring, for which Edge AI plays the role of immediate threat detection.
2. **Agriculture:** Crop health monitoring using AI-driven multispectral analysis often requires cloud resources for processing a massive dataset.
3. **Logistics:** Delivery drones use real-time route optimization using Edge AI and fleet management with Cloud AI.

6.2 Edge AI for Real-Time UAV Operations

It is important for real-time UAV operations as it enables fast processing of data and thus immediate decision-making. The local processing capability of Edge AI ensures effective working in areas with limited or no connectivity to network.

- Reduced latency: Edge AI has reduced latency and brings real-time decisions. UAVs using Edge AI can achieve latencies of below 5 ms for tasks like obstacle detection, which is very important for the time-sensitive applications such as search and rescue.
- Energy Efficiency and Limitations: Although one of the benefits of Edge AI is its energy efficiency, the UAV's limitation on batteries stands as a bottleneck. Invasive onboard AI algorithms tend to accelerate the depletion of the battery. Lightweight AI models with hardware optimizations like pruning and quantization can alleviate this issue.
- Security Implications: Edge AI data processing reduces the risk associated with cloud transmission that causes breaches in data. However, it introduces new kinds of challenges such as tampering; secure enclaves minimize some of these risks but may add extra computational overhead.
- Real-Time Edge AI Use Case: While being used on UAVs, it has already implemented a good number of remote area wildlife monitoring missions in those locations where there is identified and followed wildlife in real time and without reliable network access.

6.3 Cloud AI for Large-Scale Analysis

Cloud AI brings an opportunity wherein they can offload computationally intensive tasks onto powerful cloud servers, thus enabling large dataset processing and complex model training.

- Scalability: Cloud AI offers unmatched scalability. It can manage data from numerous UAVs operating simultaneously, making it ideal for large-scale applications like environmental monitoring, where UAVs collect extensive data that must be efficiently processed.
- Computational Power: Cloud AI utilizes the processing capability powerfully because this technology is well suitable for complex deep learning models, large-scale data analysis, etc., which are generally recommended for high-resolution applications.
- Latency vs. Energy Efficiency: Cloud AI has high computational powers, but they suffer from latency problems. Motivated solely by cloud AI, the tasks suffer from larger latencies and this could be disastrous for applications which require real time decisions. In contrast, cloud offloading significantly reduces onboard energy consumption, meaning the UAV could carry lighter processors for basic tasks.
- Fleet Management in Cloud AI: Cloud AI has been successfully utilized for the management of UAV fleets in numerous applications, including crop monitoring, where it coordinates data collection from multiple drones to provide meaningful insights for farmers.

6.4 Hybrid Edge-Cloud AI Systems

The hybrid integration of Edge AI and Cloud AI will basically distribute the best of both worlds, which are tasks based on specific requirements(S. Singh, 2017).

- Task Offloading Models Hybrid architectures- There exist nodes that selectively allocate tasks based on the latency sensitivity and computational intensity. Examples of low-latency computation include processing obstacle avoidance, among others. More computationally intensive tasks include, for example, high-resolution image analysis that offload to the cloud.
- Hybrid adaptive resource allocation: Hybrid systems dynamically adjust resources based on both network conditions and a UAV's battery lifetime so that performance is optimized and energy is conserved. In low bandwidth situations, most of the tasks are run locally. Offloading to clouds takes place as bandwidth improves.

- Data Consistency and Synchronization: The challenge in hybrid systems is to ensure data consistency between the edge devices and cloud servers. Edge devices will buffer locally, being periodically synchronized by the cloud to avoid network congestion by means of robust mechanisms that prevent data fragmentation or loss

6.5 Case Study: UAV-Enabled Mobile Edge Computing

For instance, a real-world case study demonstrates the benefits of hybrid Edge-Cloud AI systems. UAVs deployed for traffic management utilized Edge AI for real-time traffic analysis and rerouting decisions while the cloud aggregated data over time on traffic patterns in a city.

- Latency and Energy Efficiency Benefits: A characteristic of these hybrid AI systems resides in their potential to boost performance by distributing computational workloads over edge devices and the cloud. In this regard, hybrid offloads computationally intensive processes to the cloud while preferably executing latency sensitive tasks on the local device itself, thus achieving both latency and energy efficiency improvements. For what is worth, the degree of such improvements needs to be adequately supported with proper examples, taking into account the fact that specific considerations concerning system architecture, network conditions, and application requirements could lead them up in entirely different directions.
- Scalability and Performance: By offloading less time-sensitive tasks to the cloud, UAVs would be free to focus on decentralized real-time decisions at the local layer, thereby improving the effectiveness of the traffic management system. The proposed architecture demonstrates the capability of UAV-empowered Mobile Edge Computing in a city-wide traffic control scenario where multiple UAVs could collaborate in coordinated operations without much network congestion.

7. APPLICATIONS OF EDGE AI IN UAV SYSTEMS

7.1 Real-Time Surveillance and Monitoring

- This significantly enhances the real-time surveillance capabilities of the UAV since it can process video feeds locally in the Edge AI. Local processing reduces latency and minimizes reliance on external networks; hence, such systems prove to be particularly effective for real-time object detection, tracking, and anomaly detection.

- Edge AI includes advanced neural networks that can process real-time object recognition and movement tracking in sub second times, primarily made viable for applications like border patrol and disaster management operations. Being able to negate the perception of delay resulting from cloud processing, this approach thus enhances the responsiveness of surveillance operations.
- A scalable Edge AI framework with lightweight convolutional neural networks gives the platform high object detection and tracking accuracy even during computation resource constraints.

7.2 Disaster Management

Edge AI-enabled UAVs are particularly useful in disaster-affected areas where network infrastructure is damaged. Using onboard processing capability, such drones may be able to assess damage, locate survivors, and track hazards without requiring constant cloud connectivity.

- In disaster response scenarios, UAVs using Edge AI could rapidly assess damage and identify survivor locations. Such ability allows drones to work effectively in areas where network coverage is not reliable and to submit critical information as quickly as possible to first responders.
- Increased real-time detection and reporting of hazardous environments in post-disaster scenes with UAV-based systems on edge AI for better response time and safety.

7.3 Farming and Crop Monitoring

The Edge AI in UAVs can give real-time insight into crop health, soil conditions, or the need to irrigate in precision agriculture because it allows for immediate action without the time delays experienced in cloud-based analysis.

- UAV-sourced edge AI enables farmers to monitor crops in real-time to identify specific areas regarding the type of intervention necessary. Local data processing enhances real-time decisioning in precision agriculture.
- Crop health monitoring using multispectral imaging and real time analysis through Advanced Edge AI Framework for Optimizing Agriculture. Certain promising preliminary findings that have been identified so far include yield predictions and management of resources.

7.4 Logistics and Delivery

Edge AI enables a drone to autonomously navigate and optimize the delivery route in real-time for improved operation efficiency with much lesser dependence on cloud-based systems for immediate decisions.

- The delivery optimization system allows for routing calculation and adjustment by UAVs based on real-time traffic information that reduces the time of delivery effectively and overall operational efficiency while minimizing cloud dependency.
- A hybrid model integrates Edge AI for real-time navigation and Cloud AI for global coordination, which makes the entire logistics much more effective.

8. CHALLENGES IN UAV SYSTEMS

8.1 Latency and Task Offloading

Latency is huge challenge to UAV systems especially when applications require processing to be done in real time. Hybrid Edge-Cloud AI systems are gaining ground as a way to fill the gap of immediate processing with the benefits of cloud computing.

- Combining the methodologies, in this context, the time-sensitive tasks would be offloaded to Edge AI for faster processing, while other less pressing tasks are relegated to the cloud. Its structure has proven to lessen latency and improve system performance (Agha et al., 2023).
- To match the latency requirements of various tasks, the adaptive task offloading strategies should be dynamic in nature. In optimizing real-time processing and efficiency, responsiveness of UAV to immediate needs can be provided(Agha et al., 2023).

8.2 Energy Consumption and Battery Life

Energy is a critical concern for UAVs, especially when computing tasks are highly intensive and rely on the capability of Edge AI. Energy management methods play a crucial role in extending flight times.

- Dynamic offloading of tasks to cloud based on UAV's battery levels by an energy-aware task scheduling algorithm. This approach effectively minimizes the energy consumption and prolongs the time of flight, so UAVs can fly for longer duration time without recharging.

- Research has been carried out on developing energy-efficient AI algorithms tailored to UAVs. This, therefore, optimizes computational demand based on energy constraints while ensuring optimization efficiency of UAV systems in operation.

8.3 Security and Privacy Issues

Data security and privacy issues are important for a system where AI is to be integrated into UAV systems. Secure information protection and data transmission represent some of the requirements for system integrity.

- In edge device and cloud server-based task offloading, proper techniques of high-security encryptions and a secure protocol for communication must be implemented. Conservation from inherent vulnerability in AI-integrated UAV systems demands that above everything adoption of strong security measures be adopted.
- With federated learning, the approach shall ensure promising improvement in privacy within UAV networks by training the AI model without violating data confidentiality. Consequently, UAVs learn from data that may contain sensitive information. decentralized.

8.4 Scalability and Network Congestion

Scalability and network congestion have thus become two of the most important challenges arising from widespread adoption of UAVs by various sectors, ranging from agriculture, logistics, disaster management, to surveillance. The coordination and management of communication networks are highly essential for large-scale deployment of UAVs, or drone fleets/swarms, in principle.

Any kind of large-scale UAV operation typically achieves communication between drones and either the ground control systems or other drones over wireless networks. The number of drones deployed in this category clearly increases the amount of data transmitted across the network. This necessarily leads to increased congestion over the network, significant delays in data transmission, and reduction in the reliability, with a higher likelihood of system inefficiencies and possible system failures.

For instance, interference and congestion issues raise significantly in an urban or disaster area where both UAVs and smart mobile phones, IoT, etc, share the same frequency band. The massive real-time video feed, sensor data, and telemetry information to be exchanged between drones and control stations demand efficient bandwidth usage and sophisticated communication protocols(Garg et al., 2023).

9. IN-HOUSE RESEARCH

9.1 Performance Comparison

The edge AI, the hybrid UAV architecture, and cloud AI have been widely compared for their performances. Through this comparative study, it will analyze merits and demerits of each kind of architecture approach.

- Performance analysis showed that Edge AI outperforms significantly in real time and that UAVs process data, which leads to quick decisions. These are of great importance in time-sensitive operations like search rescue or surveillance in real time. This is a balanced solution where the hybrid system combines the rapid response speed of Edge AI with the processing powers of Cloud AI. In such complex applications that require comprehensive extensive processing and prompt processing, these are the right systems.
- Further comparisons between Edge AI and Cloud AI are presented on account of task complexity and latency requirement, revealing that Edge AI is relatively better suited for use cases where low latency requirements exist and scenarios that involve computationally expensive tasks, such as image analysis at high resolution or data aggregation from one or more UAVs have to be used along with Cloud AI.

9.2 Analysis of Costs

The cost of introducing and operating the systems of Edge AI and Cloud AI, respectively defines their uptake.

- While Cloud AI is likely to be costlier on the infrastructure side since it mandates robust servers and greater resources on the network, it mainly saves the operational cost in the long term. Its scalability ensures that growing volumes of data and users can easily be handled, which makes the long-run influence greater. Whereas the systems of Edge AI are cheaper at first because localized hardware necessitates constant upgrading to match emerging technologies.
- Analysis of the cost-effectiveness of Edge-Cloud AI configurations shows that edge systems, including AI, can be tuned in order to meet quite different

operational needs while also being scaled up and down according to specific applications.

9.3 Flexibility and adaptability

Scalability and adaptability are some of the important parameters that determine the adequacy of EdgeAI, Cloud AI, and hybrid systems for UAV applications.

- Cloud AI natively scales very well, supporting thousands of drones and vast datasets in applications that require voluminous data management such as in logistics and agriculture. This scalability allows extending UAVs into existing operations with minimal changes in infrastructure (Petrović &Žižović, 2019; Youvan, n.d.).
- The adaptability of Edge AI is most advantageous in dynamic environments. The particular situation of in-flight operations certainly requires rapid decision-making, and the UAV has the advantage of adapting rapidly in such conditions, enhancing operational effectiveness in challenging situations (Youvan, n.d.).

10. CONCLUSIONS AND FUTURE SCOPE

10.1 Key Findings Overview

This chapter provides an insightful overview of the trend of integrating Edge and Cloud AI in UAV systems, emphasizing applications such as real-time surveillance, disaster management, agriculture, and logistics. Key findings include:

1. Reduced Latency: Edge AI minimizes latency, enabling UAVs to perform analytics and operate independently in real-time situations.
2. Enhanced Flexibility and Power: Cloud AI offers greater flexibility and processing power, enabling the execution of more complex tasks that require substantial computational resources.
3. Optimized Task Allocation: Hybrid systems effectively allocate tasks based on their requirements, prioritizing data-intensive operations for analytics while ensuring rapid processing for time-sensitive applications.

10.2 Future Research Directions

Future research will focus on several critical areas to advance the capabilities of UAV systems:

- Enhancing Task Offloading Mechanisms: Developing novel strategies to improve task offloading between Edge and Cloud AI will enhance overall system performance and reduce latency.
- Developing Energy-Efficient Algorithms: Research into new algorithms and hardware solutions aims to increase energy efficiency in Edge AI systems, enabling UAVs to achieve longer battery life.
- Strengthening Security and Privacy Measures: There is a pressing need for advanced encryption and privacy-preserving techniques to secure data in UAV systems and protect sensitive information.
- Improving Federated Learning Models: Enhancing federated learning models will facilitate collaborative AI training while ensuring that sensitive data remains confidential.

10.3 Concluding Statements

Drones can serve as a link between smart devices and cloud data centers, to provide many services that can increase the performance of a given IoT system. Even if the integration of Edge and Cloud AI is an emerging approach that requires more effort butit sets the perfect approach for advancing UAV technology to be more autonomous, efficient, and versatile in operations. With these aims in mind, new hybrids combining processing capabilities with extensive data analysis are promised as the new standard for UAV applications. In so far as this points to the future of aerial robotics, this only accelerates innovations.

REFERENCES

Agha, S., Mohsan, H., Qasem, N., & Othman, H. (2023). Unmanned aerial vehicles (UAVs): Practical aspects, applications, open challenges, security issues, and future trends. *Intelligent Service Robotics*, *16*, 109–137. DOI: 10.1007/s11370-022-00452-4 PMID: 36687780

Borghetti, F., Caballini, C., Carboni, A., Grossato, G., Maja, R., & Barabino, B. (2022). The Use of Drones for Last-Mile Delivery: A Numerical Case Study in Milan, Italy. *Sustainability (Basel)*, *14*(3), 1766. Advance online publication. DOI: 10.3390/su14031766

Caballero-Martin, D., Lopez-Guede, J. M., Estevez, J., & Graña, M. (2024). Artificial Intelligence Applied to Drone Control: A State of the Art. *Drones (Basel)*, *8*(7), 296. DOI: 10.3390/drones8070296

Chin, R., Catal, C., & Kassahun, A. (2023). Plant disease detection using drones in precision agriculture. In *Precision Agriculture* (Vol. 24, Issue 5, pp. 1663–1682). Springer. DOI: 10.1007/s11119-023-10014-y

Edulakanti, S. R., & Ganguly, S. (2023). Review article: The emerging drone technology and the advancement of the Indian drone business industry. *The Journal of High Technology Management Research*, *34*(2), 100464. DOI: 10.1016/j.hitech.2023.100464

Erdelj, M., Natalizio, E., Chowdhury, K. R., & Akyildiz, I. F. (2017). Help from the Sky: Leveraging UAVs for Disaster Management. *IEEE Pervasive Computing*, *16*(1), 24–32. DOI: 10.1109/MPRV.2017.11

Garg, S., Ihler, A., Bentley, E. S., & Kumar, S. (2023). A Cross-Layer, Mobility, and Congestion-Aware Routing Protocol for UAV Networks. *IEEE Transactions on Aerospace and Electronic Systems*, *59*(4), 3778–3796. DOI: 10.1109/TAES.2022.3232322

Hao, T., Hwang, K., Zhan, J., Li, Y., & Cao, Y. (2023). Scenario-Based AI Benchmark Evaluation of Distributed Cloud/Edge Computing Systems. *IEEE Transactions on Computers*, *72*(3), 719–731. DOI: 10.1109/TC.2022.3176803

Iyengar, A., & Portilla, I. (2020). Analytics at the Edge. In *IBM Blog*.

Kommineni, M., & Baseer, K. K. (2024). An Architecture and Review of Intelligence Based Traffic Control System for Smart Cities. *EAI Endorsed Transactions on Energy Web*, *11*. Advance online publication. DOI: 10.4108/ew.4964

Lopatin, E., & Poikonen, P. (2023). Cost-Effective Aerial Inventory of Spruce Seedlings Using Consumer Drones and Deep Learning Techniques with Two-Stage UAV Flight Patterns. *Forests*, *14*(5), 973. Advance online publication. DOI: 10.3390/f14050973

Mohamed, N., Rao, L. S., Sharma, M., & Shukla, S. K. (2023, May). In-depth review of integration of AI in cloud computing. In 2023 3rd international conference on advance computing and innovative technologies in engineering (ICACITE) (pp. 1431-1434). IEEE.

Mohammed, S., Fang, W. C., & Ramos, C. (2023). Special issue on "“artificial intelligence in cloud computing.”". *Computing*, *105*(3), 507–511. DOI: 10.1007/s00607-021-00985-z

Naranjo, M., Fuentes, D., Muelas, E., Díez, E., Ciruelo, L., Alonso, C., Abenza, E., Gómez-Espinosa, R., & Luengo, I. (2023). Object Detection-Based System for Traffic Signs on Drone-Captured Images. *Drones (Basel)*, *7*(2), 112. Advance online publication. DOI: 10.3390/drones7020112

. Petrović, A., &Žižović, M. (2019). *Integration of Artificial Intelligence with Cloud Services*. 381–387. DOI: 10.15308/Sinteza-2019-381-387

Shen, J., & Brower, J. (2021). Access and Edge Network Architecture and Management. In *Future Networks, Services and Management: Underlay and Overlay, Edge, Applications, Slicing, Cloud, Space, AI/ML, and Quantum Computing*. DOI: 10.1007/978-3-030-81961-3_5

Singh, R., & Gill, S. S. (2023). Edge AI: A survey. *Internet of Things and Cyber-Physical Systems*, *3*, 71–92. DOI: 10.1016/j.iotcps.2023.02.004

Singh, S. (2017). Optimize cloud computations using edge computing. *2017 International Conference on Big Data, IoT and Data Science, BID 2017, 2018-January*, 49–53. DOI: 10.1109/BID.2017.8336572

Verma, A., & Verma, V. (n.d.). *Comparative Study of Cloud Computing and Edge Computing: Three Level Architecture Models and Security Challenges*. https://ssrn.com/abstract=3908915

Westheider, J., Ruckin, J., & Popovic, M. (2023). Multi-UAV Adaptive Path Planning Using Deep Reinforcement Learning. *IEEE International Conference on Intelligent Robots and Systems*, 649–656. DOI: 10.1109/IROS55552.2023.10342516

Yang, S., Bailey, E., Yang, Z., Ostrometzky, J., Zussman, G., Seskar, I., & Kostic, Z. (2020). COSMOS Smart Intersection: Edge Compute and Communications for Bird's Eye Object Tracking. *2020 IEEE International Conference on Pervasive Computing and Communications Workshops, PerCom Workshops 2020*. DOI: 10.1109/PerComWorkshops48775.2020.9156225

Yazid, Y., Ez-Zazi, I., Guerrero-González, A., El Oualkadi, A., & Arioua, M. (2021). Uav-enabled mobile edge-computing for iot based on ai: A comprehensive review. In *Drones* (Vol. 5, Issue 4). MDPI. DOI: 10.3390/drones5040148

Youvan, D. C. (n.d.). *AI at the Edge vs. AI in the Cloud: A Comparative Analysis of High-End and Edge AI Systems*. https://doi.org/DOI: 10.13140/RG.2.2.18865.80488

Chapter 10
Revolutionizing Aerial Operations With AI-Enhanced Drones

Damodharan Palaniappan
https://orcid.org/0009-0003-0721-3068
Marwadi University, India

T. Premavathi
https://orcid.org/0009-0003-0172-2021
Marwadi University, India

Rituraj Jain
https://orcid.org/0000-0002-5532-1245
Marwadi University, India

Kumar J. Parmar
https://orcid.org/0000-0002-2502-5680
Marwadi University, India

ABSTRACT

The use of AI and drones in aerial applications have come to redefine the whole concept thrown in its numerous diverse aspects. The present chapter aims at understanding the effects of AI incorporation in drone evolution from basic UAVs to highly developed complex systems. Analyses how AI is being applied in drones: computer vision for aerial real-time image analysis, autonomous navigation, edge computing for fast processing of aerial imagery. Identifying the benefits of AI enhanced drones, the chapter reflects on the technical, ethical and legal issues of their use. That deals with constraints in the algorithms, the capacities of the cell batteries as well as handling, managing, and analyzing data; privacy issues together with legal

DOI: 10.4018/979-8-3693-8497-8.ch010

implication. This chapter looks at how such cutting-edge technologies as the swarm intelligence and deep learning can be applied to realise full and total autonomy of the drones. The transformation of the aerial systems market and its potential future development through the examples of AI-based case studies of successful projects and market leaders.

1. INTRODUCTION

Aerial operation space is gradually changing because artificial intelligence and drones are receiving increasing attention. This combination is not just an evolution but a revolution offering one breakthrough after another in tackling activities across various domains. Conceptual advancements in recent years explain the immense number of opportunities that arise from this combination and how today's aerial systems break new horizons while refuting their fundamental definition. Not just programmable machines which are controlled by a remote control, these AI drones or autonomous UAVs are of a new kind of aerial vehicles. They go beyond the confines of their antecedents, introducing a new horizon of smarter, far more flexible, and highly sophistically aerial operations with enormous impact cuts across fields of endeavour. Considerably literature review indicates that AI has complex roles in rejuvenating this discipline. In doing so, it is not only necessary to include additional options, but also the very concept of aerial systems' interpretation and application (Osco et al., 2021).

1.1 Introduction to AI and Drones

This remarkable convergence signifies a significant leap forward, propelling drones far beyond the capabilities of traditional unmanned aerial vehicles (Telli et al., 2023). AI empowers these platforms with an arsenal of advanced capabilities, enabling them to - Process information: Perceive a congeries of multidimensional data originated from a diversity of sensors to assess their environment and make proper decisions. Make autonomous decisions: Use complex analysis of the environment and context to predict changes and complete tasks with as little interference from the human brain as possible. Adapt to dynamic environments: Be more able to handle contingencies and manoeuvre in such environments thus increasing organizational flexibility and resilience to change. This integration is not simply a means of extending the performance of above functionalities; it is a revolution in how we conceive and use aerospace systems. These are the early days of that new world – a world of intelligent, responsive, and highly versatile drones – ready to revolutionize a great many sectors (Telli et al., 2023).

Several factors have catalysed this emergence of AI-enhanced drones - Exponential progress in machine learning: Allowing drones to analyze information, enhance their operations with time, and adjust for novelties they encounter. Advancements in computer vision: Outfitting drones with the established skills to 'perceive' their environment, learn how to avoid objects, recognize objects and perform other duties. Growth of edge computing technologies: Enabling processing of real-time data at the drone end which helps in faster decision making in real world scenarios. That is why that in the recent past some of what was believed to be in the realms of science fiction dream is now reality. New technologies have emerged which enable drones to constantly monitor their environment as well as make decisions and perform various tasks autonomously and with high accuracy if compared to the capabilities of humans and other available technologies Xia et al., 2023.

1.2 Drone Development

The history of drones today shows how far drones have come as a marvel of technological progression and development. The first drones were almost like miniature unmanned flying machines and their main application was to spy. Lacking advanced control mechanisms and data processing tools to some extent, their applications were rather specific, and their effects were rather modest. Nonetheless, the high rate of advancement of some of the enabling technologies like sensors, computing elements and communication technology has greatly enhanced the capability of drones. Drones have now become very complex with so many solution tools such as camera, infrared, lidar, and GPS solutions. AI has played a big role in this evolution as an integration factor making the process easier and effective than before. Smart drones today can easily fly, manoeuvre, and perform manoeuvres and complex tasks in deep and challenging environments beyond the human ability to command them. Yet, with the surge of artificial intelligence as the key future innovation it became a revolutionary moment for drones (Xiao et al., 2023).

A shift from remotely piloted vehicles to highly intelligent, autonomous systems: The idea that a drone must have a human being behind every single decision it makes is far from true. They can now operate autonomously meaning that they will make decisions based on current trends as well as operation requirements set by their developers. The ability to perform complex tasks with minimal human intervention: This independence enables drones to perform elaborate tasks such as crop spraying, bridge inspection, SAR and surveillance of the environment among others with almost zero intervention from their human operators. This evolutionary leap has made amazing opportunities in drone industry, which is used in sophisticated and impactful purposes in the present days. There is an evident transformation in terms of mid-air activities due to the avionic boards and AI-supported drones that are set to premiere industries and revolutionize the future within the forthcoming years.

2. CURRENT APPLICATIONS OF ARTIFICIAL INTELLIGENCE IN DRONES

Now that drones have adopted AI, the possibilities in firms' present-day applications are vast, expanding the opportunities for a drones use in a numerous industry. In the field of computer vision, perhaps one of the most revolutionary use cases has been observed (Islam, 2023).

2.1 Real Time Analysis Based on Computer Vision for Advanced Aerial Analysis

Advanced computer vision systems equipped to the drones have enabled the drones to really 'observe' and make sense of their environment with an incredible level of speed and accuracy. This has made it possible to transform various use cases for real-time aerial analytics as explained by Osco et al., in Figure 1. highlighting the AI-Powered drone analytics cycle. Here's a closer look at how AI is shaping computer vision in drones - Such AI-based drones can now identify and categorize a rich set of objects, whether cars and structures or wild animals and risks to the environment. Environments monitoring features allow drones to detect changes in terrains, assess construction or revisit the location to address existing issues concerning deforestation or Natural calamity. Surveying and mapping involve the use of sophisticated computer vision to produce accurate, real-world 3D models of geography, structures and any other object of interest. Recent enhancements in drone technologies involve the integration of accurate sensors and deep learning techniques, which make the drones capable of soaring challenging landscapes coupled with identifying and following targets and improving the data gathering process with high precision rates (Kamarulzaman et al., 2023).

Figure 1. AI-powered drone analytics cycle

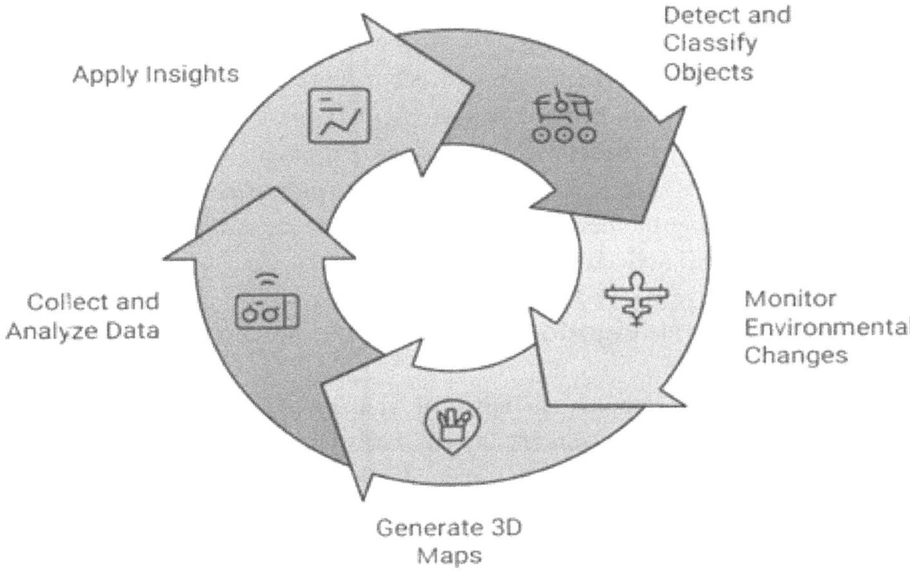

Machine learning is used to allow drones to recognise acquired objects, as well as their classification, in case if their surrounding is jam-packed and volatile. This capability is crucial for tasks such as - Surveillance and Security: Discovering possibilities of threats or abnormalities in the real time environment. Search and Rescue: Tracking lost people and things in disaster affected regions. Wildlife Monitoring: Analyzing animal's migration, and population density. AI-enabled computer vision systems mounted on UAVs can generate precise 3D topographic and facilities' contour maps. This has significant implications for - Construction and Infrastructure Development: Real-time tracking of work progress, identification of risks and integrating these into real-time planning and overall management of a construction project. Precision Agriculture: Applying computer vision technologies to monitor and evaluate the conditions of crops, detect the areas that need human involvement and management, help identify the appropriate practices for crop growth including watering or applying fertilizers. Assessing Environmental Conditions: Measuring deforestation rates, estimating glaciers shrinkage, and implementing coastal changes evaluation. With regards to the self-organising capability and decision making, AI enabled drones have shown relatively impressive performance.

Drones equipped with AI have become central to various environmental conservation activities. Through the analysis of aerial imagery, these sophisticated aerial systems can - Detect and monitor pollution levels: The tasks include finding out the contributors of air and water pollution, monitoring the distribution of pollutants, and evaluating the success of control measures. Monitor deforestation and illegal logging: Monitor alterations of forest closure, locate places with prohibited logging and establish forestation campaigns. Assess the health of coral reefs and marine ecosystems: Coral bleaching, species migration, or monitoring the marine ecosystem health. There are many ways drone FLIR systems and computer vision are used today and the utilization only broadens as time passes. From enhancing productivity in manufacturing processes to preserving our environment, these autonomous flying machines are changing the perception of how we look at things and situations in our day to day lives as noticed by Islam in 2023.

2.2 Autonomous Navigation and Obstacle Avoidance

Another important component in the use of AI in drones is the issue of the drones' capacity to move within environments in a manner that is both intelligent and safe. Here, using real-time path planning and obstacle avoidance, AI algorithms intervene successfully (Nahavandi et al., 2022). AI algorithms for real-time path planning and navigation: These algorithms enable the drones to know the best way to their destination, designated by the shortest distance, energy, the state of the weather, and the current airspace regulations. It can learn and become responsive to the surroundings by changing direction in mid-air when calibration is required to facilitate proper movement in the environment (Au, 2020). Techniques for obstacle detection and avoidance in dynamic environments: For instance, LiDAR, radar, and camera data enable drones powered by AI algorithms to detect obstacles along their flight paths even in such scenarios as dynamic and cluttered. The former can estimate the movements of an obstacle to adjust the flight path and avoid an impending collision (Gadde et al., 2021). This integration of these autonomous navigation capabilities proved to expand the function and operating envelopes of drone.

2.3 AI in Data Processing and Edge Computing

The significant data generated by drones and particularly from high-definition cameras and sensors require the availability of great processing abilities and this is where edge computing with AI comes into play (Jouhari et al., 2022). Edge computing integration for real-time data processing on drones: Whereas in traditional systems all data is sent to a central server for computing, edge computing allows drones to compute on the spot. This reduces the latency thus enabling the realization of real-time analysis and decision making especially in areas of high time sensitivity such as disaster management. AI-driven decision-making processes in remote and challenging terrains: here, making use of the local data processing function, the AI drone can make independent decisions in extra-terrestrial terrains that may not support communication infrastructure. This is very useful in tasks such as rescue missions of victims in the disaster areas or assessment of buzzard. Infrastructure in rough terrains (Lee et al., 2022). In conclusion, the enhancement of AI has made Chiropteran-enabled platforms intelligent and capable of being adaptations, or of achieving smart, self-controlled, and flexible autonomy for solving several problems in various fields.

2.4 Sensors and Data Fusion Technology

AI-driven drones employ a combination of sensors for perceiving and collecting data of the environment. These flatten sensors are the vision and auditory instruments of the drone that matching the naked raw details to artificial intelligence algorithms for direction-heading, analysis, and performing their professional tasks. High-resolution cameras capture more data, which with the help of advanced Computer Vision processing, can analyze the obtained results in detail. Figure 2. refers the following types of sensors are the most often installed in the drones based on artificial intelligence. LiDAR: LiDAR sends laser pulses and calculates the amount of time it takes to bounced back for creating a very accurate map of the surroundings. This technology is vital in pathfinding, slope mapping or survey, and infrastructure assessment. Cameras document or record visual data in different spectrums of light. RGB cameras offers normal colour imagery while multispectral and hyperspectral cameras capture data beyond the visible spectrum used in areas such as crop health assessment, mineral identification, and environment mapping. Infrared Cameras: These cameras can detect heat (through infrared rays) and hence the drones can ‘see’ at night –even locate objects depending on their temperature. It is useful in search and rescue missions, in monitoring wildlife, and physical structure examination (detection of heat loss).

Figure 2. AI-enhanced drones with sensor modalities

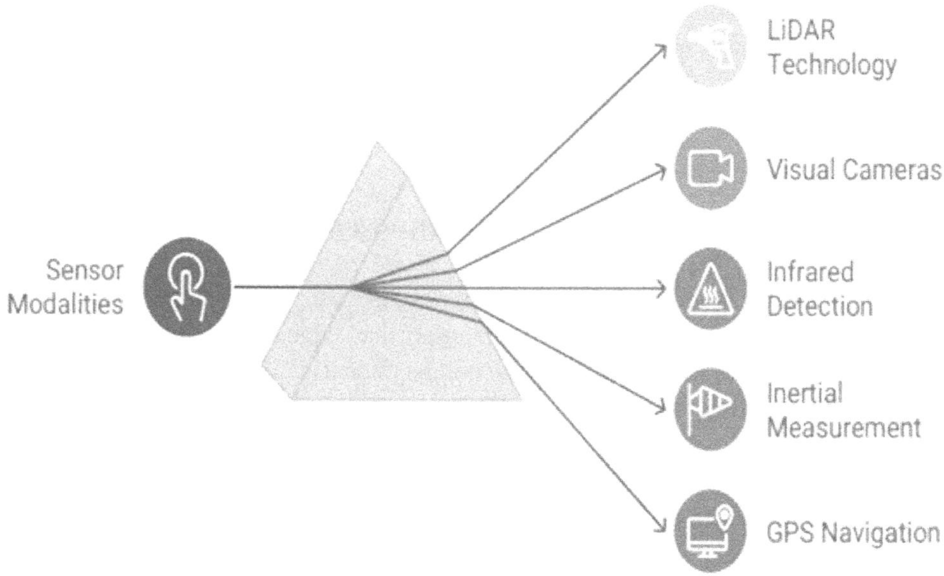

Inertial Measurement Units: IMUs are basic hardware components of drones and are made up of accelerometers and gyroscopes that determine the acceleration and rotation of the drone, which is critical for flight control and stability as well as navigating successfully. GPS: Satellite signals are used by the GPS receivers to determine the position of the drone to improve both the direction and flight plan of the drone. Data Fusion is the integration of information received from different sensors to have a more comprehensive and better picture of the surrounding world than each sensor can supply. AI algorithms play a crucial role in data fusion, using techniques like - Sensor Calibration: Coordinating data from the various sensors, this is to ensure that data being collected on one sensor is matching with the data collected by the other sensors. Data Association: Which data points in two or more sensors belong to the same object or environmental feature. State Estimation: Integrating sensor information together with a deterministic model for anticipating the drone's position, orientation, as well as velocity to high precision. Object Recognition and Tracking: Employing joined up sensor data to detect and follow objects of interest, including cars, individuals, peculiar buildings, or any sub-assemblies of a construction.

By fusing data from multiple sensors, AI-enhanced drones can - In other words, produce more accurate and reliable results in the business of controlling and managing complex environments. Create precise three-dimensional representations of the environment they are perceiving. Increase the ability to detect and follow objects, it will much more help them to make informed decisions based on the broader understanding of the situation. Data fusion technique is an important element of applying artificial intelligence in drone systems to make those systems more intelligent and independent in their operations, in performing chores as well as in delivering crucial information in different industries.

3. AI AND DRONE: REVOLUTIONIZING INDUSTRIES

The concept of AI-equipped drones is not something from a sci-fi movie anymore; instead, they are already revolutionizing markets on a global scale and providing the most possibilities and approaches. The subject area under analysis is special due to the digital transformation of the agricultural industry with the help of advanced technologies such as the use of drones with AI.

3.1 Agriculture: Implementing Precision Farming using Artificial Intelligence Based Drones for Improved Crop Management

Self-flying agricultural robots are nothing more than smart drones that are revolutionizing farming for precision. As these drones incorporate high resolution cameras and AI algorithms, these devices provide farmers even more information about fields and their crop management is beyond any comparison. While getting high resolution images from above and utilising sophisticated artificial intelligence algorithms, AI drones can offer instant information on the state of health, levels of stress, and rates of development of crops. Thus, AI-powered drones can identify initial symptoms of pest invasions, lack of nutrients or water, and address them in time (Pal et al., 2023). Precision in the monitoring and control of crops results to high yield, little or no waste, optimally utilization of resources such as water and fertilizers. Here's how AI drones are revolutionizing crop monitoring and management - Aerial Imaging and Analysis: Aerial photography can be done by drones which offer detailed images of the entire field in terms of diseases, germination and growth. The images are then perused and gone through by the Al algorithms to pick small deviations in color, texture and shape differences which would be rather difficult to discern by eye.

Early Detection of Stress and Disease: Artificial intelligence can learn to detect stress, lack of nutrients and diseases in plants early enough before they are physically noticeable. Early detection of fungal attack helps farmers to combat the situation effectively by administering appropriate measures that reduce crop damage. Targeted Application of Inputs: With specific concerns identified, the use of the AI drone allows farmers to apply fertilizers, pesticides, and herbicides a lot more efficiently. Each of these approaches helps to save input costs, preserve the environment, and increase on the yields. Irrigation Optimization: With the help of Artificial Intelligence controlled drones, field's irrigation can be evaluated, and the areas with insufficient watering can be defined. Wired Farmers, this information enable the farmers to be able to adjust the irrigation schedules to be able to water the crops appropriately to reduce wastage of water and increase their health.

Research works carried out on the use of AI integrated drone in agriculture also revealed the usefulness of the approach. Research on the utilization of the AI based drone technology for crop surveillance determined that they capable of detecting nutrient deficits 3 weeks before conventional method, overall boosting crop yields by 15% (Elbaşi et al., 2023). Precision spraying also established that the use of AI and drones in spraying chemicals cut overall pesticide use by 30% while keeping pest damage at the same level (Amarasinghe et al., 2019). The advancement of AI drones in agrarian economy is one of the best examples to certify the great impact of these amazing technologies, that are going to revolutionized not just the farmers and their crops, but also the entire agricultural sector by ensuring precision, effectiveness and sustainability. Yield Prediction: It has been found that AI drones can be employed for predicting crop yields with a high degree of success. These drones give the farmer all the necessary information about the crop such as growth stage, canopy coverage and environmental factors for planning and resource management. For instance, a study performed in the wheat filed showed that with the help of AI- dron yields could be predicted with an accuracy of more than 90% and farmers would be able to apply in decision-making processes in terms of harvesting and marketing. Talking of the benefits of using integrated with Artificial Intelligence, we have Figure 3. for benefits of AI-enhanced drones in agriculture.

Figure 3. Benefits of AI-enhanced drones in agriculture

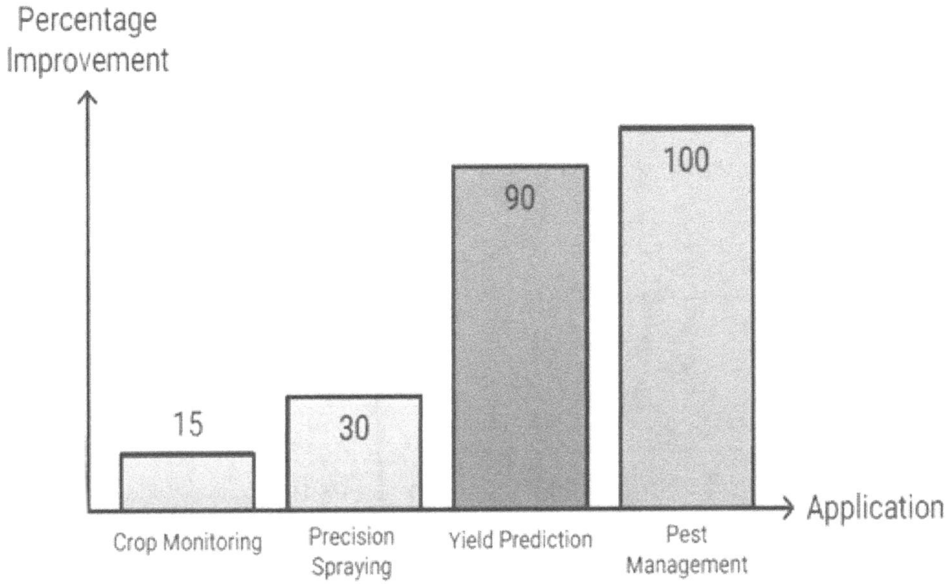

Pest Management: Multispectral or hyperspectral cameras indicate early signs of pest infestation and can be mounted on artificial intelligence drones. These drones can detect small variations of plant reflectance and locate the exact positions where pests are located and therefore, minimize the use of pesticides by spraying only on those areas. Citrus greening Disease Control: analysis of an AI drone application proved that the disease can be detected at its early stages, weeks before the visual symptoms appeared (Rajagopal & MS, 2023). According to Figure 4. AI Drones for Early Pest Detection and Pesticide Applications on Specific Areas.

Figure 4. AI drones for early pest detection and targeted pesticide application

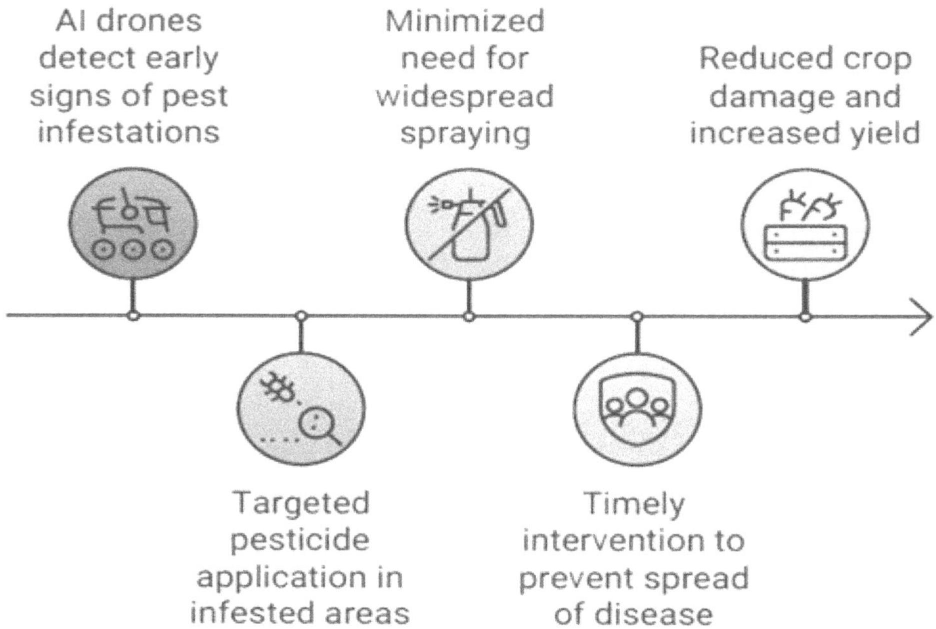

The use of AI drones in agriculture is changing the face of conventional farming by enhancing efficiency, sustainability and profitability. This trend should be expected to continually grow as development begins to unfold more complex application of the AI in agriculture hence promoting crop management and global food security.

3.2 Infrastructure: Inspection and Maintenance

To the extent that the infrastructure that require inspection and maintenance are deemed critical such as bridges, pipelines, power lines, wind turbines, it is important for safety of the public and to avoid costly occurrences. However, most of the conventional inspections require people to access the areas of interest by going through the hazardous or the hard-to-reach areas which makes them to be costly and time consuming as well as becoming risks to their own lives. This is where AI powered drones are revolutionizing the field of infrastructure inspection and management (Halder & Afsari 2023) Here in Figure 5. We described how AI drones are changing the landscape of the infrastructure inspections. Inspecting Difficult-to-Reach Areas: AI drones can also explore zones which are hard for human inspectors to reach and assess. Be it the surface under a bridge, at the top of a wind power plant or thousands of kilometres of pipeline, a drone can do the job, manoeuvring to take photos and videos from every angle. This renders expensive scaffolding, rope access methods, or other dangerous inspection practices unnecessary, as the (Xing et al., 2023). Detecting Structural Defects with AI-Powered Image Analysis: The drones' images and data can be processed by AI algorithms that will be able to point out deviations in structure as small as cracks, corrosion, vegetation growth or other misalignments. These algorithms can pick on changes in surface roughness, color and contour that might not even be noticeable to the naked eye paving way for early indications of prospective difficulties that could transform into drastic failures before being noted (Zonouz et al., 2023). Improving Safety and Efficiency: Through extending the abilities of a human inspector, AI drones minimize the dangers of working at heights or in confined spaces for workers.

Figure 5. Revolutionizing infrastructure inspections with AI drones

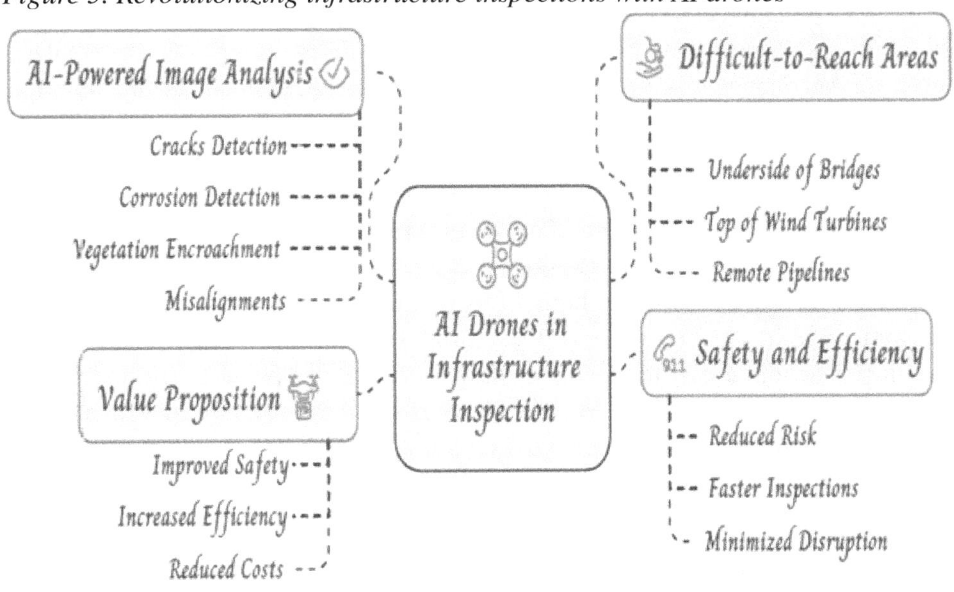

Moreover, drones can collect data from the targeted structure at a faster rate compared to other conventional techniques hence minimizing on the time taken hence reducing working interruptions. These efficiencies equal to, great reductions in cost for the various industries. The use of AI drones for infrastructure inspection offers a compelling value proposition - Improved Safety: The use of this technology will also reduce the amount of risk to human inspectors by limiting the amount of time they spend in dangerous areas. Increased Efficiency: The advantages include reductions in inspection time as well as overall downtime and better data gathering opportunities. Reduced Costs: Reduced impact of labor, lower machinery and no high capital expenditure and no chance of massive infrastructure failure. Even in the future, as the robotic technology enhances untold levels it will continue to allow the drones to deliver even better outcomes in infrastructure inspection and maintenance besides being safer, cheaper, and efficient.

3.3 Public Safety and Emergency Response

When time is of the essence, as it can often be in many crisis situations, the use of AI integrated drones is quickly becoming a critical asset for both first responders and humanitarian organizations. Police and fire departments have noted that their capacity to rapidly navigate to high risk or otherwise difficult to reach regions, collect in-person impressions, and offer invaluable and immediate visibility in critical situations is positively changing public security as well as emergency management. Here's how AI-enhanced drones are making a difference - Disaster Management: These are tiny drones that can be used after natural disasters such as earthquakes, floods, or hurricanes to give first responders a faster assessment of the situation and, therefore, help direct resource allocation. They can find people who might be stuck under debris, find out where it is safe to get to and can even drop off food and water to the victims.

Search and Rescue Operations: The use of AI in drones has enhanced the search and rescue missions especially in regions such as mountains, forest or disaster affected regions. Having infrared and high AI recognition systems that point out human heat indicators in wooded regions, these drones perform considerably decreased time in reasonable areas and find lost individuals as compared to ground search groups. Situational Awareness: AI drones are particularly used by first responders where they operate as aerial eyes giving real time view of the situation at the scene of emergency. This live feed increases the situational awareness thus allowing the incident commanders to make good decisions in terms of resource allocation and response coordination. Real-life examples of AI-enhanced drones proving their life-saving potential are becoming increasingly common: (Kyrkou et al., 2019) A case in point is the use of drones with thermal cameras used in finding people trapped in their houses after hurricane Harvey. Drones were used for enforcing social distancing, compliance with the mask and ensuring timely delivery of medical necessities amid COVID-19. With continuous development of AI technology, the possible use of drone in public safety and emergency response domains is also expected to offer greater ability in the future. It could go a long way in changing the approach and management of such vital opportunities and threats to better practice risk control and responses.

Due to the dynamic advancement in the AI technology, the possibilities of integrating drones and artificial intelligence in public safety and emergency services applications cannot be comprehensively discussed in this paper. In the aftermath of prolonged turmoil in the Middle East in 2023, two earthquakes struck Turkey and Syria. AI drones flew above the debris picking out the trapped persons as well as directing the search parties to the survivors. In California, for instance, advanced AI drones have been deployed to manage wildfires since they can offer real-time data on fire-front progression, heat sources, and safer areas for the firefighters and fire truck deployment. With increasing advancement in AI technology, there is possibility of marvel works of drones in public safety and emergency response, which in turn help to enhance capacity to save lives and property during disasters in the future.

3.4 Environmental Conservation

AI-powered drones are becoming indispensable tools for environmental conservation, offering unprecedented capabilities for monitoring and protecting our planet's fragile ecosystems. (Islam, 2023) Figure 6. refers how AI drones are making a difference in conservation efforts - Monitoring Wildlife: AI drones are changing the dynamics of wildlife tracking by offering accurate information on the number of animals, their mobility, and actions. Equipped with high-resolution cameras and AI algorithms that can identify and track individual animals, these drones enable researchers and conservationists to - Estimate population sizes and distribution: This Wikipedia data is of great importance for studying the trends of the population, as well as defining the types of animals which are endangered. Track animal movements and migration patterns: This data is used by researchers to understand the behavioural patterns of animals, find out sensitive areas of population and manage and reduce conflict between wildlife and human beings. Combat poaching and illegal wildlife trade: Surveillance drones that support thermal cameras and machine learning can identify poachers in the big woods in real time, boosting the law enforcement campaigns aimed at saving endangered species.

Figure 6. AI drones in environmental conservation

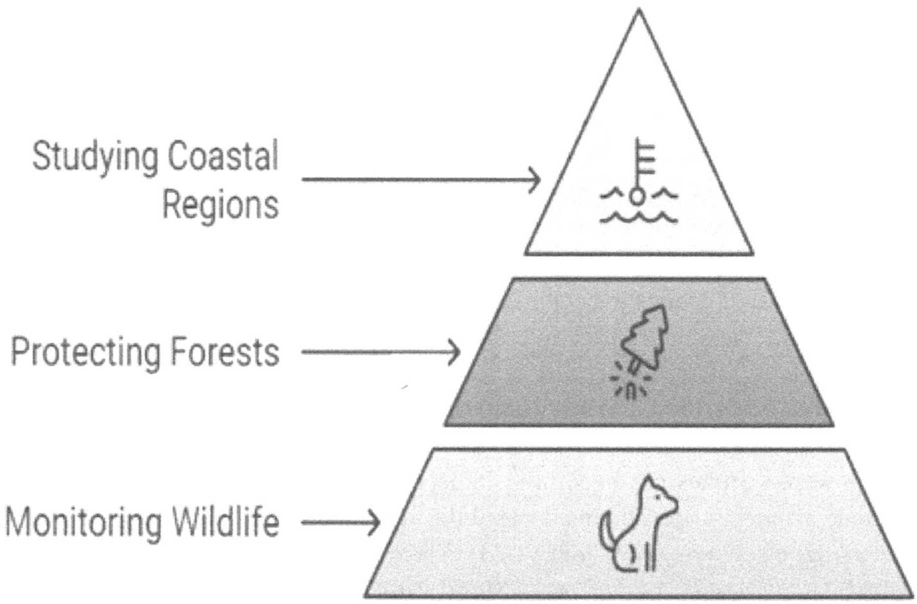

Protecting Forests: The removal of trees simply remains one of the most severe ecological problems, as it is related to climate change and the depletion of the species. AI drones are proving to be powerful tools for monitoring forests and combating illegal logging activities: Monitoring deforestation rates: Drones can provide large scale imagery of forest regions and this can help the researchers to list down the numbers of regions that are being deforested and those that need to be wiped out. Detecting illegal logging activities: It should be noted that with the help of AI algorithms it is possible to determine violations in the forest, for example, new roads or clearings for logging using drones' imagery to counteract and prevent the violations. Assessing forest health: The different technologies that can be used by drones are multispectral or hyperspectral cameras which can evaluate the status of the forest through counts such as the canopy cover, tree species density, and disease or stress indicators. Studying Coastal Regions: There is on observation noticeable that the areas close to the shores are under profound pressure from climate change, pollution and other human activities. AI drones are helping researchers and environmental managers to better understand and protect these vulnerable ecosystems: Monitoring coastlines for erosion: Coastal areas can be mapped comprehensively using drones, within the shortest time possible, observing how the coast line is changing, or the efficiency of measures put in place to protect the coast line.

Tracking changes in sea level: Through capturing images of nadiral elevation, drones can play important role in putting into identification the effects of climate change on sea level rises and small tide islands. Assessing the health of marine ecosystems: Drone technology also enables the use of specific sensors that can be used to detect water quality as well as the mapping of coral formations and the tracking of sea animals for increased understanding of the well-being of the sea. Today, data collected by AI drones is helping in conservation and climate change research and informs policies about how best to protect flora and fauna, use resources efficiently, and reduce the adverse effects of human activity on our planet. Future can see even more creative use of drones for environmental conservation as the AI technology has been identified to go a long way ahead, effectively supporting our endeavours aimed at protecting the planet for future generations.

4. CHALLENGES AND SOLUTIONS IN AI-DRIVEN AERIAL OPERATIONS

Altogether, it could be noted that the use of AI enhanced drones is quite promising; however, there is list of technical issues and constraints, which must need to overcome for realizing the full potentials of these technologies. Another problem area in aerial applications of AI is the enhancement of autonomous flight and control of unmanned aerial vehicles.

4.1 Technical Difficulties and Restrictions

Here are some key technical hurdles in AI-driven aerial operations - Limitations in AI Algorithms: Despite the progress observed in the past several years, AI algorithms are not perfect. Although, they can make decision independently they can fail in some scenarios due to certain conditions; such as in conditions that are complex or unpredictable their vision can go wrong in terms of object detection, classification or decisions made. For example, a drone intended for tracking animals might incorrectly identify a kind of animal or might not even see an animal that is well hidden in the environment. Battery Life and Flight Time: Drone autonomy is still restricted by battery power and persistence, which defines their operating distance and time. This proves even difficult for cases which would demand longer durations in the air and this includes inspections of large infrastructures as well as search and rescue missions in the wilderness. Data Processing and Transmission: Drones operating under artificial intelligence collect large volumes of data, especially in moments where they use high clarity cameras and sensors. Analysing and disseminating this data in real-time may also be a difficult proposition particularly in the case of low bandwidth/high latency operating environments. Another significant issue, therefore, pertains to the dependability and precision of the sundry sensors on which drones hinge for direction and information acquisition.

4.2 Innovations Relating to Facts Solving Technical Challenges

Despite these challenges, ongoing innovations are paving the way for more robust and capable AI-driven aerial operations - Improved AI Algorithms: Many experts are researching and trying to create even better algorithms that would be much less sensitive to certain kinds of mistakes and much better equipped to solve more intricate problems. This includes aspects such as deep learning, computer vision, and reinforcement learning that allow the drones to better capture their environment, reason and decide as well as improve their performance. Advances in Battery Technology: Lithium battery technology is advancing and the scientists are trying to gain new materials and concepts to expand the capacity and the life of the battery. There is evidence in better batteries, synergistic power supply solutions in drones, and wireless charging technologies expected to enhance both endurance and flexibility of operation.

Edge Computing and 5G Connectivity: The restrictions arising from the necessity to transmit the data to a remote centre are resolved with the help of edge computing – the data processing at the edge, on the drone itself in this case. Onboard processing aims at the minimization of frequent data transmission between drones and commands, thus increasing loop latency while conserving bandwidth. Also, with the emerging technology of the 5G network which has high bandwidth and low latency, the capabilities of data transmission will progressively improve, making it easier for drones to communicate with ground control stations. These and other developments in technology together with research ongoing are preparing the future AI-characterized aerial operations to gain more dependability and flexibility as well as expand applications into a diverse range of utility for uses such as; environmentalism, disaster relief, and many more.

4.3 Ethical and Regulatory Issues

The development of drones using artificial intelligence is progressing at a very high rate and thus brings about some critical ethical and legal questions that must be met to enable synthesis of proper and valued use of AI drones. Ethical Implications of AI in Drone Operations - Incorporation of AI in drones to enhance the capabilities of their sensors and cameras brings some of the most pressing issues of privacy and spying. Privacy Concerns: Remotely operated aerial vehicles with mounted cameras, which may have 30 times the optical zoom, as well as utilizing integrated facial recognition technology, it is an egregious invasion of privacy. It is therefore very important to set clear policies and rules regarding how data must be collected, stored and utilized to respect the privacy rights of individuals.

Bias and Discrimination: The performance of AI algorithms heavily depends on the quality of the datasets incorporated in the values of the algorithms. In case, the training data has some types of bias inbuilt in such a way, the application of AI drones may amplify the bias, therefore leading to unfairness or discrimination. For instance, an AI drone designed for surveillance with a view of protecting individuals will likely profile some ethnic groups based on the training that was given to it. Autonomous Decision-Making: With the development of new age Artificial Intelligence systems, the highly raising issue is the degree of autonomy that is to be given to the systems in critical decision-making processes, particularly those which have consequences in the life of an individual. There ought to be clearly defined structures that will help map out whom to blame in case there are incidents as such as, accidents or some equipment malfunctioning.

National and international governments and regulating authorities are still struggling to deal with the emerging threats that result from the soaring use of unmanned aerial systems propelled by Artificial intelligence (Golbin et al.,2020) (Merkert & Bushell, 2020 & Pal et al., 2023). Airspace Management: As the usage of drones rises, air traffic control is crucial since no one wants to wake up one day and find that a drone has rammed his house. Data Protection: There is increasing demand by governments in various parts of the world to enhance data protection like the GDPR in Europe. These regulations are however mandatory in drones' collection, storage and processing of personal data by the operators. Public Acceptance: It is thus necessary for the ownership and application of AI powered drones to be adopted in the public domain. To reduce concern amongst the public and increase acceptance the following should be implemented; these include; free communication, revealing to the public the data collection processes, overcome reluctance and respond to public concerns. By addressing those questions of ethics and regulation, the policymakers and industry stakeholders would be able to unlock the potential of AI-assisted fleet management and benefit from enhanced drone technologies because while on the one hand, these technologies offer a lot of potential for improving a variety of processes, it is on the other hand important to consider the risks and make sure they do not infringe on any principles of ethical operation.

4.4 Ensuring Security in AI-Enhanced Drones

AI incorporated drones are at risk of cyber threats which can interfere with the operations of the drones, tamper with collected data or even cause harm. Great care must be taken to secure these systems. As depicted on Figure 7. below highlights the Continuous Security Improvement for AI-Enhanced Drones.

Figure 7. Continuous security improvement for AI-enhanced drones

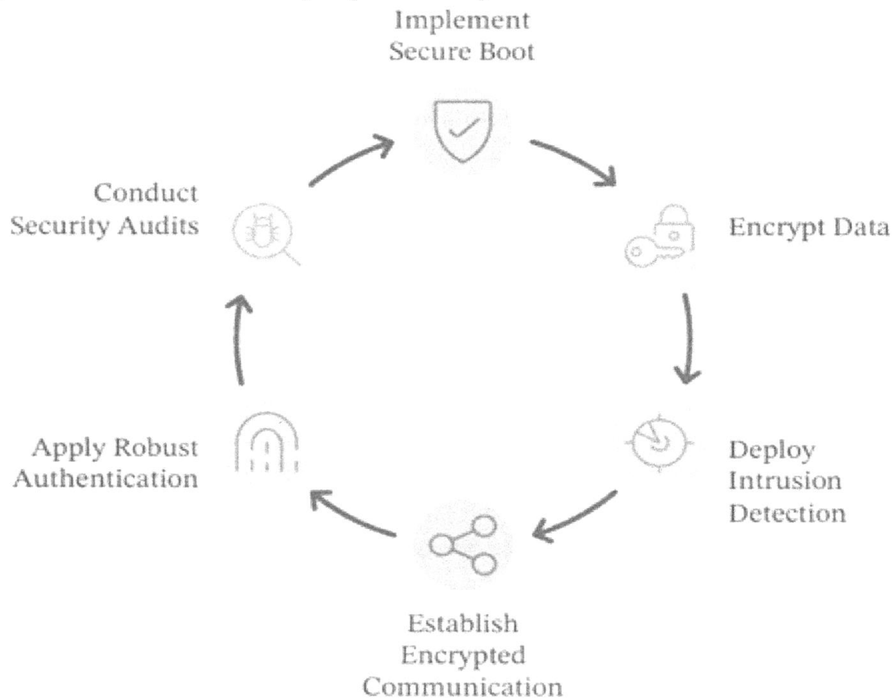

4.5 Protecting AI Systems in Drones from Cyber Threats

Cybersecurity measures for AI-driven drones must focus on - Secure Boot and Software Updates: Security procedure such as using secure boot mechanisms and guaranteeing that software upgrades aren't modified by a third-party help defend the drone OS and, correspondingly, AI algorithms against fiddling. Data Encryption: Transmitting confidential information and storing it in encrypted formats is necessary to prevent an unauthorized party's invasion. Intrusion Detection and

Prevention Systems: Use of IDS/IPS in drones makes it easier to detect and prevent cyber-attacks as they occur.

4.6 Issues and Measures for more Secure Communication and Data Transfer

The use of drones for surveillance, safety, and security is enhanced by the assurance of proper secure conversations and encryption to protect information and data being transmitted by the localized drones. The issue of secure communication and data transmission is critical especially when using AI in managing the drones. This involves the use of secure communication interfaces, stringent authentication techniques, and security assessment and vulnerability assessment. Encrypted Communication Channels: The use of appropriate encryption communication links such as VPN or TLS/SSL may also be adopted to enhance the security of the link between drones and the ground station by eliminating vulnerabilities that expose the data streaming from the drone to interception. Robust Authentication Mechanisms: Life cycle security measures should be employed to ensure that any clients or the devices that want to access the drone's systems are genuine. Regular Security Audits and Penetration Testing: Security audit and penetration testing should be conducted at least once every year to check the overall security of the drones and ensure that it has the best security measures in the current technological advancement. Though the use of AI in drones is on the advancement, there are some ethical, regulatory, and security concerns that need to be addressed to harness it fully and avoid the resulting risks. The practice of bringing together policy makers, industry professionals, academicians, and the general populace can help result in a best practice guide on the effective and socially beneficial implementation of AI operated drones in our skies and society.

5. FUTURE TRENDS IN AI AND DRONE INTEGRATION

Currently, the use of AI and drone technology, as well as their combination, are for the most part, are still in their infancy, however, the potential for ground breaking developments is incredibly high.

5.1 Emerging AI Technologies Shaping the Future of Drones

Here are some of the next-generation AI technologies poised to revolutionize drone capabilities: The advancement of technologies like computer vision, deep learning, and reinforcement learning helps the drone to identify their environment more accurately, make smarter decisions and work well in the most challenging conditions. Finally, the federated learning and edge computing are making it possible for the drones to undertake the sophisticated and intricate processing of information using AI away from centralize systems, thereby also reducing the latency and increasing the level of self-sufficiency (Pal et al., 2023).Swarm Intelligence is based on the PSO algorithm, which mimics the actions of ants or bees where several drones act together like one large group. This technology enables drones to:(Alkouz et al., 2020), perform complex tasks collaboratively: Some of the uses like search and rescue operations, huge structure inspection, or even in precision farming activities can be done much better and faster by employing a swarm of drones. Adapt to dynamic environments: From the concept of swarm intelligence, drones can transfer data and change their behaviour at these types of scenarios and thus be more robust.

Enhance coverage and redundancy: Having more than one drone reduces the amount of coverage an individual drone can provide but it makes it possible to have backup drones in case a few faces mechanical problems. Deep Learning: Machine learning specifically, deep learning it is a process of feeding a large amount of data to artificial neural networks to enable them learn and predict. In the context of drones, deep learning is driving advancements in (Kaufmann et al., 2023). Object Recognition and Tracking: Multiple objects tracking algorithms can be learned to detect objects of interest, like a vehicle or individual, wildlife, in complex environments. Autonomous Navigation and Obstacle Avoidance: Object tracking through drones makes use of deep learning; it lets the drones fly around in a determined territory, staking out objects, and even making real time decisions on what to do without human interference. Predictive Maintenance: With the help of machine learning, it can comprehensively analyse the data obtained from the sensors of drones and foresee equipment malfunctions, so that it can be promptly repaired and avoiding downtime as a result.

The Potential of AI in Enabling Fully Autonomous Drones: The primary motivation of combining AI and drones is to reach the end-to-end automation level of the drones to enable them to fly on their own (Cheng et al., 2023). This would require significant advancements in AI capabilities, including - Robust Perception and Decision-Making: A system needs to determine, with sufficient speed and precision, the surrounding environment and utilize this information to make proper decisions on its own, let alone in the situations that have not been planned beforehand. Reliable Communication and Collaboration: As with other kinds of intelligent mechanical systems, these autonomous drones must also be able to communicate and share information with other similar drones as well as ground control units. Ethical and Legal Frameworks: Specific ethical rules and legislations are necessary to regulate the use of fully autonomous drones and manage the questions of who is responsible in case of an accident, privacy infringement, etc. As fully self-driving drones are being prototyped and possibly deployed, several problems on the technical, ethical, and social levels appear. However, its diverse application in the transportation and logistics sector, agriculture or disaster management, is encouraging more scientists to continue researching on this field (Zhang and Li, 2023).

Advanced Perception and Situation Awareness are the first sub-elements under this element of EFA that have been explained as follows - Advanced sensor technologies in conjunction with the drone can achieve a level of situation awareness that has the potential of making the drone highly manoeuvrable in complex surroundings, identify threats and patterns, and learn from new conditions. Abetment of drone perception entails multispectral & hyperspectral imaging sensors which refers to imaging across a wide range of wavelengths beyond the visible spectrum forego, infrared, ultraviolet, thermal imaging Pal et al., 2023. This extended sensing capability yields useful information in several use cases like wildlife tracking, crop management, natural calamities, and search operations, as it helps in recognising objects that cannot be detected by naked eyes. Moreover, most of the SLAM methods allow drones to generate a high-resolution 3D terrain map of surrounding environment and estimate the changes in the environment over the time. As drones that are integrated with artificial intelligence top become more common, it is important to come to terms with the sort of issues arising from their use. The intelligent drones might learn from their contexts, operate and change according to the applied decrees drawn from sophisticated AI systems. Recurrent learning is subfield of machine learning which makes use of testing and improving of decisions of a system, making it able to perform in changeable conditions and making necessary corrections to achieve a given goal. Refer to the use of drones in search and rescue missions where an AI self-propelled drone can identify the target location, locate and track the victims and coordinate with other drones or ground crews to maximize the efficiency of the rescue process.

5.2 AI Enabled Drones in Urban Air Mobility

Transportation as a product of Urban Air Mobility is expected to transform transport in cities around the globe and AI will play a significant role in realizing this technology. Addressing these issues with AI incorporation into urban air mobility systems, the issues entailing the confined airspace area include airspace management, air traffic control, and dealing with several integrated vehicles flying concurrently. AI-enabled drones can transform urban transportation and logistics by providing - Autonomous Navigation and Collision Avoidance: Without implementing AI, drone taxis and delivery drones will not be able to operate in the highly congested urban space. This entails perception, spatial mapping, path planning, and self-navigation plus avoidance of obstacles. Traffic Management and Airspace Integration: With more UAVs joining the sky in particular the urban skies, air traffic control will require AI to continue managing flow and or avoid unwanted crashes with traditional aviation. Optimized Routing and Delivery Efficiency: Through various parameters such as traffic and weather conditions, time window for delivery, AI algorithms can improve the delivery routes and hence enhance delivery services. Also, incorporation of AI extends to anticipatory maintenance and other fleet operations for urban air mobility or short-range transport systems (Raivi et al., 2023).

AI-supported drones as part of transportation systems in smart cities are also a part of the overall smart city concept. With ITS system that integrated with Artificial Intelligence, the way people and goods transport within cities would change vastly symbiosis with less traffic density, pollutions, and travelling time (Wilbur et al., 2023). Seamless Integration with Transportation Networks: UAM integrated with AI can be easily connected with other transportation systems and provide an additional air transport on demand in connection with public transport, ridesharing and others. Reduced Congestion and Emissions: Through transport through the airspace, UAM could potentially solve traffic congestion in urban areas thereby shortening travelling time and consequently enhancing air quality. Enhanced Emergency Response: AI-enabled drones can become very helpful in emergency response scenarios, either for gaining quick insight into the situation, for delivering the medical equipment or for searching for the missing person in the remote area. As with any emerging industry, the further progress of safe, reliable, and effective UAM systems will depend on AI progress, regulation, and public acceptance (Altun et al., 2023). The application of aerial mobility in smart cities can have various advantages, and its impact on transforming the urban transport sector and enhancing people's quality of life should not be underestimated.

6. CASE STUDIES: REAL-WORLD APPLICATIONS OF AI-ENHANCED DRONES

Analyzing the use cases helps to look at the shift that occurs with AI integrated drone systems from a pragmatic standpoint. The use of the smart drone in advanced farming is highly transforming the practice with/octet as its profound impacts bearing benefits such as efficiency gain and productivity increments besides sustainability. For instance, the state seeks to expand the use of AI enhanced drones for disaster response and management, surveillance, mapping, search and rescue and supply delivery to disaster-stricken zones (Speth, 2022), (Altun et al., 2023). Thus, cross functions envisioned in UAV utilization for smart cities are environmental, traffic, pollution, and civil security control. In smart cities, AI integration with UAVs, however, experience several issues including but not limited to; Safety, Privacy, Ethical and Legal Issues (Dias et al., 2023).

6.1 Examples of Successful Pioneering Projects and Industry Influencers

Here are a few notable examples of projects and companies at the forefront of this technological revolution - Zipline: It is a pioneering company and it is also changing the face of healthcare delivery system in remote areas in Rwanda and Ghana. Using its artificial intelligence-based drones, it has been doing autonomous delivery of various consumables and pharmaceuticals like blood, vaccines, and others to the needy populations despite evasive terrains. This innovative approach ensures that health care services are made available where resources and network connection will be scarce (Enayati et al., 2023). Skydio: This American based drone manufacturer has recently released a line of drones which are fitted with Artificial Intelligence capabilities, which enables the drones to follow their target, avoid obstacles and take high-definition videos. Skydio has become famous for its drones that can navigate autonomously with the help of AI and come with systems for detection of obstacles. Used in industries like infrastructure industries for analysis, construction industries for surveillance, and cinematography where their ability to self-navigate through various terrains without external signals is a plus for the companies (Mohsan et al., 2023). PrecisionHawk: This firm is a leader in the use of Artificial Intelligence in creating drone solutions in the agriculture industry. Their sophisticated drone makes use of the current computer vision and machine learning techniques to accumulate illustrative information concerning numerous aspects of crop production. They can also provide accurate information on plant condition; examination of soil and evaluation of irrigation needs on large spaces of farmland. In this way with offering such a valuable set of knowledge to farmers, PrecisionHawk enables them to make smarter decisions while leveraging given amounts of data to increase the yield, decrease the use of resources, and make their businesses more sustainable. This challenge has great relevance to the future of the agricultural industry by proposing a solution that could revolutionize the ways farmers increase efficiency, minimize losses, and embrace environmentally sustainable practices.

DJI: Leading drone manufacturing company, DJI has been incorporating what AI into its products as a global manufacturer. Their AI driven home robots are used by the first responders to support search and rescue with precision assessment and immediate response. Police also use DJI's machines with AI to enhance surveillance and monitor the surroundings to avoid unfavourable outcomes. In addition, these new and improved drones are used in filmmaking by these filmmakers to get magnificent aerial shots that improve cinematography and offer interested audiences unique viewing experiences among others. These are just some of the ways in which AI enabled drones are changing industries and opening the door to smarter, better, more efficient aerial logistics.

6.2 Perspectives from Industry Pioneers and Innovators

But to have a much better picture about the effects of integrating AI with drones, it would be helpful to listen to the expertise of the key players in the industry (Pal et al., 2023). The following key points from industry pioneers and innovators highlight the transformative potential of AI-enhanced drones - Emphasis on Data: Experts of the industry underline the significance of data in creating it, as well as improving AI algorithms of drones. Availability of good, rich and varied data forms is very essential when it comes to developing effective and reliable artificial intelligence. Collaboration is Key: Most successful drones AI projects require inputs from technology firms, working alongside companies in industries, as well as government and regulatory agencies. Such reporting means exhibit free and effective flow of information as knowledge dissemination is critical in the promotion of innovation and the right employment.

Focus on Safety and Reliability: Safety is one of the main concerns when it comes to augmenting drones with AI capabilities as is reliability. The stakeholders are putting a lot of capital towards the establishment of proper safety frameworks and then putting the technologies through their paces. Analyzing these experimental programs and successful AI-drone ventures, we can derive useful lessons about the prospects for innovative intelligence-augmented UAVs and the critical success determinants.

6.3 Lessons Learned and Best Practices

There are equally a lot of lessons that could be highlighted, and they include the following - Since AI technology in drones is still relatively new, it is crucial to learn from accomplishments and issues in the process of development. Here are some key takeaways and best practices for integrating AI into aerial operations - It is essential to incorporate knowledge from different domains with AI for creating feasible and efficient solutions. One can identify the need to encourage closer cooperation between specialists in disciplines related to the subject matter and AI developers. When it comes to AI & drones, three key considerations are important to realize large scale and stability of this technology. A focus and commitment on reliability, durability and adaptability to many operating conditions can help fully realise the potential of these technologies. Legal and ethical concerns that arise must be resolved pre-emptively. Data is King: Performance of AI algorithms substantially depend upon the quality, relevance, variety and correct labelling of data. Far more time is spent on data collection, data-cleaning, and annotation to provide the best to the AI-Market. Start with a Clear Use Case: Sites specific issues that AI can tackle within the operations of the unmanned aerial vehicle again. Leveraging on clearly specified and successful use cases de-risks the entire process and yields beneficial information for future enhancement.

Iterative Development is Key: The process of AI development is cyclical. In a similar manner, use a simple plan and focus on micro deadlines before escalating the program's difficulty and moving to a macro approach as data collection and algorithms become more refined. Collaboration Fosters Innovation: For innovation and standard-setting and safe use of AI-enabled drones and associated technologies, meaningful partnerships between AI professionals, manufacturers of drones, stakeholders in industries, and other players in the regulatory sphere are critical. By leveraging these insights and best practices, organizations can navigate the complex landscape of AI-enhanced drones and harness the transformative potential of these technologies to revolutionize aerial operations (Janga et al., 2023). The following needs to be a priority and best practices for integrating AI into aerial operations - Prioritize Safety and Reliability: The testing and validation of AI enhanced drones should be strictly conducted to guarantee the safety of the drones. To reduce the risks, it is possible to apply methods such as risk diversification or redundancy and fail-safe mechanisms. Address Ethical Considerations: This requires the formulation of ethical principles that relate to privacy of data, bias confirmation during the AI algorithm creation process, and proper usage of the self-possessed drone devices. This paper finds that transparency and accountability are important in earning the trust of the people.

Invest in Training and Education: Make sure your employees are trained to use and manage the new AI enabled drones that are to be deployed in your fleet. Perhaps, continued training could alleviate this problem given technology is rapidly advancing. Stay Informed About Regulations: The government has continued to adapt policies on drones to reflect the social changes of society. It is important to remain update on current changes in the substantive law regulating the use of AI in aerial work. Therefore, by adopting these lessons learned and delights, one can effectively leverage on the use of AI-assisted drones and be able to have optimal results on different fields and or dozes throughout the whole world.

CONCLUSION

AI has been introduced into enhancing the development of drones and is making simple Remote-Controlled Equipment into complex Self-Chorusing Equipment. Advanced AI provides drone with more ability, so they can work more independently with higher efficiency and precision and especially safety. This technological advancement is having a ripple effect on many fields such as the supply chain, transportation, farming, construction, emergency response services, and surveillance of the environment. AI drones make deliveries faster and more efficient, help farmers get higher crop yields, check infrastructure health quicker, help in timely response to various emergencies, and aid in conservation and research. AI-assisted drones can expect great prospects for future development, including future developments in UAM, group intelligence, and automated operations. These drones are said to disrupt sectors, redefine transport, infrastructure, inspection, rescue, and precision farming. However, there are many safeties, privacy, security as well as ethical issues which will have to be tackled to ensure the right implementation of this revolutionary technology. Thus, collaboration of the stakeholders with AI enhanced drones is crucial for the creation of a safer, more effective, eco-friendly future.

REFERENCES

Alkouz, B., Bouguettaya, A., & Mistry, S. (2020, January 1). Swarm-based Drone-as-a-Service (SDaaS) for Delivery. Cornell University. https://doi.org//arxiv.2005.06952DOI: 10.48550

Altun, A. T., Hasanzade, M., Saldiran, E., Guner, G., Uzun, M., Fremond, R., Tang, Y., Bhundoo, P., Su, Y., Xu, Y., İnalhan, G., Hardt, M., Fransoy, A., Modha, A., Tena, J. A., Nieto, C., Vilaplana, M., Tojal, M., Gordo, V., & González, A. (2023, August 15). The Development of an Advanced Air Mobility Flight Testing and Simulation Infrastructure. *Aerospace (Basel, Switzerland)*, *10*(8), 712–712. DOI: 10.3390/aerospace10080712

Amarasinghe, A., Wijesuriya, V B., Ganepola, D., & Jayaratne, L. (2019, November 10). A swarm of crop spraying drones solution for optimising safe pesticide usage in arable lands. DOI: 10.1145/3356250.3361948

Au, T. (2020, January 1). Extending the Range of Drone-based Delivery Services by Exploration. Cornell University. https://doi.org//arxiv.2012.09367DOI: 10.48550

Bengio, Y., Hinton, G. E., Yao, A. C., Song, D., Abbeel, P., Darrell, T., Harari, Y. N., Zhang, Y., Lan, X., Shalev-Shwartz, S., Hadfield, G. K., Mouret, J., Maharaj, T., Hutter, F., Baydin, A. G., McIlraith, S. A., Gao, Q., Acharya, A., Krueger, D., & Mindermann, S. (2024, May 24). Managing extreme AI risks amid rapid progress. *Science*, *384*(6698), 842–845. DOI: 10.1126/science.adn0117 PMID: 38768279

Cheng, N., Wu, S., Wang, X., Yin, Z., Li, C., Chen, W., & Chen, F. (2023, August 15). AI for UAV-Assisted IoT Applications: A Comprehensive Review. *IEEE Internet of Things Journal*, *10*(16), 14438–14461. DOI: 10.1109/JIOT.2023.3268316

Dias, T., Fonseca, T., Vitorino, J., Martins, A., Malpique, S., & Praça, I. (2023, January 1). From Data to Action: Exploring AI and IoT-driven Solutions for Smarter Cities. Cornell University. https://doi.org/DOI: 10.48550/arXiv.2306

Elbaşi, E., Mostafa, N., Al-Arnaout, Z., Zreikat, A. I., Cina, E., Varghese, G., Shdefat, A. Y., Topcu, A. E., Abdelbaki, W., Mathew, S., & Zaki, C. (2023, January 1). Artificial Intelligence Technology in the Agricultural Sector: A Systematic Literature Review. *IEEE Access : Practical Innovations, Open Solutions*, *11*, 171–202. DOI: 10.1109/ACCESS.2022.3232485

Enayati, S., Campbell, J F., & Li, H. (2023, August 1). Vaccine distribution with drones for less developed countries: A case study in Vanuatu. Elsevier BV, 14, 100312-100312. DOI: 10.1016/j.jvacx.2023.100312

Fontaine, S., Abbadie, L., Aubert, M., Barot, S., Bloor, J., Derrien, D., Duchêne, O., Gross, N., Henneron, L., Roux, X L., Loeuille, N., Michel, J., Recous, S., Wipf, D., & Alvarez, G. (2023, December 12). Plant–soil synchrony in nutrient cycles: Learning from ecosystems to design sustainable agrosystems. Wiley, 30(1). DOI: 10.1111/gcb.17034

Gadde, C. S., Gadde, M. S., Mohanty, N., & Suresh, S. (2021, January 1). Fast Obstacle Avoidance Motion in SmallQuadcopter operation in a Cluttered Environment. Cornell University. https://doi.org//arxiv.2109.09159DOI: 10.48550

Gikunda, K. (2024, January 1). Harnessing Artificial Intelligence for Sustainable Agricultural Development in Africa: Opportunities, Challenges, and Impact. Cornell University. https://doi.org/DOI: 10.48550/arXiv.2401

Golbin, I., Rao, A S., Hadjarian, A., & Krittman, D. (2020, December 10). Responsible AI: A Primer for the Legal Community. DOI: 10.1109/BigData50022.2020.9377738

Guo, J., Jones, M. R., Djahel, S., & Wang, S. (2023, January 1). AVARS — Alleviating Unexpected Urban Road Traffic Congestion using UAVs. Cornell University. https://doi.org//arxiv.2309.04976DOI: 10.48550

Halder, S., & Afsari, K. (2023, February 10). Robots in Inspection and Monitoring of Buildings and Infrastructure: A Systematic Review. *Applied Sciences (Basel, Switzerland)*, *13*(4), 2304–2304. DOI: 10.3390/app13042304

Hirschberg, M J. (2003, September 8). American Attack Unmanned Aerial Vehicles (UAVs): A Century of Progress. DOI: 10.4271/2003-01-3064

Islam, S. M. R. (2023, January 1). Drones on the Rise: Exploring the Current and Future Potential of UAVs. Cornell University. https://doi.org//arxiv.2304.13702DOI: 10.48550

Janga, B., Asamani, G. P., Sun, Z., & Cristea, N. (2023, August 21). A Review of Practical AI for Remote Sensing in Earth Sciences. *Remote Sensing (Basel)*, *15*(16), 4112–4112. DOI: 10.3390/rs15164112

Javaid, M., Haleem, A., Khan, I H., & Suman, R. (2023, March 1). Understanding the potential applications of Artificial Intelligence in Agriculture Sector. Elsevier BV, 2(1), 15-30. DOI: 10.1016/j.aac.2022.10.001

Jouhari, M., Al-Ali, A., Baccour, E., Mohamed, A., Erbad, A., Guizani, M., & Hamdi, M. (2022, January 15). Distributed CNN Inference on Resource-Constrained UAVs for Surveillance Systems: Design and Optimization. *IEEE Internet of Things Journal*, *9*(2), 1227–1242. DOI: 10.1109/JIOT.2021.3079164

Kamarulzaman, A. M. M., Jaafar, W. S. W. M., Said, M. N. M., Saad, S. N. M., & Mohan, M. (2023, May 30). UAV Implementations in Urban Planning and Related Sectors of Rapidly Developing Nations: A Review and Future Perspectives for Malaysia. *Remote Sensing (Basel), 15*(11), 2845–2845. DOI: 10.3390/rs15112845

Kaufmann, E., Bauersfeld, L., Loquercio, A., Müller, M., Koltun, V., & Scaramuzza, D. (2023, August 30). Champion-level drone racing using deep reinforcement learning. *Nature, 620*(7976), 982–987. DOI: 10.1038/s41586-023-06419-4 PMID: 37648758

Kyrkou, C., Timotheou, S., Kolios, P., Theocharides, T., & Panayiotou, C. G. (2019, January 1). Drones: Augmenting Our Quality of Life. *IEEE Potentials, 38*(1), 30–36. DOI: 10.1109/MPOT.2018.2850386

Lee, H., Kwon, S., Jung, S., & Kim, J. (2022, January 1). Neural Myerson Auction for Truthful and Energy-Efficient Autonomous Aerial Data Delivery. Cornell University. https://doi.org//arxiv.2201.01170DOI: 10.48550

Lopes, D., & Silva, J. (2023, March 22). Urban air mobility (UAM) in the metropolitan region of São Paulo: Potential and threats. *OmniaScience, 13*(1), 1–1. DOI: 10.3926/jairm.345

Merkert, R., & Bushell, J. (2020, October 1). Managing the drone revolution: A systematic literature review into the current use of airborne drones and future strategic directions for their effective control. Elsevier BV, 89, 101929-101929. DOI: 10.1016/j.jairtraman.2020.101929

Merkle, N., Bahmanyar, R., Henry, C., Azimi, S. M., Yuan, X., Schopferer, S., Gstaiger, V., Auer, S., Schneibel, A., Wieland, M., & Kraft, T. (2023, January 1). Drones4Good: Supporting Disaster Relief Through Remote Sensing and AI. Cornell University. https://doi.org//arxiv.2308.05074DOI: 10.48550

Mohsan, S A H., Othman, N Q H., Li, Y., Alsharif, M H., & Khan, M A. (2023, January 16). Unmanned aerial vehicles (UAVs): practical aspects, applications, open challenges, security issues, and future trends. Springer Science+Business Media. DOI: 10.1007/s11370-022-00452-4

Nahavandi, S., Alizadehsani, R., Nahavandi, D., Mohamed, S., Mohajer, N., Rokonuzzaman, M., & Hossain, I. (2022, January 1). A Comprehensive Review on Autonomous Navigation. Cornell University. https://doi.org//arxiv.2212.12808DOI: 10.48550

Osco, L P., Marcato, J., Ramos, A P M., Jorge, L A D C., Fatholahi, S N., Silva, J D A., Matsubara, E T., Pistori, H., Gonçalves, W N., & Li, J. (2021, October 1). A review on deep learning in UAV remote sensing. Elsevier BV, 102, 102456-102456. DOI: 10.1016/j.jag.2021.102456

Pal, O. K., Shovon, M. S. H., Mridha, M. F., & Shin, J. (2023, January 1). A Comprehensive Review of AI-enabled Unmanned Aerial Vehicle: Trends, Vision, and Challenges. Cornell University. https://doi.org//arxiv.2310.16360DOI: 10.48550

Papyan, N., Kulhandjian, M., Kulhandjian, H., & Aslanyan, L. (2024, March 1). AI-Based Drone Assisted Human Rescue in Disaster Environments: Challenges and Opportunities. *Pattern Recognition and Image Analysis*, *34*(1), 169–186. DOI: 10.1134/S1054661824010152

Pham, Q., Ruby, R., Fang, F., Nguyen, D. C., Yang, Z., Le, M., Ding, Z., & Hwang, W. (2022, June 1). Aerial Computing: A New Computing Paradigm, Applications, and Challenges. *IEEE Internet of Things Journal*, *9*(11), 8339–8363. DOI: 10.1109/JIOT.2022.3160691

Pueyo, P., Dendarieta, J., Montijano, E., Murillo, A. C., & Schwager, M. (2024, January 1). CineMPC: A Fully Autonomous Drone Cinematography System Incorporating Zoom, Focus, Pose, and Scene Composition. *IEEE Transactions on Robotics*, *40*, 1740–1757. DOI: 10.1109/TRO.2024.3353550

Raivi, A. M., Huda, S. M. A., Alam, M. M., & Moh, S. (2023, January 28). Drone Routing for Drone-Based Delivery Systems: A Review of Trajectory Planning, Charging, and Security. *Sensors (Basel)*, *23*(3), 1463–1463. DOI: 10.3390/s23031463 PMID: 36772502

Rajagopal, M. K., & MS, B. M. (2023). Artificial Intelligence based drone for early disease detection and precision pesticide management in cashew farming. arXiv preprint arXiv:2303.08556.

Telli, K., Kraa, O., Himeur, Y., Ouamane, A., Boumehraz, M., Atalla, S., & Mansoor, W. (2023, January 1). A Comprehensive Review of Recent Research Trends on UAVs. Cornell University. https://doi.org//arxiv.2307.13691DOI: 10.48550

Wilbur, M., Sivagnanam, A., Ayman, A., Samaranayeke, S., Dubey, A., & Lászka, Á. (2023, January 1). Artificial Intelligence for Smart Transportation. Cornell University. https://doi.org/DOI: 10.48550/arXiv.2308

Wu, W., Zhou, F., Wang, B., Wu, Q., Dong, C., & Hu, R. Q. (2022, January 1). Unmanned Aerial Vehicle Swarm-Enabled Edge Computing: Potentials, Promising Technologies, and Challenges. Cornell University. https://doi.org//arxiv.2201.08517DOI: 10.48550

Xiao, J., Zhang, R., Zhang, Y., & Feroskhan, M. (2023, January 1). Vision-based Learning for Drones: A Survey. Cornell University. https://doi.org/DOI: 10.48550/arXiv.2312

Xing, J., Cioffi, G., Hidalgo-Carrió, J., & Scaramuzza, D. (2023, January 1). Autonomous Power Line Inspection with Drones via Perception-Aware MPC. Cornell University. https://doi.org//arxiv.2304.00959DOI: 10.48550

Zhang, S., & Li, L. (2023, March 23). The Multi-visits Drone-Vehicle Routing Problem with Simultaneous Pickup and Delivery Service. Springer Science+Business Media. DOI: 10.1007/s40305-023-00471-7

Zonouz, E. F., Pan, X., Hsu, Y., & Yang, T. (2023, January 1). 3D vision-based structural masonry damage detection. Cornell University. https://doi.org//arxiv.2308.16380DOI: 10.48550

Chapter 11

Security Issues and Challenges in Edge Computing Architecture for the Drone Industry

Imdad Ali Shah
https://orcid.org/0000-0003-2015-1028
Faculty of Engineering Science and Technology, Iqra University, Karachi, Pakistan

ABSTRACT

The way unmanned aerial vehicles (UAVs) function is being completely transformed by the incorporation of cutting-edge computing into the drone industry which offers notable improvements in processing information, latency reduction, and autonomous capability. This chapter explores the functional concepts and architectural frameworks essential for implementing edge computing. We outline the techniques for maximizing computational resources, boosting real-time decision-making, and increasing overall system efficiency by examining the relationship between drones and edge nodes. The implementation of edge infrastructure edge-enabled drone applications, and the difficulties with security reliability, and network connectivity are important subjects.

INTRODUCTION

In recent years the drone industry has seen exponential growth due to the quick development and implementation growth due to the quick development and implementation of unmanned aerial vehicles (UAVs). These multipurpose tools are

DOI: 10.4018/979-8-3693-8497-8.ch011

now essential to many applications, ranging from emergency response, logistics aerial photography, and agricultural monitoring(McEnroe et al., 2022, Khuwaja et al., 2018). The need for time decision-making, and autonomy has increased drones because they are more complex. (Al-Turjman and Zahmatkesh, 2020, McEnroe et al., 2022, Dong et al., 2020). This paradigm change enhances the drone system's responsiveness, lowers latency and conserves bandwidth. Edge computing allows faster data analysis for-sensitive applications, by placing processing power closer to the UAVs. We also discuss the issues of reliability, safety and network connectivity that this technological convergence. This study attempts to give a thorough grasp of how edge computing might drive the drone industry into new spheres of innovation and efficiency by looking at recent developments and upcoming ones. An overview of edge computing architecture is in Figureure 1.

Figure 1. Overview of edge computing architecture

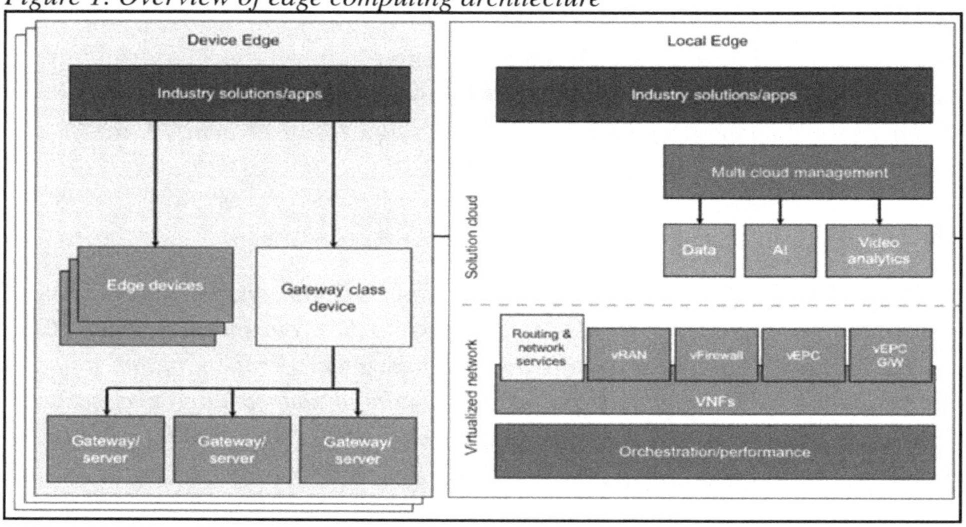

By utilizing and controlling the computing power available at distant locations, including manufacturing facilities, retail establishments, warehouses, lodging facilities, and distribution centres(Khan et al., 2019, Zhou et al., 2019). A piece of equipment with a specific function and built-in computing power is called an edge device. Edge devices, such as an assembly machine on a manufacturing floor, an ATM, an intelligent camera, or a car, can be used for interesting tasks (Zhang et al., 2019, McEnroe et al., 2022, Shakhatreh et al., 2019). While certain edge devices can host network functions or act as a restricted gateway, edge gateways are typically distinct from edge devices. IoT sensors are fixed-function devices without onboard computing, memory, or storage that gather and send data to an edge or cloud. Because of this, containers cannot be installed on them. The security issues & challenges in edge computing for the drone industry are presented in Figureure 2

Figure 2. Presents security issues & challenges in edge computing for the drone industry 2020-2024

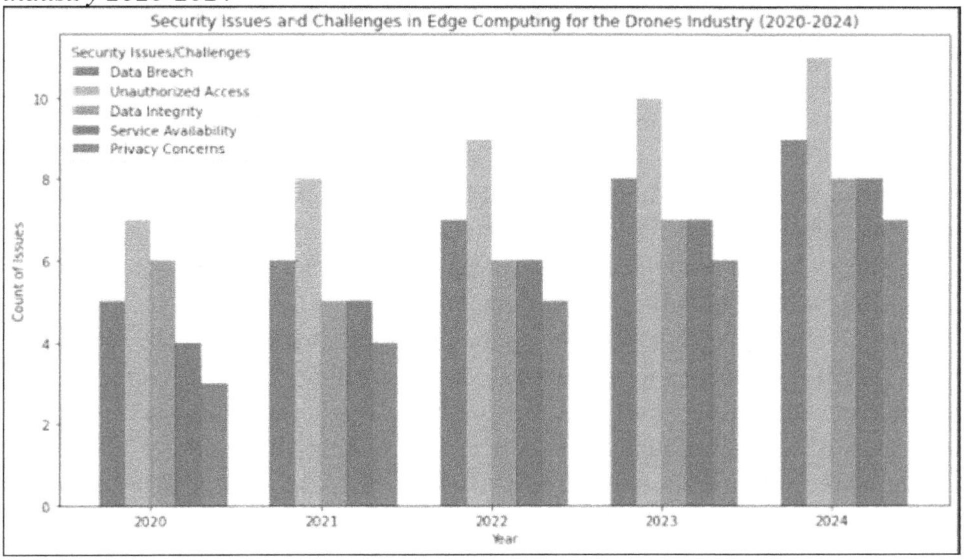

This script generates a grouped bar chart that displays the effect numbers of vulnerabilities for the issues and challenges facing the software industry. We will examine these network and application tools in further detail in subsequent articles in this series. Several problems arise in any complex setting, and there are numerous approaches to solving them. To ascertain whether it makes sense to pursue edge computing, each use case or business challenge must be evaluated individually to see if it could or should be used to address it (Cazzato et al., 2020, Shah et al., 2024a, Shah et al., 2024b). Let's talk about some possible industrial use cases—both real-world solutions and examples—for reference. All these industries have one thing in common: the network that the CSP will offer. Applications must run on networked devices, and the introduction of 5G makes it much more attractive for these sectors to begin taking edge computing seriously. Appropriate for the local edge or the device edge, while other workloads may be dynamically transferred between nodes in specific situations, either manually or automatically. Recognizing the significance of discrete workload management is crucial because the less discrete the approach, the fewer options we have for deployment and management. Network function is a crucial set of capabilities that should be included in any edge strategy and, consequently, our edge design, even if this essay has focused on application and analytics workloads(Tahir et al., 2019, Shraim et al., 2018, Skorobogatov et al., 2020). The requirement to manage application and network demands simultaneously should also be considered when adopting new tools.

The device edge, which comprises edge devices; the local edge, which comprises the application and network layer; and the cloud edge are the three main layers of the edge that are identified by the edge computing architecture. However, we have only just begun. We will delve deeper into the various layers and technologies required for developers to construct an edge-computing architecture in our upcoming post in this series(Yang et al., 2019, Simsek et al., 2016, Choy et al., 2012). Unmanned aerial vehicles (UAVs) with compute placed on them are known as edge-enabled drones. A variety of sensors, including cameras, lidar, and infrared sensors, can be added to edge-enabled drones to collect data for navigation and other specific applications. For applications that need more processing capacity, drones can send data to nearby computing facilities in addition to the computing on the device. Analysis of this data in almost real-time is made possible by the integration with local edge computing racks, which is necessary for many use cases that call for quick instantiation and action. When they lose connectivity, drones with navigation computing will typically use this as a backup, enabling them to safely finish their flight without requiring a continuous connection. Our objective is to demonstrate the revolutionary potential of edge computing by conducting a comprehensive examination of edge-enabled drone applications and infrastructure(Zhao et al., 2012, Andreu-Perez et al., 2018, Wisskirchen et al., 2017). This integration highlights the strategic significance of edge computing in the continuous development of the drone industry by enhancing the operational capabilities of UAVs and opening the door for ground-breaking innovations in several industries.

LITERATURE REVIEW

The constant issue with the abundance of data available nowadays is that a large portion of it is superfluous. Think of a medical monitoring device: it's useless to preserve days' worth of routine patient data because only the problem data matters. Short-term data that isn't retained over time makes up most of the data used in real-time analytics. After studies are completed, a corporation must determine which data to retain and which to delete. Additionally, the data that is kept needs to be safeguarded in compliance with legal and corporate regulations.These compute services may be located at adjacent edge servers or onboard. Transferring massive volumes of data is more than just a technological issue. The movement of data across national and regional borders may present extra challenges for privacy, data security, and other legal concerns. Keeping data near its source and within the parameters of existing data sovereignty regulations. IoT-based technologies have rapidly expanded and been adopted in many aspects of our lives in recent years. IoT devices are frequently observed enhancing healthcare systems. Increasing environmental monitoring, allowing smart city solutions, powering smart wearable devices, or being utilized in driverless vehicles(Shi et al., 2020, Raj and Seamans, 2019, Faniadis and Amanatiadis, 2020). In all the situations, the Internet of Things devices oversee producing enormous amounts of data, either in the form of pictures and videos from devices with cameras or numerical data for different sensor readings. It is now very difficult to process such large volumes of data at data centres, particularly when the Internet connection is unstable. A new computer paradigm known as edge computing has begun to take shape in response to these problems. Bringing processing units closer to end users and end devices is the fundamental concept of edge computing. By doing this, the data streams can be processed without having to be sent to a centralized service(Ansari et al., 2020, Shah et al., 2024c). Presents benefits of edge-enabled drones are in Figureure 3.

Figure 3. Presents benefits of edge-enabled drones

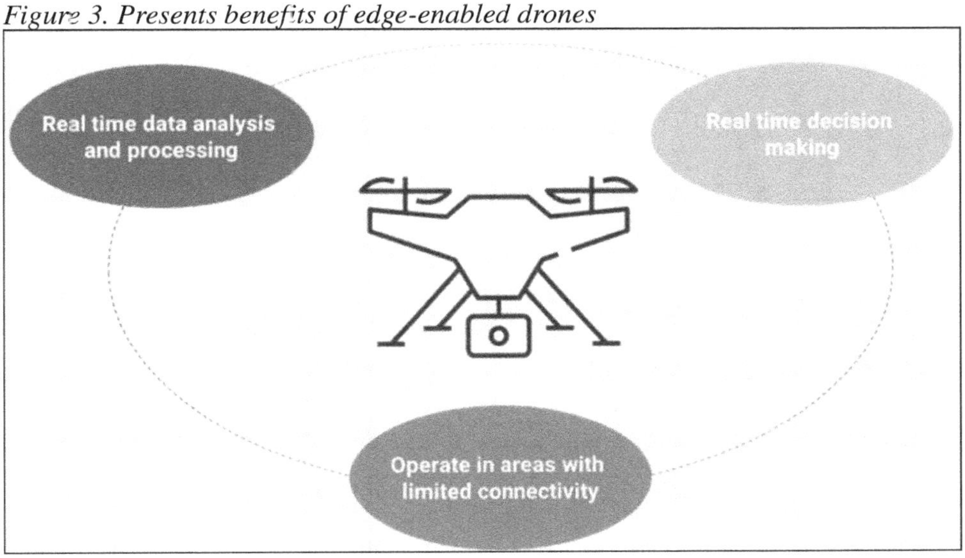

It is crucial to remember that the edge computing layer can consist of a wide range of devices, including satellites, UAVs, and dedicated edge servers, all of which carry out edge calculations for other network devices. Furthermore, in recent years, we have seen the application of machine learning (ML) and artificial intelligence (AI) techniques for an increasing number of use cases, frequently in the same context as IoT devices. A collection of technologies known as artificial intelligence (AI)

enables a computer system to think, learn, and carry out complex activities just like a human. ML, often referred to as a sub-technique of AI, builds reasoning models based on available data using a variety of self-learning algorithms. Adoption of machine learning at the periphery is more difficult, though, because many machine learning operations are still computationally costly and run on centralized servers. The same is true for deep learning methods that use multi-layered artificial neural networks (AANs), which automate feature extraction, learn from bigger data sets, and require less human involvement than standard machine learning models (Wu et al., 2021, Wu et al., 2020). The issues & challenges for the ECDE are in the Figureure 4.

Figure 4. Presents the challenges & issues for the ECDI

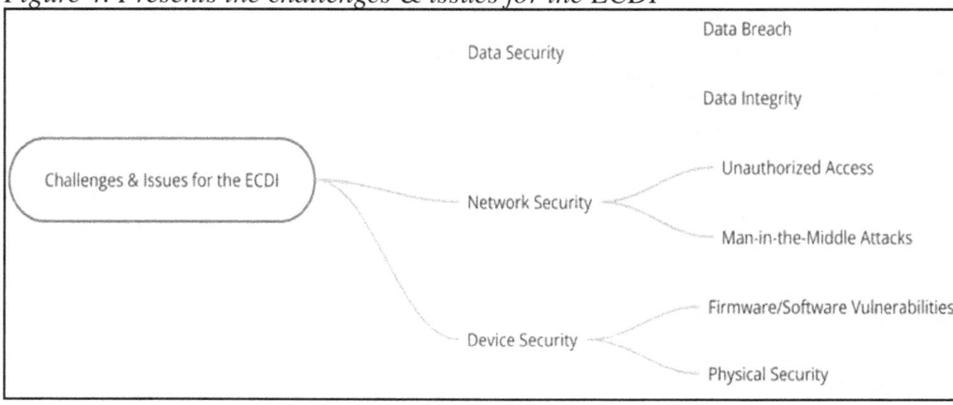

Through the employment of multipurpose tools like drones and ground robots, we firmly believe that IoT is a crucial technology in smart farming. Due to its great degree of adaptability in a variety of difficult circumstances, the Internet of Drones (IoD) has recently acquired pace. UAVs are effectively employed in a wide range of application fields, including agricultural, search and rescue operations, surveillance systems, and numerous more civic utilizations, owing to technological and practical advantages including high manoeuvrability and wireless expansion capabilities(N-guyen et al., 2021, McEnroe et al., 2022). Due to their quick, affordable, and secure deployment characteristics, these flying robots are utilized as first responders in a variety of applications. Furthermore, UAVs are becoming an essential component of airborne base stations in practically all new networks thanks to advancements in technology and installations like three-dimensional projects. Drone utilization is therefore anticipated to enhance the performance metrics of different network designs, including latency, throughput, connection, and dependability. UAVs have emerged as the most promising options for achieving significant outcomes in precision farming due to their engagement in a variety of everyday tasks. These characteristics enable UAVs to do similar tasks throughout the crop life cycle, including soil health assessments, colony, factory counting, crop health monitoring, irrigation, fertilizer and pesticide spraying, and crop yield estimation. Drones can identify poor soil health, insufficient crop nutrients, and water stress thanks to their precise and accurate sensors(Ren et al., 2019, Chen et al., 2019, Kim et al., 2021). By focusing on the regions that require the greatest attention, a drone program can make farming more inexpensive and ecologically effective. All interested agricultural experts, including farmers, agronomists, crop insurance providers, and researchers, are now interested in this drone-based technology Additionally, EC facilitates decentralized data processing and storage, removing network disruptions and bandwidth restrictions that might negatively impact crucial business choices. Edge computing doesn't need a lot of bandwidth, just like fog computing. Only the analytic findings are transmitted to an Internet-hosted cloud. Because it tackles issues with Internet connections in certain rural areas, this strategy is intriguing. Additionally, edge computing devices can offer near real-time analytics to improve uptime and optimize performance(Roghair et al., 2022). These technologies are growing rapidly, especially for facial recognition applications related to temperature measurement or mask-wear detection. However, because EC's edge nodes are already computer-equipped and employ more sophisticated technology, their deployment costs are higher than those of traditional sensors and cameras. Furthermore, because of their restricted capabilities, edge nodes with edge computing technologies are unable to process the same data stream as a computer (fog) or supercomputer (cloud). In addition to being less scalable than fog computing, EC offers limited interoperability. IoT devices may become incompatible with specific cloud services and operating systems as a result. Along with these drawbacks, edge computing does not facilitate resource pooling, and maintenance is more difficult than traditional computing because some issues can necessitate on-site physical maintenance(Huo et al., 2021, Huang and Meng, 2021).

OVERVIEW OF THE ARCHITECTURE OF EDGE COMPUTING

Edge computing is a cutting-edge paradigm that greatly improves processing speed and reaction times by moving computation and data storage closer to the data source. Examining the essential elements of edge computing architecture(Donmez et al., 2021). In order to enable applications that demand low latency, high bandwidth, and real-time decision-making, we first define edge computing and compare it to standard cloud computing, emphasizing the necessity of localized processing.

Overview Edge Servier for the Drone Industry

Strong computers known as "edge servers" are positioned at the edge of a network where data processing must take place. An edge server's physical location is frequently near the systems or applications that generate the data that the edge servers store or use. This contrasts with a more conventional server design, which often keeps data in the cloud or a data centre. A vital component of edge computing, a computing paradigm that improves response times and conserves bandwidth by bringing computation and data storage closer to the point of demand, is edge servers(Wang et al., 2021, Gopi et al., 2021). Instead of depending entirely on centralized cloud data centres, edge servers are essential for processing data at the network's edge. Edge servers are pivotal in the drone industry, offering localized processing power that enhances the capabilities and efficiency of drone operations. An overview edge servers & challenges are in Figureure 5.

Figure 5. An overview edge server & challenge

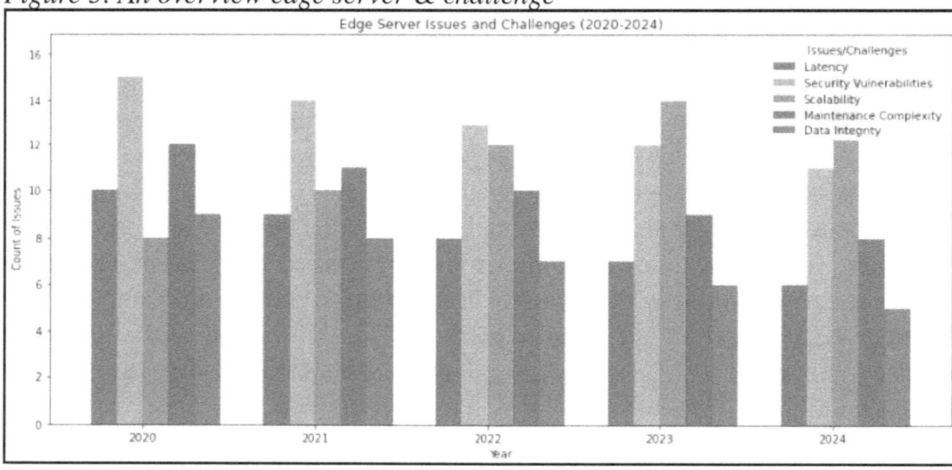

Edge computing is a technology that eliminates the need for distant servers by bringing data processing and analysis closer to the data source. The methods could significantly speed and efficiency. Drones can swiftly assess and act on information in real-time by processing data at the edge, eliminating the requirement for a continuous connection to a distant server, the application where speed is of the essence, like search and rescue, benefit greatly from this(Alsamhi et al., 2021). Through lowering the quantity of data that must be bandwidth and lower latency. Overall, edge computing has a big influence on how drone data is processed, and analysis and its significance is only going to increase as drones are utilized more frequently for a variety of purposes.

Overview Cloud Server for the Drone Industry

Cloud servers play a significant role in the drone industry by providing scalable computing resources, extension storage capabilities, and robust data management solutions. The integration of cloud computing with drone operations offers numerous benefits, ranging from enhanced data processing to improved operational efficiency. An overview of cloud servers in drones industry is in Figureure 6.

Figure 6. An overview of cloud servers in the drone's industry

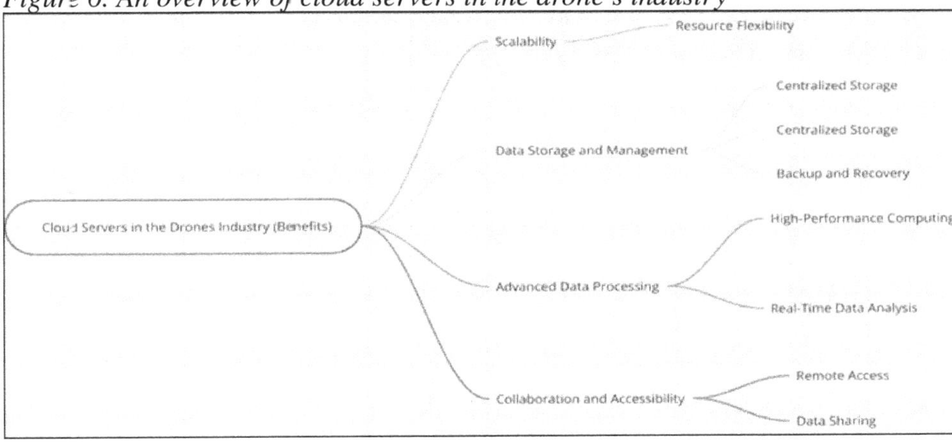

A new age of technological innovation and operational efficiency has been brought about by the drone industry's connection with cloud servers. For managing the enormous volumes of data produced by drones, cloud computing provides storage capacities and reliable data management systems, the use of drones in a variety of industries, including agriculture, surveillance, environment monitoring, infrastructure inspection, and logistics, is being revolutionized by this synergy, which makes it possible for improved real-time processing sophisticated analysis, and smooth communication(Brik et al., 2020, Zaman et al., 2023, Jabeen et al., 2023). The drone industry may increase flexibility scalability, and cost-effectiveness by utilizing the capabilities of the cloud as the door for creating applications and better operational results.

CONCLUSION

In conclusion, there are many advantages to using edge computing, additionally, it, makes real-time decision-making possible, which is essential for applications like precision agriculture, disaster response, and rescue. Even though using edge computing in drones presents certain difficulties, it greatly increases the security of the data we keep. All things considered, edge computing is a promising technology that has the potential to significantly expand drone capabilities and open new avenues for a variety of sectors. By facilitating real-time data processing boosting security and increasing operational efficiency edge servers are revolutionizing the drone sector. The incorporation of AI, 5G, and hybrid cloud architectures will significantly expand the capabilities and uses of edge computing in the drone sector as technology advances. This will result in drone operations that are more intelligent, dependable, and efficient across a range of industries.

REFERENCES

Al-Turjman, F. & Zahmatkesh, H. (2020). A comprehensive review on the use of AI in UAV communications: Enabling technologies, applications, and challenges. *Unmanned Aerial Vehicles in Smart Cities*, 1-26.

Alsamhi, S. H., Almalki, F. A., Ma, O., Ansari, M. S., & Lee, B. (2021). Predictive estimation of optimal signal strength from drones over IoT frameworks in smart cities. *IEEE Transactions on Mobile Computing*, *22*(1), 402–416. DOI: 10.1109/TMC.2021.3074442

Andreu-Perez, J., Deligianni, F., Ravi, D., & Yang, G.-Z. (2018). Artificial Intelligence and robotics. *arXiv preprint arXiv:1803.10813*.

Ansari, M. S., Alsamhi, S. H., Qiao, Y., Ye, Y. & Lee, B. (2020). Security of distributed intelligence in edge computing: Threats and countermeasures. *The Cloud-to-Thing Continuum: Opportunities and Challenges in Cloud, Fog and Edge Computing*, 95-122.

Brik, B., Ksentini, A., & Bouaziz, M. (2020). Federated learning for UAVs-enabled wireless networks: Use cases, challenges, and open problems. *IEEE Access : Practical Innovations, Open Solutions*, *8*, 53841–53849. DOI: 10.1109/ACCESS.2020.2981430

Cazzato, D., Cimarelli, C., Sanchez-Lopez, J. L., Voos, H., & Leo, M. (2020). A survey of computer vision methods for 2d object detection from unmanned aerial vehicles. *Journal of Imaging*, *6*(8), 78. DOI: 10.3390/jimaging6080078 PMID: 34460693

Chen, W., Liu, B., Huang, H., Guo, S., & Zheng, Z. (2019). When UAV swarm meets edge-cloud computing: The QoS perspective. *IEEE Network*, *33*(2), 36–43. DOI: 10.1109/MNET.2019.1800222

Choy, S., Wong, B., Simon, G., & Rosenberg, C. (2012, November). The brewing storm in cloud gaming: A measurement study on cloud to end-user latency. In 2012 11th Annual Workshop on Network and Systems Support for Games (NetGames) (pp. 1-6). IEEE.

Dong, C., Shen, Y., & Qu, Y. (2020). A survey of UAV-based edge intelligent computing. *Chinese Journal of Intelligent Science and Technology*, *2*, 227–239.

Donmez, C., Villi, O., Berberoglu, S., & Cilek, A. (2021). Computer vision-based citrus tree detection in a cultivated environment using UAV imagery. *Computers and Electronics in Agriculture*, *187*, 106273. DOI: 10.1016/j.compag.2021.106273

Faniadis, E., & Amanatiadis, A. (2020, November). Deep learning inference at the edge for mobile and aerial robotics. In *2020 IEEE International Symposium on Safety, Security, and Rescue Robotics (SSRR)* (pp. 334-340). IEEE.

Gopi, S. P., Magarini, M., Alsamhi, S. H., & Shvetsov, A. V. (2021). Machine learning-assisted adaptive modulation for optimized drone-user communication in b5g. *Drones (Basel)*, *5*(4), 128. DOI: 10.3390/drones5040128

Huang, Y., & Meng, Z. (2021). Bearing-based distributed formation control of multiple vertical take-off and landing UAVs. *IEEE Transactions on Control of Network Systems*, *8*(3), 1281–1292. DOI: 10.1109/TCNS.2021.3061941

Huo, M., Duan, H., & Fan, Y. (2021). Pigeon-inspired circular formation control for multi-UAV system with limited target information. *Guidance. Navigation and Control*, *1*(1), 2150004. DOI: 10.1142/S2737480721500047

Jabeen, T., Jabeen, I., Ashraf, H., Ullah, A., Jhanjhi, N., Ghoniem, R. M., & Ray, S. K. (2023). Smart Wireless Sensor Technology for Healthcare Monitoring System Using Cognitive Radio Networks. *Sensors (Basel)*, *23*(13), 6104. DOI: 10.3390/s23136104 PMID: 37447952

Khan, W. Z., Ahmed, E., Hakak, S., Yaqoob, I., & Ahmed, A. (2019). Edge computing: A survey. *Future Generation Computer Systems*, *97*, 219–235. DOI: 10.1016/j.future.2019.02.050

Khuwaja, A. A., Chen, Y., Zhao, N., Alouini, M.-S., & Dobbins, P. (2018). A survey of channel modeling for UAV communications. *IEEE Communications Surveys and Tutorials*, *20*(4), 2804–2821. DOI: 10.1109/COMST.2018.2856587

Kim, J., Guivant, J., Sollie, M. L., Bryne, T. H., & Johansen, T. A. (2021). Compressed pseudo-SLAM: Pseudorange-integrated compressed simultaneous localisation and mapping for unmanned aerial vehicle navigation. *Journal of Navigation*, *74*(5), 1091–1103. DOI: 10.1017/S037346332100031X

Mcenroe, P., Wang, S., & Liyanage, M. (2022). A survey on the convergence of edge computing and AI for UAVs: Opportunities and challenges. *IEEE Internet of Things Journal*, *9*(17), 15435–15459. DOI: 10.1109/JIOT.2022.3176400

Nguyen, D. C., Ding, M., Pham, Q.-V., Pathirana, P. N., Le, L. B., Seneviratne, A., Li, J., Niyato, D., & Poor, H. V. (2021). Federated learning meets blockchain in edge computing: Opportunities and challenges. *IEEE Internet of Things Journal*, *8*(16), 12806–12825. DOI: 10.1109/JIOT.2021.3072611

Raj, M., & Seamans, R. (2019). Primer on artificial intelligence and robotics. *Journal of Organization Design*, *8*(1), 11. DOI: 10.1186/s41469-019-0050-0

Ren, J., Yu, G., He, Y., & Li, G. Y. (2019). Collaborative cloud and edge computing for latency minimization. *IEEE Transactions on Vehicular Technology*, *68*(5), 5031–5044. DOI: 10.1109/TVT.2019.2904244

Roghair, J., Niaraki, A., Ko, K., & Jannesari, A. (2021, August). A vision based deep reinforcement learning algorithm for uav obstacle avoidance. In *Proceedings of SAI Intelligent Systems Conference* (pp. 115-128). Cham: Springer International Publishing.

Shah, I. A., Jhanjhi, N., & Ashraf, H. (2024a). *Logistics With the Internet of Things: Challenges, Perspectives, and Applications. Navigating Cyber Threats and Cybersecurity in the Logistics Industry*. IGI Global. DOI: 10.4018/979-8-3693-3816-2.ch006

Shah, I. A., Laraib, A., & Hussain, F. (2024b). *Logistics Industry in the Context of the Blockchain Technology. Navigating Cyber Threats and Cybersecurity in the Logistics Industry*. IGI Global.

Shah, I. A., Murugesan, R. K., & Ashraf, H. (2024c). *Advances in Health With the Help of Explainable AI. Advances in Computational Intelligence for the Healthcare Industry 4.0*. IGI Global.

Shakhatreh, H., Sawalmeh, A. H., Al-Fuqaha, A., Dou, Z., Almaita, E., Khalil, I., Othman, N. S., Khreishah, A., & Guizani, M. (2019). Unmanned aerial vehicles (UAVs): A survey on civil applications and key research challenges. *IEEE Access : Practical Innovations, Open Solutions*, *7*, 48572–48634. DOI: 10.1109/ACCESS.2019.2909530

Shi, Y., Yang, K., Jiang, T., Zhang, J., & Letaief, K. B. (2020). Communication-efficient edge AI: Algorithms and systems. *IEEE Communications Surveys and Tutorials*, *22*(4), 2167–2191. DOI: 10.1109/COMST.2020.3007787

Shraim, H., Awada, A., & Youness, R. (2018). A survey on quadrotors: ConFigureurations, modeling and identification, control, collision avoidance, fault diagnosis and tolerant control. *IEEE Aerospace and Electronic Systems Magazine*, *33*(7), 14–33. DOI: 10.1109/MAES.2018.160246

Simsek, M., Aijaz, A., Dohler, M., Sachs, J., & Fettweis, G. (2016). 5G-enabled tactile internet. *IEEE Journal on Selected Areas in Communications*, *34*(3), 460–473. DOI: 10.1109/JSAC.2016.2525398

Skorobogatov, G., Barrado, C., & Salamí, E. (2020). Multiple UAV systems: A survey. *Unmanned Systems*, *8*(2), 149–169. DOI: 10.1142/S2301385020500090

Tahir, A., Böling, J., Haghbayan, M.-H., Toivonen, H. T., & Plosila, J. (2019). Swarms of unmanned aerial vehicles—A survey. *Journal of Industrial Information Integration*, *16*, 100106. DOI: 10.1016/j.jii.2019.100106

Wang, X., Luo, B., & Zhang, Z. (2021, April). Application of UAV target tracking based on computer vision. []. IOP Publishing.]. *Journal of Physics: Conference Series*, *1881*(4), 042053.

Wisskirchen, G., Biacabe, B. T., Bormann, U., Muntz, A., Niehaus, G., Soler, G. J., & Von Brauchitsch, B. (2017). Artificial intelligence and robotics and their impact on the workplace. *IBA Global Employment Institute*, *11*, 49–67.

Wu, Q., Xu, J., Zeng, Y., Ng, D. W. K., Al-Dhahir, N., Schober, R., & Swindlehurst, A. L. (2021). A comprehensive overview on 5G-and-beyond networks with UAVs: From communications to sensing and intelligence. *IEEE Journal on Selected Areas in Communications*, *39*(10), 2912–2945. DOI: 10.1109/JSAC.2021.3088681

Wu, Y., Dai, H.-N., & Wang, H. (2020). Convergence of blockchain and edge computing for secure and scalable IIoT critical infrastructures in industry 4.0. *IEEE Internet of Things Journal*, *8*(4), 2300–2317. DOI: 10.1109/JIOT.2020.3025916

Yang, R., Yu, F. R., Si, P., Yang, Z., & Zhang, Y. (2019). Integrated blockchain and edge computing systems: A survey, some research issues and challenges. *IEEE Communications Surveys and Tutorials*, *21*(2), 1508–1532. DOI: 10.1109/COMST.2019.2894727

Zaman, N., Ghazanfar, M. A., Anwar, M., Lee, S. W., Qazi, N., Karimi, A., & Javed, A. (2023). Stock market prediction based on machine learning and social sentiment analysis. *Authorea Preprints*.

Zhang, X., Wang, Y., Lu, S., Liu, L., & Shi, W. (2019). OpenEI: An open framework for edge intelligence. 2019 IEEE 39th International Conference on Distributed Computing Systems (ICDCS),. IEEE, 1840-1851.

Zhao, Z., Hwang, K., & Villeta, J. (2012, June). Game cloud design with virtualized CPU/GPU servers and initial performance results. In *Proceedings of the 3rd workshop on Scientific Cloud Computing* (pp. 23-30).

Zhou, Z., Chen, X., Li, E., Zeng, L., Luo, K., & Zhang, J. (2019). Edge intelligence: Paving the last mile of artificial intelligence with edge computing. *Proceedings of the IEEE*, *107*(8), 1738–1762. DOI: 10.1109/JPROC.2019.2918951

Chapter 12
Swarm Intelligence and Multi-Drone Coordination With Edge AI

Siva Raja Sindiramutty
https://orcid.org/0009-0006-0310-8721
Taylor's University, Malaysia

ABSTRACT

Swarm intelligence is transforming drone tech to enable autonomous air systems to collaborate and adapt readily to real-world conditions. By flying in coordination, drones can perform sophisticated tasks that would be difficult or impossible for a single unit to accomplish. Whether for search and rescue or large-scale agricultural surveillance, coordinated systems improve speed, coverage, and decision-making. Edge AI is significant in that it allows drones to process information in real time, cutting down on reliance on remote cloud servers. This enables swarms to react quickly to changing environments, such as navigating through disaster scenes or tracking moving objects. Unlike using a central controller for all commands, drones communicate with each other and collectively decide, such as birds in a flock or ants in a colony. Swarm coordination is supplemented with advanced technologies like 5G connectivity and sensor fusion to enable the smooth sharing of data among drones.

DOI: 10.4018/979-8-3693-8497-8.ch012

1. INTRODUCTION

Definition of Swarm Intelligence in Drone Systems

Swarm intelligence in drone systems means that swarms of drones coordinate with local interactions and simple rules (Zhou et al., 2020; Alkinani et al., 2021) to effectively achieve tasks. This, following the example of ants, bees, and birds, enables drones to fly without a central controller and can therefore respond better to complex environments (Natarajan et al., 2024). In a search and rescue mission following an earthquake, instead of employing a single drone to sweep a vast region, a swarm of drones can fan out, communicate in real time, and cover a wider region in a shorter period. It has been proved in research that swarm-intelligent drones can navigate around obstacles and locate survivors in a shorter time compared to conventional means (Tan & Zheng, 2013; Hussain et al., 2024). This ability to organize in a decentralized manner makes them very applicable in agriculture, with drones flying in fields and dropping fertilizers directly to where they are needed (Lawrence et al., 2024; Jun et al., 2024).

A recent test experience shows how nature is being imitated. While birds in a flock adjust direction without bumping into one another, drones apply decentralized algorithms to avoid crashes and optimize their motion (Yedilkhan et al., 2024). The efficiency of this method is reshaping industries, from package delivery to environmental monitoring. Companies like Amazon and Zipline are testing swarm-based delivery systems to speed up operations and reduce costs (Singh et al., 2024). What stands out most is the balance between simplicity and complexity. Each drone follows a few straightforward rules, yet the group can solve problems in ways a single machine cannot. As this technology improves, swarms of drones could soon assist in disaster relief, surveillance, and construction, making everyday tasks faster and safer. Figure 1 shows the key characteristics of drones.

Figure 1. Key characteristics of drone

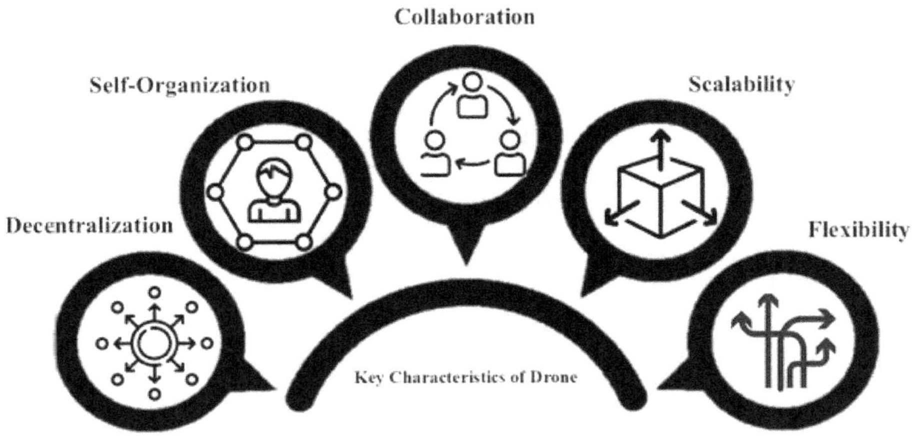

Importance of Multi-Drone Coordination in Real-World Applications

Coordinating multiple drones is increasingly important in numerous real-world applications, making operations faster, safer, and more effective. Search and rescue operations, for instance, are greatly enhanced with multiple drones working in concert. One drone can only cover a small area, but with drones travelling in coordination, more extensive areas can be swept in a shorter time, enhancing the likelihood of finding missing individuals (Singh et al., 2024; Babbar et al., 2021). Farmers use drones in agriculture for crop inspection, but one drone is limited (Aliloo et al., 2024).

With fleets flying in tandem, they can efficiently cover expansive fields, detecting problems before they develop into full-blown problems. Delivery firms are transforming with drone fleets delivering goods, saving traffic and delivery time. A drone has a weight capacity, but coordinated fleets can segment deliveries, making them more dependable (Bao et al., 2025; Kiyani et al., 2024). Disaster relief is another vital application. After earthquakes, drones can assess damage, deliver supplies, and spot survivors. A highly coordinated drone unit can cover more ground, reach priority points, and deliver information in a timely fashion (Chai et al., 2024; Brohi et al., 2020). Security and surveillance missions are being supported as well.

Instead of using a stationary camera or a single drone, synchronized groups of drones scan in real-time, reducing blind spots (Xie et al., 2024). These advancements indicate how coordination among multiple drones is transforming industries and a new method is being embraced. With technological advancements, decision-making and communication in real-time between drones will be increasingly more precise, and they will be able to perform more complex tasks.

Role of Edge AI in Enabling Real-Time Swarm Decision-Making

Drones operating in swarms need to make rapid decisions without relying on distant cloud servers. Edge AI helps by processing data on the device itself, cutting down delays and improving coordination (Sun et al., 2024). Imagine a group of drones sent into a disaster zone. Without quick responses, their effectiveness drops. By running AI models directly onboard, each drone processes environmental data and adjusts its movements accordingly (Caballero-Martin et al., 2024). One challenge is how drones share information. They must react to obstacles, weather changes, or shifting targets in real time. Traditional methods rely on centralized systems, which slow down response times. Edge AI allows drones to exchange critical data locally, improving collaboration (Gill et al., 2024; Linqiang et al., 2024). Researchers have tested this with autonomous search-and-rescue missions. A drone detects a survivor, instantly signalling others to form a coordinated search pattern (De Donato et al., 2024). This wouldn't be possible with delays from cloud-based processing. Energy efficiency also matters. Drones have limited battery life and constant communication with external networks drains power. Edge AI optimizes computing resources by processing only the most relevant data, reducing energy use (Alsadie, 2024; Manchuri et al., 2024). This means longer flight times and better overall performance. Looking ahead, edge AI could allow swarms to adapt during complex missions, such as military reconnaissance or precision agriculture. As hardware improves, real-time decision-making will become even faster, pushing drone technology into new areas (Chukwuma et al., 2024).

Chapter Contribution

This chapter is pertinent to the area of swarm intelligence and multi-drone coordination with Edge AI. First, its importance of multi-drone coordination in real-world applications and outlines an overview of swarm intelligence principles with a brief overview of key characteristics of centralized vs. Decentralized control architectures and communication models for drone swarms: peer-to-peer, leader-follower, and consensus-based approaches. The chapter then investigates the definition and benefits of edge AI in multi-drone systems and edge computing for decentralized processing and decision-making. Furthermore, the chapter presents the use of drones in search and rescue operations agriculture precision farming disaster response and environmental monitoring lastly, the section provides a glimpse of challenges and limitations in terms of energy efficiency and battery limitations, and security concerns: hacking, spoofing, and swarm takeover risks. The final content, discussed AI-driven adaptive swarm behaviour and blockchain for secure and decentralized swarm management.

Chapter Organization

In the first section, we will cover an insight into the introduction of drone navigation and its significance. In section 2, we go down to the fundamentals of swarm intelligence. In section 3, we discuss multi-drone coordination: techniques and strategies, followed by the role of edge AI in drone swarms in section 4. Section 5 delves into key technologies for swarm coordination. Section 6 will explain the applications of swarm intelligence and edge AI in drones. The final section will discuss the challenges in multi-drone coordination with Edge AI and future trends and innovations.

2. FUNDAMENTALS OF SWARM INTELLIGENCE

Overview of Swarm Intelligence Principles

Swarm intelligence means problem-solving by looking at how groups of simple agents—ants, birds, or even robots—work as a collective (McMillen & Levin, 2024). Individual agents have simple rules to follow without a central controller, and they perform complex tasks. Take ants searching for food: when one finds a good path, it deposits a chemical trail for others to follow. Over time, stronger paths build as more agents reinforce them, and it becomes effective. Bees use a similar method when searching for new hive locations, using movements in a "waggle dance" to inform each other (Palmer et al., 2024; Ravichandran et al., 2024).

It can be applied to technology as well. Researchers have implemented swarm intelligence to traffic flow in cities, with cars behaving like independent agents responding to road conditions (Agrahari et al., 2024). In robotics, engineers design swarms of small robots that can build into different shapes or search disaster zones without direct human command (Fujimoto, 2010; Chesti et al., 2020). Financial markets also have such a pattern—individual investors make small decisions using what they know, and it results in collective trends that dictate stock prices (Krishnamurthy, 2024). What's interesting is that these systems use simple rules, and they produce sophisticated outputs.

This is why scientists and engineers keep turning to nature-inspired approaches. Whether managing drone fleets, organizing warehouse logistics, or even optimizing internet searches, swarm intelligence continues to shape real-world solutions in unexpected ways (Pratihar, 2010). Figure 2 shows the fundamentals of swarm intelligence.

Figure 2. Fundamentals of swarm intelligence

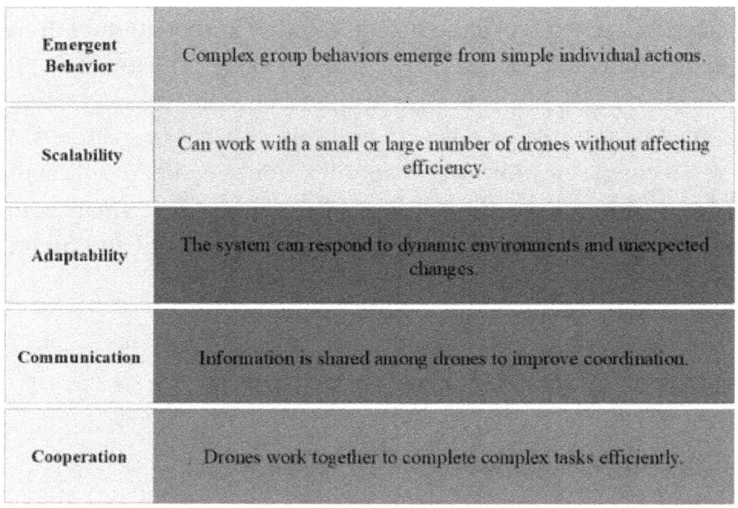

Biological Inspiration: Ant Colonies, Bird Flocking, and Fish Schooling

Ants marching along ideal paths, birds flying together in formation, and fish swimming together as one school—nature has devised brilliant mechanisms to solve complex problems without a leader. Swarm intelligence borrows from these natural phenomena to demonstrate that simple agents executing basic rules can produce complex group behaviour (Florida, 2025). Watching an ant colony, one will observe how individual ants lay down pheromone trails that guide other ants to food sources (Barrie et al., 2023; Dogra et al., 2021). With centralized control absent, the colony successfully finds food and transports it back home, akin to how modern algorithms find the best delivery routes. Bird flocking provides another beautiful illustration. Each bird has basic rules to execute: maintain spacing, orient to neighbours, and move toward the centre of the group (Barrie et al., 2023). Decentralized control causes the flocks to evade predators and cover large distances. The same concepts are employed in robotic swarms where the drones arrange themselves without orders, useful during search-and-rescue operations (Alqudsi & Makaraci, 2025)

Fish schooling behaves the same way, responding to the surroundings within near-instant synchronization. The quick response comes from the local interactions rather than one individual making the group's decision. Engineers borrow these concepts to devise self-organizing networks, from traffic control to finance modelling. Observing these natural phenomena gives birth to concepts to implement in technology, logistics, and artificial intelligence. The self-organization behaviour of ants, birds, and fish isn't simply amazing—it's a recipe to tackle modern challenges. The means nature uses to ensure efficiency through basic rules continue to inspire researchers, making swarm intelligence a valuable research subject. With each innovation, the uses expand, illustrating that sometimes the smartest solutions are the ones that emanate from the smallest.

Key Characteristics of Swarm Intelligence: Decentralization, Scalability, Self-Organization

Swarm intelligence refers to the way simple agents interact to create complex behaviours without centralized control. One of the defining characteristics is decentralization. In a swarm, individual agents follow local rules and react to their surroundings. Ant colonies are a well-known example—each ant follows pheromone trails, strengthening paths that lead to food (Dorigo & Stützle, 2019). Similarly, bird flocks and fish schools adjust their movement based on neighbours, creating synchronized patterns (Blum, 2024). These do not have a leader and therefore do not suffer from disruptions. Another feature worth noting is scalability. A swarm can expand in numbers without a decrease in efficiency, and it's because of this principle that it has applications in robotics and artificial intelligence. Swarm-inspired robots can be deployed to search for survivors in disaster zones even in uncertain environments (Hafid et al., 2024). The system can be stable whether a few or thousands of agents participate, making it flexible in different circumstances.

Self-organization enables swarms to respond to new challenges. Instead of following strict orders, agents respond to feedback from one another. Fireflies flashing in unison show this principle in action—each firefly adapts timing according to flashes around it, creating a coordinated spectacle (Buck & Buck, 1966; Gopi et al., 2021). In artificial intelligence, self-organizing algorithms improve traffic flow by dynamically adjusting signals to reduce congestion (Liu et al., 2023). This principle can be applied in man-made systems, too, such as in Wikipedia, where contributors automatically update information without a central editor.

Swarm intelligence has many applications, from logistics to drone management. Learning from these natural behaviors enables scientists to create effective, robust, and flexible systems in uncertain environments (Lenssen et al., 2025). From observations in nature, technology continues to advance in new and unexpected ways.

3. MULTI-DRONE COORDINATION: TECHNIQUES AND STRATEGIES

Centralized vs. Decentralized Control Architectures

Coordination of the various drones presents an issue that requires efficient control mechanisms. There are two basic strategies: centralized and decentralized architecture. Each has strengths and weaknesses depending on the task, the environment, and the reliability of communications.

With centralized control, one makes the entire fleet's decisions. This structuring offers precise coordination and conflict avoidance (Javed et al., 2024). It functions best under structured environments like warehouse automation or search-and-rescue operations where a command centre prescribes specific functions (Munasinghe et al., 2024; Sindiramutty, 2024). If the communications are compromised, however, the drones cannot execute autonomously, making the system vulnerable (Ficco et al., 2024). Decentralized control distributes the decision to every drone, giving them autonomy. Principles of swarm intelligence commonly control this method where the drones react to the neighbourhood and the situation locally (Ali et al., 2024; Sheng et al., 2024; Gouda et al., 2022). This technique enhances adaptability, making it best for operations like monitoring the environment and military reconnaissance where drones are to adapt to uncertain situations (Bu et al., 2024). The issue lies in the lack of central control to ensure coordination which causes inefficiencies or accidents if the drones misinterpret information (J. Sun & Hubbard, 2025). Real-world examples emphasize the trade-offs. The drone delivery system used by Amazon favours centralized control to guarantee package precision, whereas the agricultural drones that track crops use decentralized control to allow flexibility. The future will be hybrid models that integrate the best aspects of both to balance precision and autonomy. The balance has yet to be achieved but is an ongoing research topic fueled by the growth of AI and real-time communications. Figure 3 shows multi-drone coordination: techniques and strategies.

Figure 3. Multi-drone coordination: techniques and strategies

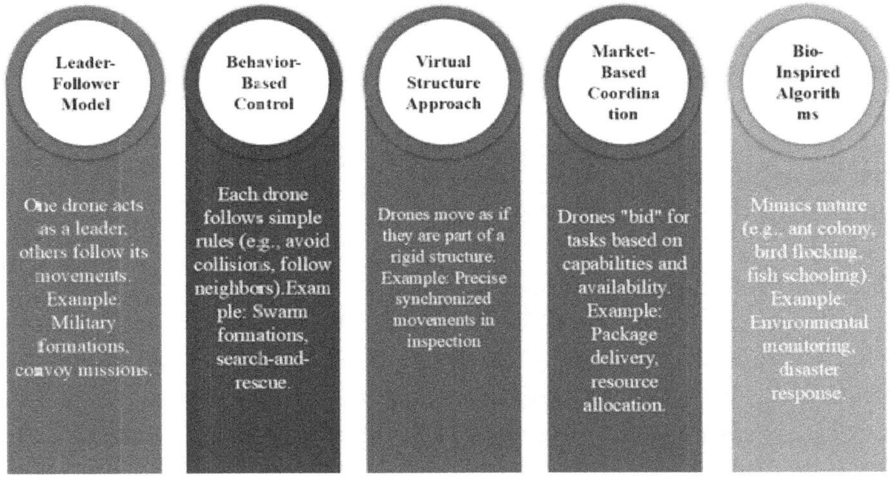

4. ROLE OF EDGE AI IN DRONE SWARMS

Definition and Benefits of Edge AI in Multi-Drone Systems

Edge AI inserts the capability to process within multi-drone frameworks, making the drones process information in real-time rather than waiting to transmit it to remote cloud servers (Banafaa et al., 2024). Real-time decision-making is essential where delays are disastrous—search-and-rescue operations, disaster relief, or warfare (Abbas et al., 2024). Take the case of a swarm of drones navigating an imploding building. If each drone waits to transmit information to a remote server to decide, seconds are lost. Edge AI ensures that these drones react in real-time, changing courses and mapping risks instantly (Tlili et al., 2024; Sindiramutty, Prabagaran, Akbar, Hussain, & Malik, 2024). Beyond speed, edge AI enhances drone synergy. If the drones are talking to each other instead of waiting for a command centre to direct them, then they are more like an orchestra than autonomous devices (Kestner, 2024). This comes especially useful in agriculture where swarms of drones monitor crops, scan for infestations, and realign irrigation cycles in real-time. A farmer spoke about how AI-powered drones enabled him to nip an infestation in the bud before it was too late to manage it—impossible using lagged analysis (Gadekallu et al., 2024; Humayun et al., 2022). Another benefit is security. Sending less information means less vulnerability to cyberattacks, making operations more secure. Military applications are especially strengthened because compromised drone communications are a serious risk. With edge AI, life-or-death decisions remain within the swarm, limiting exposure. Whether disaster relief, agriculture, or defence, edge AI-powered drones are not devices but autonomous problem-solvers learning to adapt to the situation in unimaginable ways a decade ago.

Edge Computing for Decentralized Processing and Decision-Making

Drone swarms are changing how tasks like search and rescue, agriculture, and military operations are handled. A major reason for their efficiency is Edge AI, which allows drones to process data locally instead of depending on a distant server (Maatouk et al., 2024). This reduces delays and makes quick decision-making possible, which is essential when dealing with fast-changing environments. If a drone swarm is tracking a wildfire, for instance, it cannot afford the time needed to send information to a remote data centre and wait for instructions. Instead, each drone processes the data itself and shares necessary details with others nearby (Min et al., 2025). Traditional cloud-based computing has limitations, especially in areas with weak network connectivity. By using Edge AI, drones can continue to function even without a stable internet connection (Prabhakar et al., 2025; Sindiramutty, Prabagaran, Jhanjhi, Ghazanfar, Malik, & Soomro, 2024). This makes them useful in remote areas where real-time intelligence is required, such as disaster zones or enemy territory. The ability to process sensor data instantly also helps in avoiding obstacles and ensuring coordinated movement among the swarm (Katna et al., 2024). If one drone detects an obstruction, it immediately alerts others so they can adjust their paths accordingly.

A key example is how drones equipped with Edge AI have been tested for crowd monitoring in large public events. Instead of relying on one central system, each drone evaluates patterns of movement and possible risks, helping security teams respond faster (Chamola et al., 2020). The decentralized nature of this technology reduces system failures and enhances reliability. As the technology advances, its application is expected to expand further, improving how autonomous systems handle complex operations.

6. APPLICATIONS OF SWARM INTELLIGENCE AND EDGE AI IN DRONES

Search and Rescue Operations (Distributed Target Identification)

Drones equipped with swarm intelligence and edge AI are changing how search and rescue teams locate missing persons. In large-scale disasters, centralized processing slows response times. By using edge AI, drones process data on-site without relying on remote servers (Cheng et al., 2023). This means they can quickly identify movement, heat signatures, or unusual patterns in real time. Swarm intelligence allows multiple drones to work together like a coordinated team. Instead of covering the same area, they divide tasks, reducing search time (Abdelkader et al., 2021). A drone detecting a possible target can signal others nearby to confirm or gather more details (J. Sun et al., 2016). This reduces false positives and improves accuracy. In a collapsed building, one drone might detect heat while another analyzes structural integrity, helping rescuers decide their approach (Lyu et al., 2023; Sindiramutty, Prabagaran, Jhanjhi, Murugesan, Brohi, & Masud, 2024). During a recent flood response, drones using edge AI quickly scanned vast areas and identified stranded individuals. Traditional methods, relying on satellite imagery or manual searches, took hours, while drones completed it in minutes (Witt et al., 2020). Firefighters in wildfire zones also use this technology to track missing persons and assess danger zones before sending crews. Figure 4 shows applications of swarm intelligence and edge AI in drones.

Figure 4. Applications of swarm intelligence and edge AI in drones

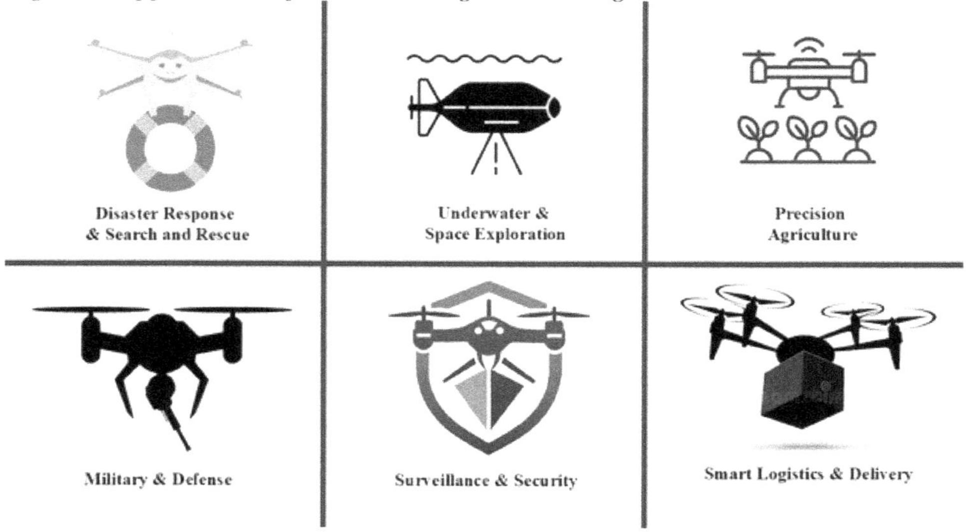

A key advantage is resilience. If one drone fails, others continue without disruption. Unlike single, high-power systems, distributed AI prevents complete failure (Alhamrouni et al., 2024). Additionally, real-time data processing reduces the risk of overloaded communication networks during emergencies (J. R. Blum et al., 2013; Jhanjhi et al., 2021). This technology is already proving its worth, and as AI models improve, future search and rescue missions will be even faster and more efficient. The combination of swarm intelligence and edge AI isn't just about making drones smarter—it's about saving lives when every second counts.

Agriculture and Precision Farming (Collaborative Crop Monitoring)

Drones with swarm intelligence and edge AI are reshaping crop monitoring (Qazi et al., 2022). Rather than a single drone flying across vast fields, a swarm of drones collaborates, processing information locally (Shakeri et al., 2019). This improves decision-making and reduces cloud-computing dependency (Gupta et al., 2023; Kumar et al., 2021). In large farms, early detection of diseases, pests, and water stress is essential. With swarm intelligence, drones report in real-time, adjusting course to scan high-priority areas (Saeed et al., 2022). Edge AI allows for onboard processing to let drones scan plant health without delay (Sathya et al., 2024).

Farmers have long struggled with actionable information in a timely fashion. Satellite imagery and visual inspection once took days or weeks. Now, drones equipped with multispectral cameras and AI software can spot early warning signals in minutes (Raniga et al., 2024). Take a vineyard with thousands of grapevines. Instead of waiting for visible symptoms to appear, drones detect subtle differences in leaf colour and temperature, alerting farmers before damage can be widespread (A. Abbas et al., 2023). In one instance, a wheat farm in India reduced pesticide use by 30% by spraying affected spots instead of fields in general (Yadav et al., 2014).

Drones can be used in collaboration to enhance efficiency. A drone can scan a section and another can use fertilizers or pesticides in targeted areas. This targeted application enhances production while minimizing costs and environmental damage.

With climate change affecting growing conditions, real-time adaptability becomes crucial. Farmers are no longer reacting to problems—they are preventing them before they escalate. The impact of these technologies extends beyond efficiency. It shapes a future where food production is more resilient, sustainable, and responsive to changing conditions.

Disaster Response and Environmental Monitoring

Swarm intelligence-powered drones and edge AI are transforming disaster relief and environmental monitoring. During wildfires, the drones cover large swaths quickly, providing real-time information to rescue teams. AI-powered drones monitored the 2019 Australian bushfires' progression, aiding fast evacuations and resource allocation (Advanced Navigation, 2025). Compared to using individual drones or waiting for delayed satellite imagery, swarms collaborate to cover more areas effectively (Cao et al., 2024; Lim et al., 2019). During flood zones, edge AI-powered drones scan water heights and predict breaches that are likely to happen. After the 2017 Hurricane Harvey, autonomous drones surveyed flood damage and located stranded survivors earlier than teams made up of humans (McNabb, 2024). The technology reduces the need to rely on the central server, making it more efficient where there are weak networks.

Drone swarms track deforestation and poaching in real-time during environmental monitoring. Conservationists use AI-powered drones to track vulnerable species and detect illicit forestry, prompting timely interventions (Ullah et al., 2024; Sindiramutty, Prabagaran, Jhanjhi, Murugesan, Brohi, & Wei, 2024). Marine research uses drones to explore the health of the coral reef without disturbing the ecosystem, an added advantage over divers or larger monitoring devices (Piñeros et al., 2024). The precision and speed of the drones speed emergency decision-making and ecological research. With lower human exposure to risks and higher efficiency, drones are an intelligent tool to manage crises and natural threats. Refinement of AI models and integration with existing emergency frameworks will further unlock its full potential (Bibri et al., 2024). With the advancement of technology, the uses will be more efficient in disaster control and natural habitat protection

Military and Defense Applications (Autonomous Surveillance and Reconnaissance)

Military drones are becoming smarter using the concepts of swarm intelligence and edge AI. The two technologies allow drones to collaborate autonomously, making them more efficient during surveillance and reconnaissance. Swarm intelligence enables multiple drones to communicate and coordinate in real-time, like birds or insects moving in groups (Wang et al., 2024). This improves coverage and reduces blind zones, ensuring operations are faster and more accurate. Edge AI enhances drone decision-making by processing data locally rather than relying on remote servers (Tlili et al., 2024b; Nayyar et al., 2021). This reduces delays, which is crucial when tracking fast-moving targets or detecting threats. During battlefield operations, edge AI-powered drones can quickly analyze images and detect unusual behaviour without needing a constant internet connection (Laghari et al., 2024; Sindiramutty, Prabagaran, Jhanjhi, Murugesan, Malik, & Hussain, 2024). This improves security because sensitive information does not have to be transmitted over networks, reducing the risk of interception.

An example is the U.S. military's tests using drone swarms for reconnaissance missions. These drones communicate and adapt their movements based on changing conditions (L. Sun et al., 2024). If one detects enemy movement, the others adjust to gain a clearer view or track targets more accurately. Such coordination improves reliability and reduces the need for human involvement. Another case involved AI-powered drones monitoring conflict zones. In one instance, drones identified enemy hideouts by analyzing movement patterns in real-time, leading to a successful operation without risking soldier lives. This illustrates how combining swarm intelligence and edge AI can revolutionize modern warfare by providing smarter, faster, and more adaptable reconnaissance tools. Figure 5 shows challenges in multi-drone coordination with edge AI.

Figure 5. Challenges in multi-drone coordination with edge AI

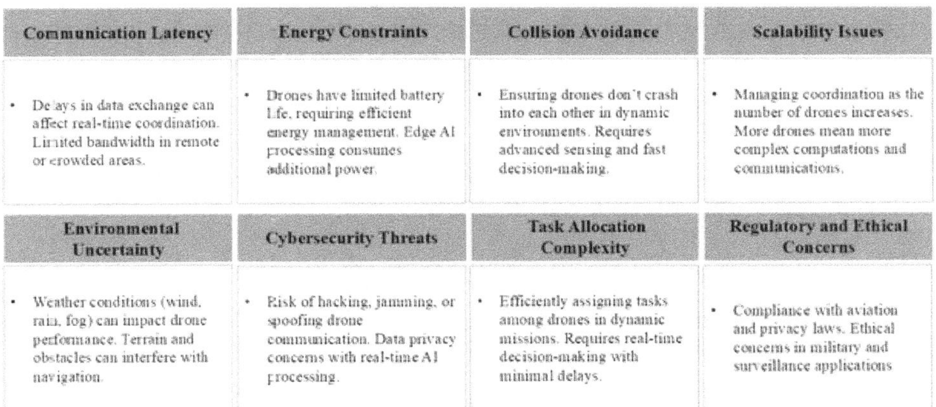

7. CHALLENGES IN MULTI-DRONE COORDINATION WITH EDGE AI

Energy Efficiency and Battery Limitations

Coordinating several drones using edge AI poses serious challenges, not least energy efficiency and battery life. Considering that each drone can only be power-limited, having an efficient hardware and software solution becomes a need as well. Since the past few years, a lot has been concentrated on edge-AI, which means that the drones are performing the data processing on a local level rather than sending data to cloud servers for data processing, but the local data processing increases strain on the batteries (Rocha et al., 2024; Weiqi et al., 2024). While efficient routing and load balancing strategies can reduce energy use real-world conditions, like unanticipated wind patterns or late task changes, often spoil the idyllic plans. Other research indicates that pruning lightweight AI models could reduce power drain, but finding the sweet spot for making intelligent algorithms without burning out the energy bank per se is challenging. I once saw a test flight where a flock of drones that had been programmed to survey an agricultural field had their batteries fail in a staggered fashion. Although the AI altered their paths in real-time, some drones were barely able to return before shutting down. This necessitates the development of novel power-aware algorithms capable of estimating depletion rates and adjusting plans accordingly. A proposed solution includes energy-sharing strategies, where a drone offloads energy wirelessly or recharges at a station mid-mission. Those solutions demand extra hardware, increasing weight and minimizing efficiency. A second possible solution is swarm intelligence models, inspired by principles of nature, enabling drones to coordinate and collaborate their movements in a manner to minimize net power loss (Cetinsaya et al., 2024). The trick is getting these systems responsive enough for real-time applications. If not done properly, AI-based coordination can burn through resources more quickly than anticipated, making multi-drone operations less realistically useful.

Security Concerns: Hacking, Spoofing, and Swarm Takeover Risks

Having multiple drones together to fly is exciting but raises security concerns. Consider an artificial intelligence group of drones making real-time decisions without the need to be connected to a centralized system. This makes the system efficient but leaves the system open to hackers. If a hacker gains access to one drone, the entire group can be taken over, causing widespread damage Shah et al., 2024; (Shah et al., 2022). Spoofing is another issue. Hackers can transmit spoof commands to trick the drone into thinking that it is receiving real commands. The drone can be tricked to land in the wrong place, go off track, or crash. Researchers demonstrated how an entire drone mission was diverted using a simple GPS spoofing attack (Joeaneke et al., 2024; Xun et al., 2025). If someone takes control of a swarm, it's not just a technical failure—it's a potential disaster. Drones also communicate wirelessly, making them vulnerable to data interception. Sensitive information, like video footage or target locations, could be stolen. Encrypting communication helps, but it's not foolproof. A study showed that even encrypted drone transmissions could be disrupted by interference attacks (Ahmad et al., 2024). Beyond technical risks, there's a real-world impact. In 2018, suspected drone hacks disrupted airport operations, stranding thousands of passengers (Topham et al., 2018). If criminals or hostile groups exploit these weaknesses, critical infrastructure could be at risk. Governments and companies are now racing to develop stronger defences, like AI-driven anomaly detection, but staying ahead of attackers is a constant challenge (Sartika, 2025).

8. FUTURE TRENDS AND INNOVATIONS

AI-Driven Adaptive Swarm Behaviour

Drones are getting smarter, working together like never before. AI-driven swarm behaviour is turning them into adaptable groups that adjust to changing conditions. This is no longer just about pre-programmed flight paths. Swarms can now react to real-time data, making them useful for search and rescue, disaster response, and surveillance (Papyan et al., 2024). Imagine a swarm of drones searching for survivors in an earthquake-hit area. Instead of waiting for commands, they spread out, share data instantly, and change their paths based on obstacles or new information.

This shift is powered by machine learning and decentralized decision-making. Each drone processes information independently while staying connected to the group. The result? Faster reactions, better coordination, and fewer failures due to lost communication with a central system. Military applications are also expanding. Drones can now identify threats, relay information, and execute tactical manoeuvres without human input (Ceviz et al., 2024). There's an urgency to refine this technology because it's not just militaries that are interested. Industries like agriculture use swarms to monitor crops and optimize pesticide use, reducing waste and increasing efficiency (Lochan et al., 2024; Ying et al., 2024). The challenge is making swarms work in unpredictable environments. Natural obstacles, signal interference, and security threats all pose risks. Researchers are exploring bio-inspired algorithms, learning from birds and fish to make drone movements more fluid and resilient (Budholiya et al., 2021). With AI continuing to improve, drone swarms will become even more independent, opening possibilities we are only beginning to understand (Andreoni et al., 2024). Figure 6 shows future trends and innovations.

Figure 6. Future trends and innovations

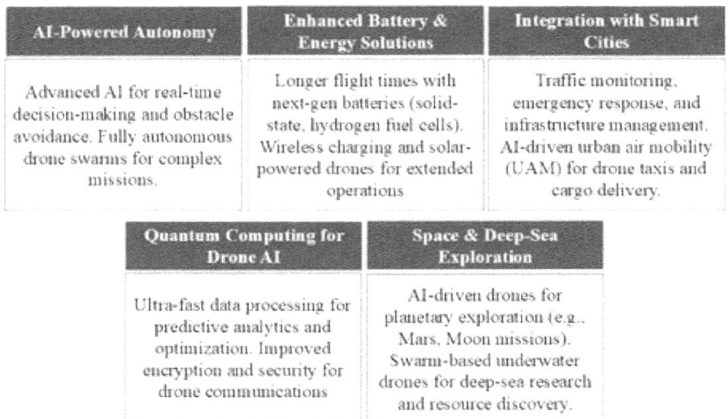

Blockchain for Secure and Decentralized Swarm Management

Drone technologies are transforming industries from delivery to agriculture. With more drones operating in swarms, security concerns arise. Blockchain is being regarded as a solution to manage swarms securely without a central authority. By having each drone document what it does on a shared ledger, blockchain ensures data integrity and device-to-device trust and prevents cyberattacks that can manipulate drone behaviour (Dildar et al., 2025). One such promising innovation is using smart contracts in drone operations. Self-enforcing contracts enable drones to adjust according to pre-set rules, for instance, to change routes if one unit is not working (Van Groesen & Pauwels, 2022; Alex et al., 2022). This reduces human intervention and optimizes efficiency in operations. Another innovation is integrating blockchain with artificial intelligence, which can enable drones to adapt to changing environments in real time. AI-powered drones can detect obstacles, adjust routes, and record decisions on a blockchain, which is a transparent and safe system (Singh, Kumar, et al., 2024; Alferidah & Jhanjhi, 2020). Energy use is still a problem. Lightweight blockchain designs are being designed to minimize battery use from high computational demands (M. K. Singh, Pippal, et al., 2024). Instead of all drones verifying all transactions, sharding is utilized where smaller groups verify different parts of the ledger to enable smooth operation without overwhelming the system. Blockchain is enhancing communication security in that drones can communicate with one another without being intercepted (Jagatheesaperumal et al., 2025). This is particularly crucial for military and disaster response missions, where coordination security is critical. With all this innovation, blockchain integration with drone swarms is paving the way for a future with greater security, efficiency, and autonomy.

9. CONCLUSION

Summary of Key Concepts

Use of swarm intelligence and edge AI to command multiple drones is transforming industries with autonomous systems being faster, more intelligent, and responsive. Unlike a central controller, drones in a swarm communicate with one another, exchanging real-time information to navigate complex environments. This has been useful in disaster relief, where decision-making can be a matter of life and death, and in agriculture, with targeted crop monitoring.

Edge AI has a central role in that it runs directly in drones, processing information in a localized fashion to avoid latency and offer real-time coordination. This allows drones to fly autonomously in low or zero-cloud environments. Perhaps one of the most surprising aspects of this technology is that it can organize itself. Like ants, birds, or a school of fish, drones can autonomously organize tasks without human direction. This is useful in defence, logistics, and environmental monitoring, where flexibility is required. But there are challenges. Short battery life limits mission duration, and security threats such as hacking and spoofing pose a challenge to operation. Solutions are in progress, however, with developers working on blockchain-based security and AI-based swarm behaviour optimisation for drones. With this technology set to continue to improve, multi-drone use in conjunction will continue to get more sophisticated and pervasive in industries. Whether in emergency response, increasing food production, or security in public spaces, these technologies will dictate automation's future.

Future Directions for Research and Development

The trend for future multi-drone coordination is towards more autonomy, efficiency, and security. Research directions include making energy efficiency a priority. Current limitations in batteries restrict flight time, and intelligent power management becomes a necessity. Researchers are developing energy-aware algorithms that can predict battery drain and dynamically allocate drones in real time. Wireless charging stations and in-flight energy transfer from drone to drone can further enhance mission time. Security comes into play. Increased autonomy means increased vulnerability to cyber threats such as hacking and spoofing. Future research will have to address swarm security enhancement using blockchain-based authentication and AI-based anomaly detection. Protection against interference and flying in hostile environments will be critical, particularly for defence and emergency services missions.

Swarm algorithms will further enhance coordination. While current models borrow from nature, real-world adaptability can still be increased. With machine learning, drones can learn to optimize their flying patterns from previous missions and, over time, enhance their efficiency. AI and predictive analytics integration can help drones predict environmental changes such as unforeseen weather patterns and modify their operations accordingly. One area with great promise is human-swarm interaction. Developing intuitive control systems to allow humans to control swarms with ease—by gesture, voice, or brain-machine interface—could bring the tech within reach. With drones being used more and more in disaster response, agriculture, and infrastructure inspection, enhancing control systems will reduce the automation-human decision-making gap. With such developments, multi-drone systems will continue to advance, opening new avenues in sectors. Future research will define how seamlessly they can be incorporated into regular operations, making them a basic tool for automation in the next generation.

REFERENCES

Abbas, A., Zhang, Z., Zheng, H., Alami, M. M., Alrefaei, A. F., Abbas, Q., Naqvi, S. H., Rao, M. J., Mosa, W. F. A., Abbas, Q., Hussain, A., Hassan, M. Z., & Zhou, L. (2023). Drones in plant Disease Assessment, Efficient Monitoring, and Detection: A way forward to smart agriculture. *Agronomy (Basel)*, *13*(6), 1524. DOI: 10.3390/agronomy13061524

Abbas, S., Talib, M. A., Ahmed, I., & Belal, O. (2024). Integration of UAVs and FANETs in Disaster Management: A review on applications, challenges and future directions. *Transactions on Emerging Telecommunications Technologies*, *35*(12), e70023. Advance online publication. DOI: 10.1002/ett.70023

Abdelkader, M., Güler, S., Jaleel, H., & Shamma, J. S. (2021). Aerial swarms: Recent applications and challenges. *Current Robotics Reports*, *2*(3), 309–320. DOI: 10.1007/s43154-021-00063-4 PMID: 34977595

Advanced Navigation. (2025, January 9). *Inertial navigation systems for sea, land & air*. https://www.advancednavigation.com/

Agrahari, A., Dhabu, M. M., Deshpande, P. S., Tiwari, A., Baig, M. A., & Sawarkar, A. D. (2024). Artificial Intelligence-Based Adaptive Traffic Signal Control System: A Comprehensive Review. *Electronics (Basel)*, *13*(19), 3875. DOI: 10.3390/electronics13193875

Ahmad, I., Rodriguez, F., Kumar, T., Suomalainen, J., Jagatheesaperumal, S. K., Walter, S., Asghar, M. Z., Li, G., Papakonstantinou, N., Ylianttila, M., Huusko, J., Sauter, T., & Harjula, E. (2024). Communications Security in Industry X: A survey. *IEEE Open Journal of the Communications Society*, *5*, 982–1025. DOI: 10.1109/OJCOMS.2024.3356076

Alhamrouni, I., Kahar, N. H. A., Salem, M., Swadi, M., Zahroui, Y., Kadhim, D. J., Mohamed, F. A., & Nazari, M. A. (2024). A Comprehensive review on the role of Artificial intelligence in power system stability, control, and Protection: Insights and Future Directions. *Applied Sciences (Basel, Switzerland)*, *14*(14), 6214. DOI: 10.3390/app14146214

Ali, H. M. S., Jalal, S. K., Saab, M. W., Sulaiman, S. K., Ghno, G. S. N., Mustafa, S. I., & Azimov, B. (2024). Deciphering the Implications of Swarm Intelligence Algorithms in Efficiently Managing Drone Swarms. *2024 35th Conference of Open Innovations Association (FRUCT)*, 112–123. DOI: 10.23919/FRUCT61870.2024.10516408

Aliloo, J., Abbasi, E., Karamidehkordi, E., Parmehr, E. G., & Canavari, M. (2024). Dos and Don'ts of using drone technology in the crop fields. *Technology in Society*, *76*, 102456. DOI: 10.1016/j.techsoc.2024.102456

Alkinani, M. H., Almazroi, A. A., Jhanjhi, N., & Khan, N. A. (2021). 5G and IoT Based Reporting and Accident Detection (RAD) System to Deliver First Aid Box Using Unmanned Aerial Vehicle. *Sensors (Basel)*, *21*(20), 6905. DOI: 10.3390/s21206905 PMID: 34696118

Alqudsi, Y., & Makaraci, M. (2025). UAV swarms: Research, challenges, and future directions. *Journal of Engineering and Applied Sciences (Asian Research Publishing Network)*, *72*(1), 12. Advance online publication. DOI: 10.1186/s44147-025-00582-3

Alsadie, D. (2024). A comprehensive review of AI techniques for resource management in Fog Computing: Trends, challenges, and future directions. *IEEE Access : Practical Innovations, Open Solutions*, *12*, 118007–118059. DOI: 10.1109/ACCESS.2024.3447097

Andreoni, M., Lunardi, W. T., Lawton, G., & Thakkar, S. (2024). Enhancing autonomous system security and resilience with Generative AI: A comprehensive survey. *IEEE Access : Practical Innovations, Open Solutions*, *12*, 109470–109493. DOI: 10.1109/ACCESS.2024.3439363

Babbar, H., Rani, S., Masud, M., Verma, S., Anand, D., & Jhanjhi, N. (2021). Load balancing algorithm for migrating switches in software-defined vehicular networks. *Computers, Materials & Continua/Computers. Computers, Materials & Continua*, *67*(1), 1301–1316. DOI: 10.32604/cmc.2021.014627

Banafaa, M. K., Pepeoğlu, Ö., Shayea, I., Alhammadi, A., Shamsan, Z. A., Razaz, M. A., Alsagabi, M., & Al-Sowayan, S. (2024). A comprehensive survey on 5G-and-Beyond networks with UAVs: Applications, emerging technologies, regulatory aspects, research trends and challenges. *IEEE Access : Practical Innovations, Open Solutions*, *12*, 7786–7826. DOI: 10.1109/ACCESS.2023.3349208

Bao, D., Yan, Y., Li, Y., & Chu, J. (2025). The Future of Last-Mile Delivery: Lifecycle environmental and economic impacts of Drone-Truck Parallel Systems. *Drones (Basel)*, *9*(1), 54. DOI: 10.3390/drones9010054

Barrie, R., Haalck, L., Risse, B., Nowotny, T., Graham, P., & Buehlmann, C. (2023). Trail using ants follow idiosyncratic routes in complex landscapes. *Learning & Behavior*, *52*(1), 105–113. DOI: 10.3758/s13420-023-00615-y PMID: 37993707

Bibri, S. E., Huang, J., Jagatheesaperumal, S. K., & Krogstie, J. (2024). The synergistic interplay of artificial intelligence and digital twin in environmentally planning sustainable smart cities: A comprehensive systematic review. *Environmental Science and Ecotechnology*, *20*, 100433. DOI: 10.1016/j.ese.2024.100433 PMID: 38831974

Blum, C. (2024). Ant Colony Optimization: A Bibliometric Review. *Physics of Life Reviews*, *51*, 87–95. DOI: 10.1016/j.plrev.2024.09.014 PMID: 39341089

Blum, J. R., Eichhorn, A., Smith, S., Sterle-Contala, M., & Cooperstock, J. R. (2013). Real-time emergency response: Improved management of real-time information during crisis situations. *Journal on Multimodal User Interfaces*, *8*(2), 161–173. DOI: 10.1007/s12193-013-0139-7

Brohi, S. N., Jhanjhi, N., Brohi, N. N., & Brohi, M. N. (2020). Key Applications of State-of-the-Art Technologies to Mitigate and Eliminate COVID-19.pdf. *TECHRxiv*. DOI: 10.36227/techrxiv.12115596.v1

Bu, Y., Yan, Y., & Yang, Y. (2024). Advancement Challenges in UAV Swarm Formation Control: A Comprehensive review. *Drones (Basel)*, *8*(7), 320. DOI: 10.3390/drones8070320

Buck, J., & Buck, E. (1966). Biology of synchronous flashing of fireflies. *Nature*, *211*(5049), 562–564. DOI: 10.1038/211562a0

Budholiya, S., Bhat, A., Raj, S. A., Sultan, M. T. H., Shah, A. U. M., & Basri, A. A. (2021). State of the Art Review about Bio-Inspired Design and Applications: An Aerospace Perspective. *Applied Sciences (Basel, Switzerland)*, *11*(11), 5054. DOI: 10.3390/app11115054

Caballero-Martin, D., Lopez-Guede, J. M., Estevez, J., & Graña, M. (2024). Artificial intelligence applied to drone control: A state of the art. *Drones (Basel)*, *8*(7), 296. DOI: 10.3390/drones8070296

Cao, P., Lei, L., Cai, S., Shen, G., Liu, X., Wang, X., Zhang, L., Zhou, L., & Guizani, M. (2024). Computational Intelligence Algorithms for UAV swarm Networking and Collaboration: A Comprehensive survey and Future Directions. *IEEE Communications Surveys and Tutorials*, *26*(4), 2684–2728. DOI: 10.1109/COMST.2024.3395358

Cetinsaya, B., Reiners, D., & Cruz-Neira, C. (2024). From PID to swarms: A decade of advancements in drone control and path planning - A systematic review (2013–2023). *Swarm and Evolutionary Computation*, *89*, 101626. DOI: 10.1016/j.swevo.2024.101626

Ceviz, O., Sen, S., & Sadioglu, P. (2024). A survey of security in UAVs and FANETs: Issues, threats, analysis of attacks, and solutions. *IEEE Communications Surveys and Tutorials*, *1*, 1. Advance online publication. DOI: 10.1109/COMST.2024.3515051

Chai, R., Guo, Y., Zuo, Z., Chen, K., Shin, H., & Tsourdos, A. (2024). Cooperative motion planning and control for aerial-ground autonomous systems: Methods and applications. *Progress in Aerospace Sciences*, *146*, 101005. DOI: 10.1016/j.paerosci.2024.101005

Chamola, V., Hassija, V., Gupta, V., & Guizani, M. (2020). A Comprehensive Review of the COVID-19 Pandemic and the Role of IoT, Drones, AI, Blockchain, and 5G in Managing its Impact. *IEEE Access : Practical Innovations, Open Solutions*, *8*, 90225–90265. DOI: 10.1109/ACCESS.2020.2992341

Cheng, N., Wu, S., Wang, X., Yin, Z., Li, C., Chen, W., & Chen, F. (2023). AI for UAV-Assisted IoT Applications: A Comprehensive Review. *IEEE Internet of Things Journal*, *10*(16), 14438–14461. DOI: 10.1109/JIOT.2023.3268316

Chesti, I. A., Humayun, M., Sama, N. U., & Jhanjhi, N. (2020). Evolution, Mitigation, and Prevention of Ransomware. *2020 2nd International Conference on Computer and Information Sciences (ICCIS)*. DOI: 10.1109/ICCIS49240.2020.9257708

Chukwuma, U., Gebremedhin, K. G., & Uyeh, D. D. (2024). Imagining AI-driven decision making for managing farming in developing and emerging economies. *Computers and Electronics in Agriculture*, *221*, 108946. DOI: 10.1016/j.compag.2024.108946

De Donato, M. C., Corradini, F., Fornari, F., & Re, B. (2024). SAFE: An ICT platform for supporting monitoring, localization and rescue operations in case of earthquake. *Internet of Things : Engineering Cyber Physical Human Systems*, *27*, 101273. DOI: 10.1016/j.iot.2024.101273

Dildar, M. S., Khan, A. S., Abbasi, I. A., Shaheen, R., Ruqaishi, K. A., & Ahmed, S. (2025). End-to-end security mechanism using blockchain for Industrial Internet of Things. *IEEE Access : Practical Innovations, Open Solutions*, *1*, 20584–20598. Advance online publication. DOI: 10.1109/ACCESS.2025.3535821

Dogra, V., Singh, A., Verma, S., Kavita, N., Jhanjhi, N. Z., & Talib, M. N. (2021). Analyzing DistilBERT for Sentiment Classification of Banking Financial News. In *Lecture notes in networks and systems* (pp. 501–510). DOI: 10.1007/978-981-16-3153-5_53

Ficco, M., Granata, D., Palmieri, F., & Rak, M. (2024). A systematic approach for threat and vulnerability analysis of unmanned aerial vehicles. *Internet of Things : Engineering Cyber Physical Human Systems*, *26*, 101180. DOI: 10.1016/j.iot.2024.101180

Floridi, L. (2025). AI as Agency without Intelligence: On Artificial Intelligence as a New Form of Artificial Agency and the Multiple Realisability of Agency Thesis. *Philosophy & Technology, 38*(1). DOI: 10.1007/s13347-025-00858-9

Fujimoto, H. (2010). *Emerging trends in mobile robotics:Proceedings of the 13th International Conference on Climbing and Walking Robots and the Support Technologies for Mobile Machines :31 August-3 September 2010,Nagoya Institute of Technology,Japan*. World Scientific.

Gadekallu, T. R., Maddikunta, P. K. R., Boopathy, P., Deepa, N., Chengoden, R., Victor, N., Wang, W., Wang, W., Zhu, Y., & Dev, K. (2024). XAI for Industry 5.0 -Concepts, Opportunities, Challenges and Future Directions. *IEEE Open Journal of the Communications Society*, *1*, 1. Advance online publication. DOI: 10.1109/OJCOMS.2024.3473891

Gill, S. S., Golec, M., Hu, J., Xu, M., Du, J., Wu, H., Walia, G. K., Murugesan, S. S., Ali, B., Kumar, M., Ye, K., Verma, P., Kumar, S., Cuadrado, F., & Uhlig, S. (2024). Edge AI: A Taxonomy, Systematic review and Future Directions. *Cluster Computing*, *28*(1), 18. Advance online publication. DOI: 10.1007/s10586-024-04686-y

Gopi, R., Sathiyamoorthi, V., Selvakumar, S., Manikandan, R., Chatterjee, P., Jhanjhi, N. Z., & Luhach, A. K. (2021). Enhanced method of ANN based model for detection of DDoS attacks on multimedia internet of things. *Multimedia Tools and Applications*, *81*(19), 26739–26757. DOI: 10.1007/s11042-021-10640-6

Gouda, W., Almurafeh, M., Humayun, M., & Jhanjhi, N. Z. (2022). Detection of COVID-19 based on chest x-rays using deep learning. *Health Care*, *10*(2), 343. DOI: 10.3390/healthcare10020343 PMID: 35206957

Hafid, A., Hocine, R., & Guezouli, L. (2024). Analyzing Swarm Robotics Approaches in Natural Disaster Scenarios: A Comparative Study. *2024 1st International Conference on Innovative and Intelligent Information Technologies (IC3IT)*, 1–6. DOI: 10.1109/IC3IT63743.2024.10869410

Humayun, M., Sujatha, R., Almuayqil, S. N., & Jhanjhi, N. Z. (2022). A Transfer Learning Approach with a Convolutional Neural Network for the Classification of Lung Carcinoma. *Health Care*, *10*(6), 1058. DOI: 10.3390/healthcare10061058 PMID: 35742109

Hussain, K., Rahmatyar, A. R., Riskhan, B., Sheikh, M. U., & Sindiramutty, S. R. (2024). Threats and Vulnerabilities of Wireless Networks in the Internet of Things (IoT). *IEEE 1st Karachi Section Humanitarian Technology Conference (KHI-HTC), 2*, 1–8. DOI: 10.1109/KHI-HTC60760.2024.10482197

Jagatheesaperumal, S. K., Rahouti, M., Chehri, A., Xiong, K., & Bieniek, J. (2025). Blockchain-Based Security Architecture for unmanned aerial vehicles in B5G/6G Services and Beyond: A Comprehensive approach. *IEEE Open Journal of the Communications Society, 1*, 1042–1069. Advance online publication. DOI: 10.1109/OJCOMS.2025.3528220

Javed, S., Hassan, A., Ahmad, R., Ahmed, W., Ahmed, R., Saadat, A., & Guizani, M. (2024). State-of-the-Art and future research challenges in UAV swarms. *IEEE Internet of Things Journal, 11*(11), 19023–19045. DOI: 10.1109/JIOT.2024.3364230

Jhanjhi, N., Humayun, M., & Almuayqil, S. N. (2021). Cyber security and privacy issues in industrial internet of things. *Computer Systems Science and Engineering, 37*(3), 361–380. DOI: 10.32604/csse.2021.015206

Joeaneke, P., Val, O. O., Olaniyi, O. O., Ogungbemi, O. S., Olisa, A. O., & Akinola, O. I. (2024). Protecting Autonomous UAVs from GPS Spoofing and Jamming: A Comparative Analysis of Detection and Mitigation Techniques. SSRN *Electronic Journal*. DOI: 10.2139/ssrn.4995938

Katna, S., Singh, S. K., Kumar, S., Manro, D., Chhabra, A., & Sharma, S. K. (2024). Communication systems for drone swarms and remote operations. In *Advances in computational intelligence and robotics book series* (pp. 115–136). DOI: 10.4018/979-8-3693-2707-4.ch006

Kestner, P. (2024). Techniques and tactics of cyber warfare. In *Springer eBooks* (pp. 215–242). DOI: 10.1007/978-3-658-43879-1_7

Kiyani, F. F., Hamid, B., Humayun, M., Sindiramutty, S. R. L., & Chowdhury, S. (2024). Discovery of Influential Publications Using Research Article's Usage Context. *2024 International Conference on Emerging Trends in Networks and Computer Communications (ETNCC)*, 1–7. DOI: 10.1109/ETNCC63262.2024.10767576

Krishnamurthy, A. (2024). Beyond Rationality. *Wiley*, 409–427. DOI: 10.1002/9781394214334.ch17

Kumar, M. S., Vimal, S., Jhanjhi, N., Dhanabalan, S. S., & Alhumyani, H. A. (2021). Blockchain based peer to peer communication in autonomous drone operation. *Energy Reports, 7*, 7925–7939. DOI: 10.1016/j.egyr.2021.08.073

Laghari, A. A., Jumani, A. K., Laghari, R. A., Li, H., Karim, S., & Khan, A. A. (2024). Unmanned Aerial Vehicles Advances in Object Detection and Communication Security review. *Cognitive Robotics*, *4*, 128–141. DOI: 10.1016/j.cogr.2024.07.002

Lawrence, I. D., Pavitra, A. R. R., Karu, R., & Saravanan, M. (2024). An Advanced Application of UAV – Drone Technologies in Precision Agriculture for Seed Dropping, Fertilizers and Pesticides Spraying and Field Monitoring. *Wiley*, 139–157. DOI: 10.1002/9781394186686.ch7

Lenssen, E., Nagtegaal, I., Van Oostveen, C., Sieben, A., Van Rijssen, L., & Weggelaar, A. M. (2025). Exploring nurses' leadership and resilience in a complex daily work environment: A qualitative study. *BMC Nursing*, *24*(1), 173. Advance online publication. DOI: 10.1186/s12912-025-02761-2 PMID: 39953459

Lim, M., Abdullah, A., Jhanjhi, N., Khan, M. K., & Supramaniam, M. (2019). Link Prediction in Time-Evolving Criminal Network with deep Reinforcement learning technique. *IEEE Access : Practical Innovations, Open Solutions*, *7*, 184797–184807. DOI: 10.1109/ACCESS.2019.2958873

Linqiang, Y., Sindiramutty, S. R. L., Ashraf, H., Muzammal, S. M., Balakrishnan, S. P., Gupta, S., & Kavita, N. (2024). Intelligent Household Waste Classification System Based on Machine Learning. *2024 International Conference on Emerging Trends in Networks and Computer Communications (ETNCC)*, 760–768. DOI: 10.1109/ETNCC63262.2024.10767563

Liu, Y., Huo, L., Wu, J., & Bashir, A. K. (2023). SWarm Learning-Based Dynamic optimal management for traffic congestion in 6G-Driven intelligent Transportation System. *IEEE Transactions on Intelligent Transportation Systems*, *24*(7), 7831–7846. DOI: 10.1109/TITS.2023.3234444

Lochan, K., Khan, A., Elsayed, I., Suthar, B., Seneviratne, L., & Hussain, I. (2024). Advancements in precision Spraying of Agricultural Robots: A Comprehensive review. *IEEE Access : Practical Innovations, Open Solutions*, *12*, 129447–129483. DOI: 10.1109/ACCESS.2024.3450904

Lyu, M., Zhao, Y., Huang, C., & Huang, H. (2023). Unmanned aerial vehicles for search and rescue: A survey. *Remote Sensing (Basel)*, *15*(13), 3266. DOI: 10.3390/rs15133266

Maatouk, N., Letaifa, A. B., & Rachedi, A. (2024). AI empowered data offloading in MEC enabled UAV networks. *Multimedia Tools and Applications*. Advance online publication. DOI: 10.1007/s11042-024-19859-5

Manchuri, A., Kakera, A., Saleh, A., & Raja, S. (2024). pplication of Supervised Machine Learning Models in Biodiesel Production Research - A Short Review. *Borneo Journal of Sciences and Technology*. DOI: 10.35370/bjost.2024.6.1-10

McMillen, P., & Levin, M. (2024). Collective intelligence: A unifying concept for integrating biology across scales and substrates. *Communications Biology*, *7*(1), 378. Advance online publication. DOI: 10.1038/s42003-024-06037-4 PMID: 38548821

McNabb, M. (2024, January 16). Drones in Disaster Response: Harnessing the potential of unmanned aerial systems. *DRONELIFE*. https://dronelife.com/2024/01/16/drones-in-disaster-response-harnessing-the-potential-of-unmanned-aerial-systems/

Min, W., Muthanna, M. S. A., Ibrahim, M., Alkanhel, R., Muthanna, A., & Laouid, A. (2025). Privacy-preserving federated UAV data collection framework for autonomous path optimization in maritime operations. *Applied Soft Computing*, *112906*, 112906. Advance online publication. DOI: 10.1016/j.asoc.2025.112906

Munasinghe, I., Perera, A., & Deo, R. C. (2024). A Comprehensive review of UAV-UGV Collaboration: Advancements and challenges. *Journal of Sensor and Actuator Networks*, *13*(6), 81. DOI: 10.3390/jsan13060081

Natarajan, G., Elango, E., Sundaravadivazhagan, B., & Rethinam, S. (2024). Artificial Intelligence Algorithms and Models for Embodied Agents: Enhancing Autonomy in Drones and Robots. In *Building Embodied AI Systems: The Agents, the Architecture Principles, Challenges, and Application Domains* (pp. 103–132). DOI: 10.1007/978-3-031-68256-8_5

Nayyar, A., Gadhavi, L., & Zaman, N. (2021). Machine learning in healthcare: review, opportunities and challenges. In *Elsevier eBooks* (pp. 23–45). DOI: 10.1016/B978-0-12-821229-5.00011-2

Palmer, J., Samuelson, A. E., Gill, R. J., Leadbeater, E., & Jansen, V. A. (2024). Foraging distance distributions reveal how honeybee waggle dance recruitment varies with landscape. *Communications Biology*, *7*(1), 1306. Advance online publication. DOI: 10.1038/s42003-024-06987-9 PMID: 39394388

Papyan, N., Kulhandjian, M., Kulhandjian, H., & Aslanyan, L. (2024). AI-Based drone Assisted Human rescue in disaster Environments: Challenges and opportunities. *Pattern Recognition and Image Analysis*, *34*(1), 169–186. DOI: 10.1134/S1054661824010152

Piñeros, V. J., Reveles-Espinoza, A. M., & Monroy, J. A. (2024). From remote sensing to artificial intelligence in coral reef monitoring. *Machines (Basel)*, *12*(10), 693. DOI: 10.3390/machines12100693

Prabhakar, P., Yokesh, V., Aruchamy, P., & Nanthakumar, S. (2025). Artificial Intelligence-Enabled fully echoed Q-Routing and adaptive directional medium access control protocol for flying Ad-Hoc networks. *International Journal of Communication Systems*, *38*(4), e6138. Advance online publication. DOI: 10.1002/dac.6138

Pratihar, D. K. (2010). *Intelligent Autonomous Systems: Foundations and Applications*. Springer Science & Business Media. DOI: 10.1007/978-3-642-11676-6

Qazi, S., Khawaja, B. A., & Farooq, Q. U. (2022). IoT-Equipped and AI-Enabled Next Generation Smart Agriculture: A Critical review, current challenges and future trends. *IEEE Access : Practical Innovations, Open Solutions*, *10*, 21219–21235. DOI: 10.1109/ACCESS.2022.3152544

Raniga, D., Amarasingam, N., Sandino, J., Doshi, A., Barthelemy, J., Randall, K., Robinson, S. A., Gonzalez, F., & Bollard, B. (2024). Monitoring of Antarctica's fragile vegetation using Drone-Based remote sensing, multispectral imagery and AI. *Sensors (Basel)*, *24*(4), 1063. DOI: 10.3390/s24041063 PMID: 38400222

Ravichandran, N., Tewaraja, T., Rajasegaran, V., Kumar, S. S., Gunasekar, S. K. L., & Sindiramutty, S. R. (2024). Comprehensive Review Analysis and Countermeasures for Cybersecurity Threats: DDoS, Ransomware, and Trojan Horse Attacks. *Preprints.org*. DOI: 10.20944/preprints202409.1369.v1

Rocha, A., Monteiro, M., Mattos, C., Dias, M., Soares, J., Magalhães, R., & Macedo, J. (2024). Edge AI for Internet of Medical Things: A literature review. *Computers & Electrical Engineering*, *116*, 109202. DOI: 10.1016/j.compeleceng.2024.109202

Saeed, R. A., Omri, M., Abdel-Khalek, S., Ali, E. S., & Alotaibi, M. F. (2022). Optimal path planning for drones based on swarm intelligence algorithm. *Neural Computing & Applications*, *34*(12), 10133–10155. DOI: 10.1007/s00521-022-06998-9

Şahin, M. C., & Tarhan, A. K. (2025). Evaluation and selection of hardware and AI models for edge applications: A method and a case study on UAVs. *Applied Sciences (Basel, Switzerland)*, *15*(3), 1026. DOI: 10.3390/app15031026

Sartika, D. (2025). Critical phishing defense strategies and digital asset protection. In *IGI Global eBooks* (pp. 213–220). DOI: 10.4018/979-8-3693-8784-9.ch010

Sathya, D., Thangamani, R., & Balaji, B. S. (2024). The revolution of edge computing in smart farming. In *Signals and communication technology* (pp. 351–389). DOI: 10.1007/978-3-031-51195-0_17

Shah, I. A., Jhanjhi, N. Z., & Laraib, A. (2022). Cybersecurity and blockchain usage in contemporary business. In *Advances in information security, privacy, and ethics book series* (pp. 49–64). DOI: 10.4018/978-1-6684-5284-4.ch003

Shakeri, R., Al-Garadi, M. A., Badawy, A., Mohamed, A., Khattab, T., Al-Ali, A. K., Harras, K. A., & Guizani, M. (2019). Design Challenges of Multi-UAV Systems in Cyber-Physical Applications: A Comprehensive Survey and Future Directions. *IEEE Communications Surveys and Tutorials*, *21*(4), 3340–3385. DOI: 10.1109/COMST.2019.2924143

Sharma, R., Singh, A., Kavita, N., Jhanjhi, N. Z., Masud, M., Jaha, E. S., & Verma, S. (2021). Plant disease diagnosis and image classification using deep learning. *Computers, Materials & Continua/Computers. Materials & Continua (Print)*, *71*(2), 2125–2140. DOI: 10.32604/cmc.2022.020017

Sheng, L. T., Azman, A. B., Khan, N. A., Sindiramutty, S. R., Singh, S., & Verma, S. (2024). Reinforcement learning-based traffic signal control. *2024 International Conference on Emerging Trends in Networks and Computer Communications (ETNCC)*, 792–798. DOI: 10.1109/ETNCC63262.2024.10767501

Sindiramutty, S. R. (2024). Autonomous Threat Hunting: a future paradigm for AI-Driven Threat intelligence. *arXiv (Cornell University)*. https://doi.org//arxiv.2401.00286DOI: 10.48550

Sindiramutty, S. R., Jhanjhi, N. Z., Tan, C. E., Yun, K. J., Manchuri, A. R., Ashraf, H., Murugesan, R. K., Tee, W. J., & Hussain, M. (2024). Data security and privacy concerns in drone operations. In *Advances in information security, privacy, and ethics book series* (pp. 236–290). DOI: 10.4018/979-8-3693-0774-8.ch010

Sindiramutty, S. R., Prabagaran, K. R. V., Akbar, R., Hussain, M., & Malik, N. A. (2024). Overview of Generative AI techniques for cybersecurity. In *Advances in information security, privacy, and ethics book series* (pp. 1–52). DOI: 10.4018/979-8-3693-5415-5.ch001

Sindiramutty, S. R., Prabagaran, K. R. V., Jhanjhi, N. Z., Ghazanfar, M. A., Malik, N. A., & Soomro, T. R. (2024). Security Considerations in Generative AI for web Applications. In *Advances in information security, privacy, and ethics book series* (pp. 281–332). DOI: 10.4018/979-8-3693-5415-5.ch009

Sindiramutty, S. R., Prabagaran, K. R. V., Jhanjhi, N. Z., Murugesan, R. K., Brohi, S. N., & Masud, M. (2024). Generative AI in network security and intrusion detection. In *Advances in information security, privacy, and ethics book series* (pp. 77–124). DOI: 10.4018/979-8-3693-5415-5.ch003

Sindiramutty, S. R., Prabagaran, K. R. V., Jhanjhi, N. Z., Murugesan, R. K., Brohi, S. N., & Wei, G. W. (2024). Generative AI for threat intelligence and information sharing. In *Advances in digital crime, forensics, and cyber terrorism book series* (pp. 191–234). DOI: 10.4018/979-8-3693-8944-7.ch006

Sindiramutty, S. R., Prabagaran, K. R. V., Jhanjhi, N. Z., Murugesan, R. K., Malik, N. A., & Hussain, M. (2024). Ethics and transparency in secure web model generation. In *Advances in information security, privacy, and ethics book series* (pp. 411–464). DOI: 10.4018/979-8-3693-5415-5.ch012

Singh, M. K., Pippal, S. K., & Sharma, V. (2024). Lightweight blockchain mechanism for secure data transmission in healthcare system. *Biomedical Signal Processing and Control, 102*, 107411. DOI: 10.1016/j.bspc.2024.107411

Singh, T., Kumar, S., Singh, S. K., Priyanshu, N., Gupta, B. B., Wu, J., & Castiglione, A. (2024). Enhancing autonomous system security with AI and secure computation technologies. In *Advances in computational intelligence and robotics book series* (pp. 159–186). DOI: 10.4018/979-8-3693-2707-4.ch008

Singh, T., Sinjana, Y., & Kapoor, A. P. (2024). Integrating Mobile Edge Computing in Drone Delivery: A Holistic Service Provision Strategy for Drone-Based Healthcare Logistics. *SSRN*. DOI: 10.2139/ssrn.4994845

Sun, C., Fontanesi, G., Canberk, B., Mohajerzadeh, A., Chatzinotas, S., Grace, D., & Ahmadi, H. (2024). Advancing UAV Communications: A Comprehensive survey of Cutting-Edge Machine Learning Techniques. *IEEE Open Journal of Vehicular Technology, 5*, 825–854. DOI: 10.1109/OJVT.2024.3401024

Sun, J., & Hubbard, S. (2025). An examination of UAS incidents: Characteristics and safety considerations. *Drones (Basel), 9*(2), 112. DOI: 10.3390/drones9020112

Sun, J., Li, B., Jiang, Y., & Wen, C. (2016). A Camera-Based Target Detection and Positioning UAV system for search and rescue (SAR) purposes. *Sensors (Basel), 16*(11), 1778. DOI: 10.3390/s16111778 PMID: 27792156

Sun, L., Wang, J., Wan, L., Li, K., Wang, X., & Lin, Y. (2024). Human-UAV interaction assisted heterogeneous UAV swarm scheduling for target searching in communication denial environment. *IEEE Transactions on Automation Science and Engineering*, •••, 1–16. DOI: 10.1109/TASE.2024.3412005

Tan, Y., & Zheng, Z. (2013). Research advance in swarm robotics. *Defence Technology, 9*(1), 18–39. DOI: 10.1016/j.dt.2013.03.001

Tlili, F., Ayed, S., & Fourati, L. C. (2024a). Advancing UAV security with artificial intelligence: A comprehensive survey of techniques and future directions. *Internet of Things : Engineering Cyber Physical Human Systems, 27*, 101281. DOI: 10.1016/j.iot.2024.101281

Tlili, F., Ayed, S., & Fourati, L. C. (2024b). Advancing UAV security with artificial intelligence: A comprehensive survey of techniques and future directions. *Internet of Things : Engineering Cyber Physical Human Systems*, *27*, 101281. DOI: 10.1016/j.iot.2024.101281

Topham, G., Weaver, M., & Siddique, H. (2018, December 21). Runway reopens after days of drone disruption at Gatwick. *The Guardian*. https://www.theguardian.com/uk-news/2018/dec/20/tens-of-thousands-of-passengers-stranded-by-gatwick-airport-drones

Ullah, F., Saqib, S., & Xiong, Y. (2024). Integrating artificial intelligence in biodiversity conservation: Bridging classical and modern approaches. *Biodiversity and Conservation*. Advance online publication. DOI: 10.1007/s10531-024-02977-9

Van Groesen, W., & Pauwels, P. (2022). Tracking prefabricated assets and compliance using quick response (QR) codes, blockchain and smart contract technology. *Automation in Construction*, *141*, 104420. DOI: 10.1016/j.autcon.2022.104420

Wang, C., Zhang, S., Ma, T., Xiao, Y., Chen, M. Z., & Wang, L. (2024). Swarm intelligence: A survey of model classification and applications. *Chinese Journal of Aeronautics*, 102982. Advance online publication. DOI: 10.1016/j.cja.2024.03.019

Weiqi, X., Hooi, S. T. C., Sindiramutty, S. R. L., Asirvatham, D. L., Kumar, D., & Verma, S. (2024). Surface Anomaly Detection Using Machine Learning Technique. *2024 International Conference on Emerging Trends in Networks and Computer Communications (ETNCC)*, 1–7. DOI: 10.1109/ETNCC63262.2024.10767562

Witt, R. R., Beranek, C. T., Howell, L. G., Ryan, S. A., Clulow, J., Jordan, N. R., Denholm, B., & Roff, A. (2020). Real-time drone derived thermal imagery outperforms traditional survey methods for an arboreal forest mammal. *PLoS One*, *15*(11), e0242204. DOI: 10.1371/journal.pone.0242204 PMID: 33196649

Xie, W., Wan, Y., Wu, G., Li, Y., Zhou, F., & Wu, Q. (2024). A RF-Visual Directional fusion framework for precise UAV positioning. *IEEE Internet of Things Journal*, *11*(22), 36736–36747. DOI: 10.1109/JIOT.2024.3424271

Yadav, I. C., Devi, N. L., Syed, J. H., Cheng, Z., Li, J., Zhang, G., & Jones, K. C. (2014). Current status of persistent organic pesticides residues in air, water, and soil, and their possible effect on neighboring countries: A comprehensive review of India. *The Science of the Total Environment*, *511*, 123–137. DOI: 10.1016/j.scitotenv.2014.12.041 PMID: 25540847

Yedilkhan, D., Kyzyrkanov, A. E., Kutpanova, Z. A., Aljawarneh, S., & Atanov, S. K. (2024). Intelligent obstacle avoidance algorithm for safe urban monitoring with autonomous mobile drones. *Journal of Electronic Science and Technology*, *22*(4), 100277. DOI: 10.1016/j.jnlest.2024.100277

Ying, X., Murugesan, R. K., Sindiramutty, S. R., Wei, G. W., Balakrishnan, S., Kumar, D., & Verma, S. (2024). Scene Text Recognition using Deep Learning Techniques. *2024 International Conference on Emerging Trends in Networks and Computer Communications (ETNCC)*, 1–9. DOI: 10.1109/ETNCC63262.2024.10767484

Zhou, Y., Rao, B., & Wang, W. (2020). UAV swarm intelligence: Recent advances and future trends. *IEEE Access : Practical Innovations, Open Solutions*, *8*, 183856–183878. DOI: 10.1109/ACCESS.2020.3028865

Compilation of References

. Petrović, A., &Žižović, M. (2019). *Integration of Artificial Intelligence with Cloud Services*. 381–387. DOI: 10.15308/Sinteza-2019-381-387

<jrn>Niu, K., & Yan, Y. (2023). A small-object-detection model based on improved yolov8 for uav aerial images. 2023 2nd International Conference on Artificial Intelligence and Intelligent Information Processing (AIIIP)

Abadade, Y., Temouden, A., Bamoumen, H., Benamar, N., Chtouki, Y., & Hafid, A. S. (2023). A comprehensive survey on tinyml. *IEEE Access : Practical Innovations, Open Solutions*, *11*, 96892–96922. DOI: 10.1109/ACCESS.2023.3294111

Abbas, A., Zhang, Z., Zheng, H., Alami, M. M., Alrefaei, A. F., Abbas, Q., Naqvi, S. H., Rao, M. J., Mosa, W. F. A., Abbas, Q., Hussain, A., Hassan, M. Z., & Zhou, L. (2023). Drones in plant Disease Assessment, Efficient Monitoring, and Detection: A way forward to smart agriculture. *Agronomy (Basel)*, *13*(6), 1524. DOI: 10.3390/agronomy13061524

Abbas, S., Talib, M. A., Ahmed, I., & Belal, O. (2024). Integration of UAVs and FANETs in Disaster Management: A review on applications, challenges and future directions. *Transactions on Emerging Telecommunications Technologies*, *35*(12), e70023. Advance online publication. DOI: 10.1002/ett.70023

Abdelkader, M., Güler, S., Jaleel, H., & Shamma, J. S. (2021). Aerial swarms: Recent applications and challenges. *Current Robotics Reports*, *2*(3), 309–320. DOI: 10.1007/s43154-021-00063-4 PMID: 34977595

Abdelnabi, A. A. B., & Rabadi, G. (2024). Human detection from unmanned aerial vehicles' images for search and rescue missions: A state-of-the-art review. *IEEE Access : Practical Innovations, Open Solutions*, *12*, 152009–152035. DOI: 10.1109/ACCESS.2024.3479988

Abdullah, A. S., AB, A. A., & Geetha, S. (2024). DecentraliDrone: A decentralized, fully autonomous drone delivery system for reliable, efficient transport of goods. *Alexandria Engineering Journal*, *88*, 1–30.

Abiodun, T. F. (2020). Usage of drones or unmanned aerial vehicles (uavs) for effective aerial surveillance, mapping system and intelligence gathering in combating insecurity in nigeria. *African Journal of Social Sciences and Humanities Research*, *3*(2), 29–44.

Abro, G., Zulkifli, S., Masood, R., Asirvadam, V. & Laouti, A. (2022). Comprehensive Review of UAV Detection, Security, and Communication Advancements to Prevent Threats. Drones 2022, 6, 284. *Advances in UAV Detection, Classification and Tracking*, 63.

Adil, M., Song, H., Mastorakis, S., Abulkasim, H., Farouk, A., & Jin, Z. (2023). UAV-assisted IoT applications, cybersecurity threats, AI-enabled solutions, open challenges with future research directions. *IEEE Transactions on Intelligent Vehicles*, *9*(4), 4583–4605. DOI: 10.1109/TIV.2023.3309548

Advanced Navigation. (2025, January 9). *Inertial navigation systems for sea, land & air*. https://www.advancednavigation.com/

Agelli, M., Corona, N., Maggio, F., & Moi, P. V. (2024a). Unmanned ground vehicles for continuous crop monitoring in agriculture: Assessing the readiness of current ICT technology. *Machines (Basel)*, *12*(11), 750. DOI: 10.3390/machines12110750

Agrahari, A., Dhabu, M. M., Deshpande, P. S., Tiwari, A., Baig, M. A., & Sawarkar, A. D. (2024). Artificial Intelligence-Based Adaptive Traffic Signal Control System: A Comprehensive Review. *Electronics (Basel)*, *13*(19), 3875. DOI: 10.3390/electronics13193875

Ahmad, I., Rodriguez, F., Kumar, T., Suomalainen, J., Jagatheesaperumal, S. K., Walter, S., Asghar, M. Z., Li, G., Papakonstantinou, N., Ylianttila, M., Huusko, J., Sauter, T., & Harjula, E. (2024). Communications Security in Industry X: A survey. *IEEE Open Journal of the Communications Society*, *5*, 982–1025. DOI: 10.1109/OJCOMS.2024.3356076

Ahmed, A. A., & Echi, M. (2021). Hawk-eye: An AI-powered threat detector for intelligent surveillance cameras. *IEEE Access : Practical Innovations, Open Solutions*, *9*, 63283–63293. DOI: 10.1109/ACCESS.2021.3074319

Ahn, H., Choi, H.-L., Kang, M., & Moon, S. (2019). Learning-based anomaly detection and monitoring for swarm drone flights. *Applied Sciences (Basel, Switzerland)*, *9*(24), 5477. DOI: 10.3390/app9245477

Aibibu, T., Lan, J., Zeng, Y., Lu, W., & Gu, N. (2023). An efficient rep-style gaussian–wasserstein network: Improved uav infrared small object detection for urban road surveillance and safety. *Remote Sensing (Basel)*, *16*(1), 25. DOI: 10.3390/rs16010025

Akbari, Y., Almaadeed, N., Al-Maadeed, S., & Elharrouss, O. (2021). Applications, databases and open computer vision research from drone videos and images: A survey. *Artificial Intelligence Review*, *54*(5), 3887–3938. DOI: 10.1007/s10462-020-09943-1

Al-Ahmadi, A. (2024). Drone attitude and position prediction via stacked hybrid deep learning model for massive MIMO applications. *IEEE Access : Practical Innovations, Open Solutions*, *1*, 1. Advance online publication. DOI: 10.1109/ACCESS.2024.3516126

Alanhdi, A., & Toka, L. (2024). A survey on integrating edge computing with ai and blockchain in maritime domain, aerial systems, iot, and industry 4.0. *IEEE Access : Practical Innovations, Open Solutions*, *12*, 28684–28709. DOI: 10.1109/ACCESS.2024.3367118

Albano, M., Ferreira, L L., Orio, G D., Maló, P., Webers, G., Jantunen, E., Gabilondo, I., Viguera, M., & Papa, G. (2020, July 21). Advanced sensor-based maintenance in real-world exemplary cases. Taylor & Francis, 61(4), 537-553. DOI: 10.1080/00051144.2020.1794192

Aldao, E., González-De Santos, L. M., & González-Jorge, H. (2022). LiDAR based detect and avoid system for UAV navigation in UAM corridors. *Drones (Basel)*, *6*(8), 185. DOI: 10.3390/drones6080185

Aldossri, R., Aljughaiman, A., & Albuali, A. (2024). Advancing Drone Operations through Lightweight Blockchain and Fog Computing Integration: A Systematic Review. *Drones (Basel)*, *8*(4), 153. DOI: 10.3390/drones8040153

Aldughayfiq, B., Ashfaq, F., Jhanjhi, N. Z., & Humayun, M. (2023). YOLOv5-FPN: A robust framework for multi-sized cell counting in fluorescence images. *Diagnostics (Basel)*, *13*(13), 2280. DOI: 10.3390/diagnostics13132280 PMID: 37443674

Alex, S. A., Jhanjhi, N., Humayun, M., Ibrahim, A. O., & Abulfaraj, A. W. (2022). Deep LSTM Model for Diabetes Prediction with Class Balancing by SMOTE. *Electronics (Basel)*, *11*(17), 2737. DOI: 10.3390/electronics11172737

Alferidah, D. K., & Jhanjhi, N. (2020). Cybersecurity Impact over Bigdata and IoT Growth. *2020 International Conference on Computational Intelligence (ICCI)*. DOI: 10.1109/ICCI51257.2020.9247722

Al-Fuqaha, A., Guizani, M., Mohammadi, M., Aledhari, M., & Ayyash, M. (2015). Internet of things: A survey on enabling technologies, protocols, and applications. *IEEE Communications Surveys and Tutorials*, *17*(4), 2347–2376. DOI: 10.1109/COMST.2015.2444095

Alhamrouni, I., Kahar, N. H. A., Salem, M., Swadi, M., Zahroui, Y., Kadhim, D. J., Mohamed, F. A., & Nazari, M. A. (2024). A Comprehensive review on the role of Artificial intelligence in power system stability, control, and Protection: Insights and Future Directions. *Applied Sciences (Basel, Switzerland)*, *14*(14), 6214. DOI: 10.3390/app14146214

Ali, H. M. S., Jalal, S. K., Saab, M. W., Sulaiman, S. K., Ghno, G. S. N., Mustafa, S. I., & Azimov, B. (2024). Deciphering the Implications of Swarm Intelligence Algorithms in Efficiently Managing Drone Swarms. *2024 35th Conference of Open Innovations Association (FRUCT)*, 112–123. DOI: 10.23919/FRUCT61870.2024.10516408

Ali, S S., Kaur, R., & Khan, S. (2023, January 21). Identification of innovative technology enablers and drone technology determinants adoption: a graph theory matrix analysis framework. Springer Science+Business Media, 16(2), 830-852. DOI: 10.1007/s12063-023-00346-3

Ali, S. (2024). *Sensor fusion-based indoor navigation in GNSS-denied environments - Espace ETS*. https://espace.etsmtl.ca/id/eprint/3442/

Aliloo, J., Abbasi, E., Karamidehkordi, E., Parmehr, E. G., & Canavari, M. (2024). Dos and Don’ts of using drone technology in the crop fields. *Technology in Society*, *76*, 102456. DOI: 10.1016/j.techsoc.2024.102456

Al-Jarrah, O. Y., Shatnawi, A. S., Shurman, M. M., Ramadan, O. A., & Muhaidat, S. (2024). Exploring deep learning-based visual localization techniques for UAVs in GPS-Denied environments. *IEEE Access : Practical Innovations, Open Solutions*, *12*, 113049–113071. DOI: 10.1109/ACCESS.2024.3440064

Alkaabi, K., Raza, M., Qasemi, E., Alderei, H., Alderei, M., & Almheiri, S. (2024). Using crowd-sourced traffic data and open-source tools for urban congestion analysis. *Transportation Research Interdisciplinary Perspectives*, *28*, 101261. DOI: 10.1016/j.trip.2024.101261

Al-kfairy, M., Mustafa, D., Kshetri, N., Insiew, M., & Alfandi, O. (2024). Ethical challenges and solutions of generative ai: An interdisciplinary perspective. [MDPI.]. *Informatics (MDPI)*, *11*(3), 58. DOI: 10.3390/informatics11030058

Alkinani, M. H., Almazroi, A. A., Jhanjhi, N., & Khan, N. A. (2021). 5G and IoT Based Reporting and Accident Detection (RAD) System to Deliver First Aid Box Using Unmanned Aerial Vehicle. *Sensors (Basel), 21*(20), 6905. DOI: 10.3390/s21206905 PMID: 34696118

Al-lQubaydhi, N., Alenezi, A., Alanazi, T., Senyor, A., Alanezi, N., Alotaibi, B., Alotaibi, M., Razaque, A., & Hariri, S. (2024). Deep learning for unmanned aerial vehicles detection: A review. *Computer Science Review, 51*, 100614. DOI: 10.1016/j.cosrev.2023.100614

Alqudsi, Y., & Makaraci, M. (2025). UAV swarms: Research, challenges, and future directions. *Journal of Engineering and Applied Sciences (Asian Research Publishing Network), 72*(1), 12. Advance online publication. DOI: 10.1186/s44147-025-00582-3

Alquwayzani, A. A., & Albuali, A. A. (2024). A Systematic Literature review of zero trust architecture for military UAV security systems. *IEEE Access : Practical Innovations, Open Solutions, 1*, 176033–176056. Advance online publication. DOI: 10.1109/ACCESS.2024.3503587

Alsadie, D. (2024). A comprehensive review of AI techniques for resource management in Fog Computing: Trends, challenges, and future directions. *IEEE Access : Practical Innovations, Open Solutions, 12*, 118007–118059. DOI: 10.1109/ACCESS.2024.3447097

Alsamhi, S. H., Almalki, F. A., Ma, O., Ansari, M. S., & Lee, B. (2021). Predictive estimation of optimal signal strength from drones over IoT frameworks in smart cities. *IEEE Transactions on Mobile Computing, 22*(1), 402–416. DOI: 10.1109/TMC.2021.3074442

Alsayed, A., Yunusa-Kaltungo, A., Quinn, M. K., Arvin, F., & Nabawy, M. R. A. (2021, August 24). Drone-Assisted Confined Space Inspection and Stockpile Volume Estimation. *Remote Sensing (Basel), 13*(17), 3356–3356. DOI: 10.3390/rs13173356

Althati, C., Tomar, M., & Malaiyappan, J. N. A. (2024). Scalable machine learning solutions for heterogeneous data in distributed data platform. *Journal of Artificial Intelli- gence General science (JAIGS) ISSN: 3006-4023*, 4(1):299–309.

Altun, A. T., Hasanzade, M., Saldiran, E., Guner, G., Uzun, M., Fremond, R., Tang, Y., Bhundoo, P., Su, Y., Xu, Y., İnalhan, G., Hardt, M., Fransoy, A., Modha, A., Tena, J. A., Nieto, C., Vilaplana, M., Tojal, M., Gordo, V., & González, A. (2023, August 15). The Development of an Advanced Air Mobility Flight Testing and Simulation Infrastructure. *Aerospace (Basel, Switzerland), 10*(8), 712–712. DOI: 10.3390/aerospace10080712

Al-Turjman, F. & Zahmatkesh, H. (2020). A comprehensive review on the use of AI in UAV communications: Enabling technologies, applications, and challenges. *Unmanned Aerial Vehicles in Smart Cities*, 1-26.

Amarasinghe, A., Wijesuriya, V B., Ganepola, D., & Jayaratne, L. (2019, November 10). A swarm of crop spraying drones solution for optimising safe pesticide usage in arable lands. DOI: 10.1145/3356250.3361948

Amirgaliyev, B., Mussabek, M., Rakhimzhanova, T., & Zhumadillayeva, A. (2025). A Review of Machine Learning and Deep Learning Methods for Person Detection, Tracking and Identification, and Face Recognition with Applications. *Sensors (Basel), 25*(5), 1410. DOI: 10.3390/s25051410

Andersen, T. S., Andersen, N. A., Ravn, O., & Fumagalli, M. (2024). A long distance mono optical localization system for unmanned aerial vehicles. *2022 17th International Conference on Control, Automation, Robotics and Vision (ICARCV)*, 947–953. DOI: 10.1109/ICARCV63323.2024.10821626

Andreoni, M., Lunardi, W. T., Lawton, G., & Thakkar, S. (2024). Enhancing autonomous system security and resilience with Generative AI: A comprehensive survey. *IEEE Access : Practical Innovations, Open Solutions, 12*, 109470–109493. DOI: 10.1109/ACCESS.2024.3439363

Andreu-Perez, J., Deligianni, F., Ravi, D., & Yang, G.-Z. (2018). Artificial Intelligence and robotics. *arXiv preprint arXiv:1803.10813.*

An, J., Putro, M. D., Priadana, A., & Jo, K.-H. (2023). Improved yolov5 network with cbam for object detection vision drone. 2023 IEEE International Conference on Industrial Technology (ICIT), Beheshti, N., & Johnsson, L. (2020). Squeeze u-net: A memory and energy efficient image segmentation network. Proceedings of the IEEE/CVF conference on computer vision and pattern recognition workshops, Bhat, S. (2024). Leveraging 5g network capabilities for smart grid communication. *Journal of Electrical Systems, 20*(2), 2272–2283.

Ansari, M. S., Alsamhi, S. H., Qiao, Y., Ye, Y. & Lee, B. (2020). Security of distributed intelligence in edge computing: Threats and countermeasures. *The Cloud-to-Thing Continuum: Opportunities and Challenges in Cloud, Fog and Edge Computing*, 95-122.

Apeland, J. (2021). Use of fuel cells to extend multirotor drone endurance.

Arafat, M. Y., Alam, M. M., & Moh, S. (2023). Vision-Based Navigation Techniques for Unmanned Aerial Vehicles: Review and challenges. *Drones (Basel), 7*(2), 89. DOI: 10.3390/drones7020089

Aragon, R. V., Castaño, C A R., & Cerón, A. (2020, September 30). Impact and Technological Innovation of UAS/Drones in the World Economy: *VI International Conference on Innovation and Trends in Engineering (CONIITI 2020)*. DOI: 10.1109/CONIITI51147.2020.9240228

Ariante, G., Papa, U., Ponte, S., & Del Core, G. (2019, June). UAS for positioning and field mapping using LIDAR and IMU sensors data: Kalman filtering and integration. In 2019 IEEE 5th International Workshop on Metrology for AeroSpace (MetroAeroSpace) (pp. 522-527). IEEE.

Ariante, G., Papa, U., Ponte, S., & Del Core, G. (2020, June). Velocity and attitude estimation of a small unmanned aircraft with micro Pitot tube and Inertial Measurement Unit (IMU). In 2020 IEEE 7th International Workshop on Metrology for AeroSpace (MetroAeroSpace) (pp. 314-318). IEEE.

Ariante, G., Ponte, S., Papa, U., & Del Core, G. (2021, June). Safe landing area determination (slad) for unmanned aircraft systems by using rotary lidar. In 2021 IEEE 8th international workshop on metrology for AeroSpace (MetroAeroSpace) (pp. 110-115). IEEE.

Ariante, G., & Papa, U. (2024). *Embedded Platforms for UAS Landing Path and Obstacle Detection II: Navigation Support Systems for Urban Air Mobility Operations*. Springer Nature. DOI: 10.1007/978-3-031-59754-1

Ariante, G., Ponte, S., Papa, U., & Del Core, G. (2021). 2021a. Estimation of Airspeed, Angle of Attack, and Sideslip for Small Unmanned Aerial Vehicles (UAVs) Using a Micro-Pitot Tube. [s Note: MDPI stays neutral with regard to jurisdictional claims in published....]. *Electronics (Basel)*, *10*(19), 2325. DOI: 10.3390/electronics10192325

Ariante, G., Ponte, S., Papa, U., Greco, A., & Del Core, G. (2022). Ground control system for UAS safe landing area determination (SLAD) in urban air mobility operations. *Sensors (Basel)*, *22*(9), 3226. DOI: 10.3390/s22093226 PMID: 35590916

Arshad, M. A., Khan, S. H., Qamar, S., Khan, M. W., Murtza, I., Gwak, J., & Khan, A. (2022). Drone navigation using region and Edge Exploitation-Based Deep CNN. *IEEE Access : Practical Innovations, Open Solutions*, *10*, 95441–95450. DOI: 10.1109/ACCESS.2022.3204876

Arya, L., Sharma, Y. K., Padmanaban, H., Devi, S., & Kumar, R. (2023, October). Eyes in the sky: Safeguarding borders security with AI-powered aerial monitoring and IoT integration. In *The International Conference on Recent Innovations in Computing* (pp. 863-873). Springer Nature Singapore.

Ashfaq, F., Ghoniem, R. M., Jhanjhi, N. Z., Khan, N. A., & Algarni, A. D. (2023). Using Dual Attention BiLSTM to Predict Vehicle Lane Changing Maneuvers on Highway Dataset. *Systems*, *11*(4), 196. DOI: 10.3390/systems11040196

Awasthi, S., Gramse, N., Reining, C., & Roidl, M. (2022, January 1). UAVs for Industries and Supply Chain Management. Cornell University. https://doi.org//arxiv.2212.03346DOI: 10.48550

Ayaz, M., Ammad-Uddin, M., Sharif, Z., Mansour, A., & Aggoune, E.-H. M. (2019). Internet-of-things (iot)-based smart agriculture: Toward making the fields talk. *IEEE Access : Practical Innovations, Open Solutions*, *7*, 129551–129583. DOI: 10.1109/ACCESS.2019.2932609

Babbar, H., Rani, S., Masud, M., Verma, S., Anand, D., & Jhanjhi, N. (2021). Load balancing algorithm for migrating switches in software-defined vehicular networks. *Computers, Materials & Continua/Computers. Computers, Materials & Continua*, *67*(1), 1301–1316. DOI: 10.32604/cmc.2021.014627

Bain, G., & Blaney, M. (2025). *Managing safety in the drone industry: A Practical Guide*.

Bakirci, M. (2024). Enhancing vehicle detection in intelligent transportation systems via autonomous uav platform and yolov8 integration. *Applied Soft Computing*, *164*, 112015. DOI: 10.1016/j.asoc.2024.112015

Balestrieri, E., Daponte, P., De Vito, L., Picariello, F., & Tudosa, I. (2021). Sensors and Measurements for UAV safety: An overview. *Sensors (Basel)*, *21*(24), 8253. DOI: 10.3390/s21248253 PMID: 34960347

Banafaa, M. K., Pepeoğlu, Ö., Shayea, I., Alhammadi, A., Shamsan, Z. A., Razaz, M. A., Alsagabi, M., & Al-Sowayan, S. (2024). A comprehensive survey on 5G-and-Beyond networks with UAVs: Applications, emerging technologies, regulatory aspects, research trends and challenges. *IEEE Access : Practical Innovations, Open Solutions*, *12*, 7786–7826. DOI: 10.1109/ACCESS.2023.3349208

Bao, D., Yan, Y., Li, Y., & Chu, J. (2025). The Future of Last-Mile Delivery: Lifecycle environmental and economic impacts of Drone-Truck Parallel Systems. *Drones (Basel)*, *9*(1), 54. DOI: 10.3390/drones9010054

Barrie, R., Haalck, L., Risse, B., Nowotny, T., Graham, P., & Buehlmann, C. (2023). Trail using ants follow idiosyncratic routes in complex landscapes. *Learning & Behavior*, *52*(1), 105–113. DOI: 10.3758/s13420-023-00615-y PMID: 37993707

Bayomi, N., & Fernandez, J. E. (2023). Eyes in the Sky: Drones Applications in the Built Environment under Climate Change Challenges. *Drones (Basel)*, *7*(10), 637. DOI: 10.3390/drones7100637

Bekmurzaeva, R., Kalimullin, R., & Aguzarova, F. (2024). Application of drones and artificial intelligence to monitor and protect natural ecosystems. In *BIO Web of Conferences* (Vol. 140, p. 01008). EDP Sciences. DOI: 10.1051/bioconf/202414001008

Bengio, Y., Hinton, G. E., Yao, A. C., Song, D., Abbeel, P., Darrell, T., Harari, Y. N., Zhang, Y., Lan, X., Shalev-Shwartz, S., Hadfield, G. K., Mouret, J., Maharaj, T., Hutter, F., Baydin, A. G., McIlraith, S. A., Gao, Q., Acharya, A., Krueger, D., & Mindermann, S. (2024, May 24). Managing extreme AI risks amid rapid progress. *Science*, *384*(6698), 842–845. DOI: 10.1126/science.adn0117 PMID: 38768279

Besada, J. A., Campaña, I., Carramiñana, D., Bergesio, L., & De Miguel, G. (2021). Review and simulation of counter-UAS sensors for unmanned traffic management. *Sensors (Basel)*, *22*(1), 189. DOI: 10.3390/s22010189 PMID: 35009730

Bhat, S., & Kavasseri, A. (2024). Multi-source data integration for navigation in gps-denied autonomous driving environments. *International Journal of Electrical and Electronics Research*, *12*(3), 863–869. DOI: 10.37391/ijeer.120317

Bhoi, S. K., Chakraborty, S., Verbrugge, B., Helsen, S., Robyns, S., Baghdadi, M. E., & Hegazy, O. (2024). Intelligent data-driven condition monitoring of power electronics systems using smart edge–cloud framework. *Internet of Things : Engineering Cyber Physical Human Systems*, *26*, 101158. DOI: 10.1016/j.iot.2024.101158

Bibri, S. E., Huang, J., Jagatheesaperumal, S. K., & Krogstie, J. (2024). The synergistic interplay of artificial intelligence and digital twin in environmentally planning sustainable smart cities: A comprehensive systematic review. *Environmental Science and Ecotechnology*, *20*, 100433. DOI: 10.1016/j.ese.2024.100433 PMID: 38831974

Bilal, K., Khalid, O., Erbad, A., & Khan, S. U. (2018). Potentials, trends, and prospects in edge technologies: Fog, cloudlet, mobile edge, and micro data centers. *Computer Networks*, *130*, 94–120. DOI: 10.1016/j.comnet.2017.10.002

Blum, C. (2024). Ant Colony Optimization: A Bibliometric Review. *Physics of Life Reviews*, *51*, 87–95. DOI: 10.1016/j.plrev.2024.09.014 PMID: 39341089

Blum, J. R., Eichhorn, A., Smith, S., Sterle-Contala, M., & Cooperstock, J. R. (2013). Real-time emergency response: Improved management of real-time information during crisis situations. *Journal on Multimodal User Interfaces*, *8*(2), 161–173. DOI: 10.1007/s12193-013-0139-7

Boiteau, S., Vanegas, F., & Gonzalez, F. (2024). Framework for autonomous uav navigation and target detection in global-navigation-satellite-system-denied and visually degraded environments. *Remote Sensing (Basel)*, *16*(3), 471. DOI: 10.3390/rs16030471

Borghetti, F., Caballini, C., Carboni, A., Grossato, G., Maja, R., & Barabino, B. (2022). The Use of Drones for Last-Mile Delivery: A Numerical Case Study in Milan, Italy. *Sustainability (Basel)*, *14*(3), 1766. Advance online publication. DOI: 10.3390/su14031766

Bousdekis, A., Lepenioti, K., Apostolou, D., & Mentzas, G. (2021, March 31). A Review of Data-Driven Decision-Making Methods for Industry 4.0 Maintenance Applications. *Electronics (Basel)*, *10*(7), 828–828. DOI: 10.3390/electronics10070828

Brik, B., Ksentini, A., & Bouaziz, M. (2020). Federated learning for UAVs-enabled wireless networks: Use cases, challenges, and open problems. *IEEE Access : Practical Innovations, Open Solutions*, *8*, 53841–53849. DOI: 10.1109/ACCESS.2020.2981430

Brohi, S. N., Jhanjhi, N., Brohi, N. N., & Brohi, M. N. (2020). Key Applications of State-of-the-Art Technologies to Mitigate and Eliminate COVID-19.pdf. *TECHRxiv*. DOI: 10.36227/techrxiv.12115596.v1

Buck, J., & Buck, E. (1966). Biology of synchronous flashing of fireflies. *Nature*, *211*(5049), 562–564. DOI: 10.1038/211562a0

Budholiya, S., Bhat, A., Raj, S. A., Sultan, M. T. H., Shah, A. U. M., & Basri, A. A. (2021). State of the Art Review about Bio-Inspired Design and Applications: An Aerospace Perspective. *Applied Sciences (Basel, Switzerland)*, *11*(11), 5054. DOI: 10.3390/app11115054

Bu, Y., Yan, Y., & Yang, Y. (2024). Advancement Challenges in UAV Swarm Formation Control: A Comprehensive review. *Drones (Basel)*, *8*(7), 320. DOI: 10.3390/drones8070320

Caballero-Martin, D., Lopez-Guede, J. M., Estevez, J., & Gran˜a, M. (2024). Artificial intelligence applied to drone control: A state of the art. *Drones (Basel)*, *8*(7), 296. DOI: 10.3390/drones8070296

Cadet, E., Osundare, O. S., Ekpobimi, H. O., Samira, Z., & Wondaferew, Y. (2024). AI-powered threat detection in surveillance systems: A real-time data processing framework.

Cai, Y., Gao, F., Huang, J., & Qiao, G. (2024). GPS-aided Stereo Inertial Navigation Localization Algorithm for outdoor scenarios. *IEEE Sensors Journal*, *1*. Advance online publication. DOI: 10.1109/JSEN.2024.3519880

Caminero, G., Lopez-Martin, M., & Carro, B. (2019). Adversarial environ- ment reinforcement learning algorithm for intrusion detection. *Computer Networks*, *159*, 96–109. DOI: 10.1016/j.comnet.2019.05.013

Campini, M. (2024). *GPS-Driven Autonomous Navigation for Precision Agriculture with AWS - Webthesis*. https://webthesis.biblio.polito.it/31580/

Cao, K., Liu, Y., Meng, G., & Sun, Q. (2020). An overview on edge computing research. *IEEE Access : Practical Innovations, Open Solutions*, *8*, 85714–85728. DOI: 10.1109/ACCESS.2020.2991734

Cao, P., Lei, L., Cai, S., Shen, G., Liu, X., Wang, X., Zhang, L., Zhou, L., & Guizani, M. (2024). Computational Intelligence Algorithms for UAV swarm Networking and Collaboration: A Comprehensive survey and Future Directions. *IEEE Communications Surveys and Tutorials*, *26*(4), 2684–2728. DOI: 10.1109/COMST.2024.3395358

Capitán, C., Capitán, J., Castano, A. R., & Ollero, A. (2019, June). Risk assessment based on SORA methodology for a UAS media production application. In *2019 International Conference on Unmanned Aircraft Systems (ICUAS)* (pp. 451-459). IEEE.

Carvalho, G., Cabral, B., Pereira, V., & Bernardino, J. (2021). Edge computing: Current trends, research challenges and future directions. *Computing*, *103*(5), 993–1023. DOI: 10.1007/s00607-020-00896-5

Castrillo, V. U., Manco, A., Pascarella, D., & Gigante, G. (2022). A review of counter-UAS technologies for cooperative defensive teams of drones. [CIOFFI, J. R., CERQUEIRA, C. S. & DOS SANTOS LIMA, J. S. A brief review of concepts and technological solutions to perfom DAA functions in the context of SIMUA Project.]. *Drones (Basel)*, *6*(3), 65. DOI: 10.3390/drones6030065

Cazzato, D., Cimarelli, C., Sanchez-Lopez, J. L., Voos, H., & Leo, M. (2020). A survey of computer vision methods for 2d object detection from unmanned aerial vehicles. *Journal of Imaging*, *6*(8), 78. DOI: 10.3390/jimaging6080078 PMID: 34460693

Cerutti, A. (2024). *Enhancing Drone Detection and Tracking Using LiDAR Sensor with Adaptive Motion Strategies - Webthesis*. https://webthesis.biblio.polito.it/33086/

Cetinsaya, B., Reiners, D., & Cruz-Neira, C. (2024). From PID to swarms: A decade of advancements in drone control and path planning - A systematic review (2013–2023). *Swarm and Evolutionary Computation*, *89*, 101626. DOI: 10.1016/j.swevo.2024.101626

Ceviz, O., Sen, S., & Sadioglu, P. (2024). A survey of security in UAVs and FANETs: Issues, threats, analysis of attacks, and solutions. *IEEE Communications Surveys and Tutorials*, *1*, 1. Advance online publication. DOI: 10.1109/COMST.2024.3515051

Chai, R., Guo, Y., Zuo, Z., Chen, K., Shin, H., & Tsourdos, A. (2024). Cooperative motion planning and control for aerial-ground autonomous systems: Methods and applications. *Progress in Aerospace Sciences*, *146*, 101005. DOI: 10.1016/j.paerosci.2024.101005

Chamola, V., Hassija, V., Gupta, V., & Guizani, M. (2020). A Comprehensive Review of the COVID-19 Pandemic and the Role of IoT, Drones, AI, Blockchain, and 5G in Managing its Impact. *IEEE Access : Practical Innovations, Open Solutions*, *8*, 90225–90265. DOI: 10.1109/ACCESS.2020.2992341

Chang, Z., Liu, S., Xiong, X., Cai, Z., & Tu, G. (2021). A survey of recent advances in Edge-Computing-Powered artificial intelligence of Things. *IEEE Internet of Things Journal*, *8*(18), 13849–13875. DOI: 10.1109/JIOT.2021.3088875

Chen, C., Jiao, S., Zhang, S., Liu, W., Feng, L., & Wang, Y. (2018). TripImputor: Real-time imputing taxi trip purpose leveraging multi-sourced urban data. *IEEE Transactions on Intelligent Transportation Systems*, *19*(10), 3292–3304. DOI: 10.1109/TITS.2017.2771231

Cheng, N., Wu, S., Wang, X., Yin, Z., Li, C., Chen, W., & Chen, F. (2023, August 15). AI for UAV-Assisted IoT Applications: A Comprehensive Review. *IEEE Internet of Things Journal*, *10*(16), 14438–14461. DOI: 10.1109/JIOT.2023.3268316

Chen, W., Liu, B., Huang, H., Guo, S., & Zheng, Z. (2019). When UAV swarm meets edge-cloud computing: The QoS perspective. *IEEE Network*, *33*(2), 36–43. DOI: 10.1109/MNET.2019.1800222

Chen, Y., Zhang, H., & Wu, J. (2023). Real-time traffic monitoring using YOLOv4: A comprehensive study. *IEEE Transactions on Intelligent Transportation Systems*, *24*(3), 1123–1135.

Chesti, I. A., Humayun, M., Sama, N. U., & Jhanjhi, N. (2020). Evolution, Mitigation, and Prevention of Ransomware. *2020 2nd International Conference on Computer and Information Sciences (ICCIS)*. DOI: 10.1109/ICCIS49240.2020.9257708

Chin, R., Catal, C., & Kassahun, A. (2023). Plant disease detection using drones in precision agriculture. In *Precision Agriculture* (Vol. 24, Issue 5, pp. 1663–1682). Springer. DOI: 10.1007/s11119-023-10014-y

Choy, S., Wong, B., Simon, G., & Rosenberg, C. (2012, November). The brewing storm in cloud gaming: A measurement study on cloud to end-user latency. In 2012 11th Annual Workshop on Network and Systems Support for Games (NetGames) (pp. 1-6). IEEE.

Chukwuma, U., Gebremedhin, K. G., & Uyeh, D. D. (2024). Imagining AI-driven decision making for managing farming in developing and emerging economies. *Computers and Electronics in Agriculture*, *221*, 108946. DOI: 10.1016/j.compag.2024.108946

Chung, H. J., Kang, B., & Yang, Y. S. (2024). N-DriverMotion: Driver motion learning and prediction using an event-based camera and directly trained spiking neural networks on Loihi 2. *IEEE Open Journal of Vehicular Technology*, •••, 1–14. DOI: 10.1109/OJVT.2024.3504481

Cisco, U. 2020. Cisco annual internet report (2018–2023) white paper, March 2020.

Clarke, R., & Moses, L. B. (2014). The regulation of civilian drones' im- pacts on public safety. *Computer Law & Security Report*, *30*(3), 263–285. DOI: 10.1016/j.clsr.2014.03.007

Cohen, A. P., Shaheen, S. A., & Farrar, E. M. (2021). Urban air mobility: History, ecosystem, market potential, and challenges. *IEEE Transactions on Intelligent Transportation Systems*, *22*(9), 6074–6087. DOI: 10.1109/TITS.2021.3082767

Coifman, B., McCord, M., Mishalani, R. G., & Redmill, K. (2004). Surface transportation surveillance from unmanned aerial vehicles. *Proc. of the 83rd Annual Meeting of the Transportation Research Board*

Couturier, A., & Akhloufi, M. A. (2024). A review on deep learning for UAV Absolute Visual Localization. *Drones (Basel)*, *8*(11), 622. DOI: 10.3390/drones8110622

Criollo, L., Mena-Arciniega, C., & Xing, S. (2024). Classification, military applications, and opportunities of unmanned aerial vehicles. *Aviation*, *28*(2), 115–127. DOI: 10.3846/aviation.2024.21672

Crupi, L., Butera, L., Ferrante, A., & Palossi, D. (2024). A deep learning-based pest insect monitoring system for ultra-low power pocket-sized drones. In *2024 20th International Conference on Distributed Computing in Smart Systems and the Internet of Things (DCOSS-IoT)*, pages 323–330. IEEE. DOI: 10.1109/DCOSS-IoT61029.2024.00056

Das, D., Njilla, L., Ghosh, U., Shetty, S., & Levin, E. (2024). An edge-cloud approach for dynamic resource allocation in drone communication networks. In *IEEE INFOCOM 2024-IEEE Conference on Computer Communications Workshops (INFOCOM WKSHPS)*, pages 1–6. IEEE. DOI: 10.1109/INFOCOMWKSHPS61880.2024.10620670

Dass, S. M., Mahalingam, S. R., Hang, P. L., Kelvin, K., Wingsky, J., Mohamed, A. M. M. A., Rebecca, A. J., Hatmal, N. I. Y. B. A., Ali, A. R., Tjahyosedjati, E., Wei, T., & Sindiramutty, S. R. (2025). The Role of Cryptography and AI in Enhancing Blockchain Security. *preprints.org*. DOI: 10.20944/preprints202501.0386.v1

De Donato, M. C., Corradini, F., Fornari, F., & Re, B. (2024). SAFE: An ICT platform for supporting monitoring, localization and rescue operations in case of earthquake. *Internet of Things : Engineering Cyber Physical Human Systems*, *27*, 101273. DOI: 10.1016/j.iot.2024.101273

Dhami, H. S. (2024). *Planning and Perception for Unmanned Aerial Vehicles in Object and Environmental Monitoring*. PhD thesis, University of Maryland, College Park.

Dias, T., Fonseca, T., Vitorino, J., Martins, A., Malpique, S., & Praça, I. (2023, January 1). From Data to Action: Exploring AI and IoT-driven Solutions for Smarter Cities. Cornell University. https://doi.org/DOI: 10.48550/arXiv.2306

Dildar, M. S., Khan, A. S., Abbasi, I. A., Shaheen, R., Ruqaishi, K. A., & Ahmed, S. (2025). End-to-end security mechanism using blockchain for Industrial Internet of Things. *IEEE Access : Practical Innovations, Open Solutions*, *1*, 20584–20598. Advance online publication. DOI: 10.1109/ACCESS.2025.3535821

Dilshad, N., Hwang, J., Song, J., & Sung, N. (2020, October 21). Applications and Challenges in Video Surveillance via Drone: A Brief Survey. DOI: 10.1109/ICTC49870.2020.9289536

Dimitrova, M., Aminzadeh, A., Meiabadi, M. S., Sattarpanah Karganroudi, S., Taheri, H., & Ibrahim, H. (2022). A survey on non-destructive smart inspection of wind turbine blades based on industry 4.0 strategy. *Applied Mechanics*, *3*(4), 1299–1326. DOI: 10.3390/applmech3040075

Dinelli, C., Racette, J., Escarcega, M., Lotero, S., Gordon, J., Montoya, J., Dunaway, C., Androulakis, V., Khaniani, H., Shao, S., Roghanchi, P., & Hassanalian, M. (2023). Configurations and Applications of Multi-Agent Hybrid Drone/Unmanned Ground Vehicle for Underground Environments: A review. *Drones (Basel)*, *7*(2), 136. DOI: 10.3390/drones7020136

Dogra, V., Singh, A., Verma, S., Kavita, N., Jhanjhi, N. Z., & Talib, M. N. (2021). Analyzing DistilBERT for Sentiment Classification of Banking Financial News. In *Lecture notes in networks and systems* (pp. 501–510). DOI: 10.1007/978-981-16-3153-5_53

Dong, C., Shen, Y., & Qu, Y. (2020). A survey of UAV-based edge intelligent computing. *Chinese Journal of Intelligent Science and Technology*, *2*, 227–239.

Donmez, C., Villi, O., Berberoglu, S., & Cilek, A. (2021). Computer vision-based citrus tree detection in a cultivated environment using UAV imagery. *Computers and Electronics in Agriculture*, *187*, 106273. DOI: 10.1016/j.compag.2021.106273

Dosovitskiy, A., Beyer, L., Kolesnikov, A., Weissenborn, D., Zhai, X., Unterthiner, T., Dehghani, M., Minderer, M., Heigold, G., & Gelly, S. (2020). An image is worth 16x16 words. arXiv preprint arXiv:2010.11929, 7.

Edulakanti, S. R., & Ganguly, S. (2023). Review article: The emerging drone technology and the advancement of the Indian drone business industry. *The Journal of High Technology Management Research*, *34*(2), 100464. DOI: 10.1016/j.hitech.2023.100464

Elbaşi, E., Mostafa, N., Al-Arnaout, Z., Zreikat, A. I., Cina, E., Varghese, G., Shdefat, A. Y., Topcu, A. E., Abdelbaki, W., Mathew, S., & Zaki, C. (2023, January 1). Artificial Intelligence Technology in the Agricultural Sector: A Systematic Literature Review. *IEEE Access : Practical Innovations, Open Solutions*, *11*, 171–202. DOI: 10.1109/ACCESS.2022.3232485

Elfwing, S., Uchibe, E., & Doya, K. (2018). Sigmoid-weighted linear units for neural network function approximation in reinforcement learning. *Neural Networks*, *107*, 3–11. DOI: 10.1016/j.neunet.2017.12.012 PMID: 29395652

Enayati, S., Campbell, J F., & Li, H. (2023, August 1). Vaccine distribution with drones for less developed countries: A case study in Vanuatu. Elsevier BV, 14, 100312-100312. DOI: 10.1016/j.jvacx.2023.100312

Erdelj, M., Natalizio, E., Chowdhury, K. R., & Akyildiz, I. F. (2017). Help from the Sky: Leveraging UAVs for Disaster Management. *IEEE Pervasive Computing*, *16*(1), 24–32. DOI: 10.1109/MPRV.2017.11

ESCAP, U. (2017). Innovations in disaster rapid assessment: a framework for early recovery in asean countries.

Fagundes-Junior, L. A., De Carvalho, K. B., Ferreira, R. S., & Brandão, A. S. (2024). Machine Learning for Unmanned Aerial Vehicles Navigation: An Overview. *SN Computer Science*, *5*(2), 256. Advance online publication. DOI: 10.1007/s42979-023-02592-5

Fahlstrom, P. G., & Gleason, T. J. (2013). Introduction to UAV systems. *Choice (Chicago, Ill.)*, *50*(11), 50–6201. DOI: 10.5860/CHOICE.50-6201

Faniadis, E., & Amanatiadis, A. (2020, November). Deep learning inference at the edge for mobile and aerial robotics. In *2020 IEEE International Symposium on Safety, Security, and Rescue Robotics (SSRR)* (pp. 334-340). IEEE.

Farsath, K. R., Jitha, K., Marwan, V. M., Jouhar, A. M. A., Farseen, K. M., & Musrifa, K. A. (2024, March). AI-enhanced unmanned aerial vehicles for search and rescue operations. In *2024 5th International Conference on Innovative Trends in Information Technology (ICITIIT)* (pp. 1-10). IEEE.

Fatima, N., Saxena, P., & Gupta, M. (2022). Integration of multi access edge computing with unmanned aerial vehicles: Current techniques, open issues and research directions. *Physical Communication*, *52*, 101641. DOI: 10.1016/j.phycom.2022.101641

Fatima-Tuz-Zahra, N., Jhanjhi, N., Brohi, S. N., Malik, N. A., & Humayun, M. (2020). Proposing a Hybrid RPL Protocol for Rank and Wormhole Attack Mitigation using Machine Learning. *2020 2nd International Conference on Computer and Information Sciences (ICCIS)*. DOI: 10.1109/ICCIS49240.2020.9257607

Fawakherji, M., Potena, C., Pretto, A., Bloisi, D. D., & Nardi, D. (2021). Multispectral image synthesis for crop/weed segmentation in precision farming. *Robotics and Autonomous Systems*, *146*, 103861. DOI: 10.1016/j.robot.2021.103861

Fellner, R. (2021). *Unmanned Aircraft Systems Risk Assessment Based on SORA for First Responders and Disaster Management. Janik, P., Zawistowski, M., Fellner, R., & Zawistowski, G.* Unmanned Aircraft Systems Risk Assessment Based on SORA for First Responders and Disaster Management.

Feroz, S., & Dabous, S. A. (2021, May 6). UAV-Based Remote Sensing Applications for Bridge Condition Assessment. *Remote Sensing (Basel)*, *13*(9), 1809–1809. DOI: 10.3390/rs13091809

Ficco, M., Granata, D., Palmieri, F., & Rak, M. (2024). A systematic approach for threat and vulnerability analysis of unmanned aerial vehicles. *Internet of Things : Engineering Cyber Physical Human Systems*, *26*, 101180. DOI: 10.1016/j.iot.2024.101180

Firdaus, M., & Rhee, K.-H. (2021). On blockchain-enhanced secure data storage and sharing in vehicular edge computing networks. *Applied Sciences (Basel, Switzerland), 11*(1), 414. DOI: 10.3390/app11010414

Fitwi, A., Chen, Y., & Zhu, S. A lightweight blockchain-based privacy protection for smart surveillance at the edge. *2019 IEEE International Conference on Blockchain (Blockchain)*, 2019. IEEE, 552-555. DOI: 10.1109/Blockchain.2019.00080

Floridi, L. (2025). AI as Agency without Intelligence: On Artificial Intelligence as a New Form of Artificial Agency and the Multiple Realisability of Agency Thesis. *Philosophy & Technology, 38*(1). DOI: 10.1007/s13347-025-00858-9

Fontaine, S., Abbadie, L., Aubert, M., Barot, S., Bloor, J., Derrien, D., Duchêne, O., Gross, N., Henneron, L., Roux, X L., Loeuille, N., Michel, J., Recous, S., Wipf, D., & Alvarez, G. (2023, December 12). Plant–soil synchrony in nutrient cycles: Learning from ecosystems to design sustainable agrosystems. Wiley, 30(1). DOI: 10.1111/gcb.17034

Foudeh, H. A., Luk, P. C., & Whidborne, J. F. (2021). An Advanced Unmanned Aerial Vehicle (UAV) Approach via Learning-Based Control for Overhead Power Line Monitoring: A Comprehensive Review. *IEEE Access : Practical Innovations, Open Solutions*, *9*, 130410–130433. DOI: 10.1109/ACCESS.2021.3110159

Fujimoto, H. (2010). *Emerging trends in mobile robotics:Proceedings of the 13th International Conference on Climbing and Walking Robots and the Support Technologies for Mobile Machines :31 August-3 September 2010,Nagoya Institute of Technology,Japan*. World Scientific.

Gadekallu, T. R., Maddikunta, P. K. R., Boopathy, P., Deepa, N., Chengoden, R., Victor, N., Wang, W., Wang, W., Zhu, Y., & Dev, K. (2024). XAI for Industry 5.0 -Concepts, Opportunities, Challenges and Future Directions. *IEEE Open Journal of the Communications Society*, *1*, 1. Advance online publication. DOI: 10.1109/OJCOMS.2024.3473891

Gao, L., Luan, T. H., Gu, B., Qu, Y., & Xiang, Y. (2021). *Privacy-preserving in edge computing*. Springer. DOI: 10.1007/978-981-16-2199-4

Garg, S., Ihler, A., Bentley, E. S., & Kumar, S. (2023). A Cross-Layer, Mobility, and Congestion-Aware Routing Protocol for UAV Networks. *IEEE Transactions on Aerospace and Electronic Systems, 59*(4), 3778–3796. DOI: 10.1109/TAES.2022.3232322

Gerstein, D M., & Leidy, E. (2024, January 1). Emerging Technology and Risk Analysis: Unmanned Aerial Systems Intelligent Swarm Technology. DOI: 10.7249/RRA2380-1

Gikunda, K. (2024, January 1). Harnessing Artificial Intelligence for Sustainable Agricultural Development in Africa: Opportunities, Challenges, and Impact. Cornell University. https://doi.org/DOI: 10.48550/arXiv.2401

Gill, S. S., Golec, M., Hu, J., Xu, M., Du, J., Wu, H., Walia, G. K., Murugesan, S. S., Ali, B., Kumar, M., Ye, K., Verma, P., Kumar, S., Cuadrado, F., & Uhlig, S. (2024). Edge AI: A Taxonomy, Systematic review and Future Directions. *Cluster Computing*, *28*(1), 18. Advance online publication. DOI: 10.1007/s10586-024-04686-y

Giordan, D., Adams, M., Aicardi, I., Alicandro, M., Allasia, P., Baldo, M., Berardinis, P D., Dominici, D., Godone, D., Hobbs, P., Lechner, V., Niedzielski, T., Piras, M., Rotilio, M., Salvini, R., Segor, V., Sotier, B., & Troilo, F. (2020, April 1). The use of unmanned aerial vehicles (UAVs) for engineering geology applications. Springer Science+Business Media, 79(7), 3437-3481

Giordan, D., Adams, M., Aicardi, I., Alicandro, M., Allasia, P., Baldo, M., Berardinis, P D., Dominici, D., Godone, D., Hobbs, P., Lechner, V., Niedzielski, T., Piras, M., Rotilio, M., Salvini, R., Segor, V., Sotier, B., & Troilo, F. (2020, April 1). The use of unmanned aerial vehicles (UAVs) for engineering geology applications. Springer Science+Business Media, 79(7), 3437-3481. DOI: 10.1007/s10064-020-01766-2

Gohari, A., Ahmad, A. B., Rahim, R. B. A., Supa'at, A. S. M., Abd Razak, S., & Gismalla, M. S. M. (2022). Involvement of surveillance drones in smart cities: A systematic review. *IEEE Access : Practical Innovations, Open Solutions*, *10*, 56611–56628. DOI: 10.1109/ACCESS.2022.3177904

Golbin, I., Rao, A S., Hadjarian, A., & Krittman, D. (2020, December 10). Responsible AI: A Primer for the Legal Community. DOI: 10.1109/BigData50022.2020.9377738

Gonçalves, R. S., Oliveira, M. D., Rocioli, M., Souza, F. D., Gallo, C. A., Sudbrack, D. E. T., Trautmann, P. V., Clasen, B., & Homma, R. Z. (2023, June 13). Drone–Robot to Clean Power Line Insulators. *Sensors (Basel)*, *23*(12), 5529–5529. DOI: 10.3390/s23125529 PMID: 37420696

Gopi, R., Sathiyamoorthi, V., Selvakumar, S., Manikandan, R., Chatterjee, P., Jhanjhi, N. Z., & Luhach, A. K. (2021). Enhanced method of ANN based model for detection of DDoS attacks on multimedia internet of things. *Multimedia Tools and Applications*, *81*(19), 26739–26757. DOI: 10.1007/s11042-021-10640-6

Gopi, S. P., Magarini, M., Alsamhi, S. H., & Shvetsov, A. V. (2021). Machine learning-assisted adaptive modulation for optimized drone-user communication in b5g. *Drones (Basel)*, *5*(4), 128. DOI: 10.3390/drones5040128

Goruganthu, T. S., Marepalli, H., Bheemineni, S., Bobba, R. C., Wang, D., Richhariya, S., & Liu, J. (2024). Building an AI-based surveillance drone cloud platform. In *Proc. of the IEEE International Workshop on Future Intelligent Technologies for Young Researchers (FITYR)* (pp. 1-10). DOI: 10.1109/FITYR63263.2024.00008

Gouda, W., Almurafeh, M., Humayun, M., & Jhanjhi, N. Z. (2022). Detection of COVID-19 based on chest x-rays using deep learning. *Health Care*, *10*(2), 343. DOI: 10.3390/healthcare10020343 PMID: 35206957

Goyal, R., Reiche, C., Fernando, C., Serrao, J., Kimmel, S., Cohen, A. & Shaheen, S. (2018). Urban air mobility (UAM) market study.

Goyal, R., Reiche, C., Fernando, C., & Cohen, A. (2021). Advanced air mobility: Demand analysis and market potential of the airport shuttle and air taxi markets. *Sustainability (Basel)*, *13*(13), 7421. DOI: 10.3390/su13137421

Green, D. R., Hagon, J. J., G'omez, C., & Gregory, B. J. (2019). Using low-cost uavs for environmental monitoring, mapping, and modelling: Examples from the coastal zone. In *Coastal management* (pp. 465–501). Elsevier. DOI: 10.1016/B978-0-12-810473-6.00022-4

Grenier, A., Lohan, E. S., Ometov, A., & Nurmi, J. (2023). Towards Smarter Positioning through Analyzing Raw GNSS and Multi-Sensor Data from Android Devices: A Dataset and an Open-Source Application. *Electronics (Basel)*, *12*(23), 4781. DOI: 10.3390/electronics12234781

Guglielmo, D. C. (2024, September 30). *Refining forest mapping drones for swift movement and reliable precision via deep learning*. https://aaltodoc.aalto.fi/items/9d5f9e0d-ef2f-4df2-9be0-e4be137349d8

Guimarães, N., Pádua, L., Marques, P., Silva, N., Peres, E., & Sousa, J. J. (2020, March 24). Forestry Remote Sensing from Unmanned Aerial Vehicles: A Review Focusing on the Data, Processing and Potentialities. *Remote Sensing (Basel)*, *12*(6), 1046–1046. DOI: 10.3390/rs12061046

Günal, M M. (2019, January 1). Data Collection Inside Industrial Facilities with Autonomous Drones. Springer International Publishing, 141-151. DOI: 10.1007/978-3-030-04137-3_9

Guo, F., Tang, B., Tang, M., Zhao, H., & Liang, W. (2021, June). Microservice selection in edge-cloud collaborative environment: A deep reinforcement learning approach. In 2021 8th IEEE International Conference on Cyber Security and Cloud Computing (CSCloud)/2021 7th IEEE International Conference on Edge Computing and Scalable Cloud (EdgeCom) (pp. 24-29). IEEE.

Guo, H., Li, W., Nejad, M., & Shen, C.-C. (2019). *Access control for electronic health records with hybrid blockchain-edge architecture. 2019 IEEE international conference on blockchain (Blockchain)*. IEEE.

Guo, Q., Liu, Z., Yang, Z., Jiang, Y., Sun, Y., Xu, J., Zhao, W., Wang, W., Wang, W., Ren, Q., & Shu, C. (2024). Development, challenges and future trends on the fabrication of micro-textured surfaces using milling technology. *Journal of Manufacturing Processes*, *126*, 285–331. DOI: 10.1016/j.jmapro.2024.07.112

Gupta, S., Gupta, S. P., Kumar, S. A., and Wasim, M. (2022). Robust multi- sensor facial recognition in real time using nvidia deepstream.

Gupta, A., Jain, S., Choudhary, P., & Parida, M. (2024). Dynamic object detection using sparse LiDAR data for autonomous machine driving and road safety applications. *Expert Systems with Applications*, *255*, 124636. DOI: 10.1016/j.eswa.2024.124636

Gyagenda, N., Hatilima, J. V., Roth, H., & Zhmud, V. (2022). A review of GNSS-independent UAV navigation techniques. *Robotics and Autonomous Systems*, *152*, 104069. DOI: 10.1016/j.robot.2022.104069

Hafid, A., Hocine, R., & Guezouli, L. (2024). Analyzing Swarm Robotics Approaches in Natural Disaster Scenarios: A Comparative Study. *2024 1st International Conference on Innovative and Intelligent Information Technologies (IC3IT)*, 1–6. DOI: 10.1109/IC3IT63743.2024.10869410

Halder, S., & Afsari, K. (2023, February 10). Robots in Inspection and Monitoring of Buildings and Infrastructure: A Systematic Review. *Applied Sciences (Basel, Switzerland)*, *13*(4), 2304–2304. DOI: 10.3390/app13042304

Hansen, J. G., & De Figueiredo, R. P. (2024). Active object detection and tracking using gimbal mechanisms for autonomous drone applications. *Drones (Basel)*, *8*(2), 55. DOI: 10.3390/drones8020055

Hao, C., Zhang, H., Song, W., Liu, F., & Wu, E. (2024). Slinet: Slicing-aided learning for small object detection. *IEEE Signal Processing Letters*, *31*, 790–794. DOI: 10.1109/LSP.2024.3373261

Hao, T., Hwang, K., Zhan, J., Li, Y., & Cao, Y. (2023). Scenario-Based AI Benchmark Evaluation of Distributed Cloud/Edge Computing Systems. *IEEE Transactions on Computers*, *72*(3), 719–731. DOI: 10.1109/TC.2022.3176803

Hasan, S. (2019). Urban air mobility (UAM) market study.

He, B., Ji, X., Li, G., & Cheng, B. (2024). Key Technologies and Applications of UAVs in Underground Space: A review. *IEEE Transactions on Cognitive Communications and Networking*, *10*(3), 1026–1049. DOI: 10.1109/TCCN.2024.3358545

Heidari, A., Navimipour, N. J., Unal, M., & Zhang, G. (2022). Machine Learning Applications in Internet-of-Drones: Systematic review, recent deployments, and open Issues. *ACM Computing Surveys*, *55*(12), 1–45. DOI: 10.1145/3571728

Hemmati, A., Raoufi, P., & Rahmani, A. M. (2024). Edge artificial intelligence for big data: A systematic review. *Neural Computing & Applications*, *36*(19), 11461–11494. DOI: 10.1007/s00521-024-09723-w

Hern'andez Vicente, D. (2023). *Analysis design and implementation of ar- tificial intelligence techniques in edge computing environments*. PhD thesis, Universitat Polit`ecnica de Val`encia.

Himeur, Y., Rimal, B., Tiwary, A., & Amira, A. (2022). Using artificial intelli- gence and data fusion for environmental monitoring: A review and future perspectives. *Information Fusion*, *86*, 44–75. DOI: 10.1016/j.inffus.2022.06.003

Hirschberg, M J. (2003, September 8). American Attack Unmanned Aerial Vehicles (UAVs): A Century of Progress. DOI: 10.4271/2003-01-3064

Hong, Z., Chen, W., Huang, H., Guo, S., & Zheng, Z. (2019). Multi-hop cooperative computation offloading for industrial IoT–edge–cloud computing environments. *IEEE Transactions on Parallel and Distributed Systems*, *30*(12), 2759–2774. DOI: 10.1109/TPDS.2019.2926979

Hosain, M. T., Zaman, A., Abir, M. R., Akter, S., Mursalin, S., & Khan, S. S. (2024). Synchronizing Object Detection: Applications, advancements and existing challenges. *IEEE Access : Practical Innovations, Open Solutions*, *12*, 54129–54167. DOI: 10.1109/ACCESS.2024.3388889

Hossain, M. A., Hossain, A. R., & Ansari, N. (2022). Numerology-capable UAV-MEC for future generation massive IoT networks. *IEEE Internet of Things Journal*, *9*(23), 23860–23868. DOI: 10.1109/JIOT.2022.3189945

Huang, Y., & Meng, Z. (2021). Bearing-based distributed formation control of multiple vertical take-off and landing UAVs. *IEEE Transactions on Control of Network Systems*, *8*(3), 1281–1292. DOI: 10.1109/TCNS.2021.3061941

Humayun, M., Almufareh, M. F., & Jhanjhi, N. Z. (2022). Autonomous traffic system for emergency vehicles. *Electronics (Basel)*, *11*(4), 510. DOI: 10.3390/electronics11040510

Humayun, M., Ashfaq, F., Jhanjhi, N. Z., & Alsadun, M. K. (2022). Traffic management: Multi-scale vehicle detection in varying weather conditions using yolov4 and spatial pyramid pooling network. *Electronics (Basel)*, *11*(17), 2748. DOI: 10.3390/electronics11172748

Humayun, M., Sujatha, R., Almuayqil, S. N., & Jhanjhi, N. Z. (2022). A Transfer Learning Approach with a Convolutional Neural Network for the Classification of Lung Carcinoma. *Health Care*, *10*(6), 1058. DOI: 10.3390/healthcare10061058 PMID: 35742109

Huo, M., Duan, H., & Fan, Y. (2021). Pigeon-inspired circular formation control for multi-UAV system with limited target information. *Guidance. Navigation and Control*, *1*(1), 2150004. DOI: 10.1142/S2737480721500047

Husain, B. H., & Askar, S. (2021). Survey on edge computing security. *International Journal of Science and Business*, *5*, 52–60.

Hussain, K., Rahmatyar, A. R., Riskhan, B., Sheikh, M. U., & Sindiramutty, S. R. (2024). Threats and Vulnerabilities of Wireless Networks in the Internet of Things (IoT). *IEEE 1st Karachi Section Humanitarian Technology Conference (KHI-HTC)*, *2*, 1–8. DOI: 10.1109/KHI-HTC60760.2024.10482197

Hussain, A., Li, S., Hussain, T., Lin, X., Ali, F., & AlZubi, A. A. (2024). Computing Challenges of UAV Networks: A Comprehensive survey. *Computers, Materials & Continua/Computers. Computers, Materials & Continua*, *81*(2), 1999–2051. DOI: 10.32604/cmc.2024.056183

Hüttenrauch, M., Šošić, A., & Neumann, G. (2019). Deep reinforcement learning for swarm systems. *Journal of Machine Learning Research*, *20*(54), 1–31.

Hymel, S., Banbury, C., Situnayake, D., Elium, A., Ward, C., & Kelcey, M. (2022). Edge impulse: An mlops platform for tiny machine learning. *arXiv preprint arXiv:2212.03332*.

Illiashenko, O., Kharchenko, V., Babeshko, I., Fesenko, H., & Di Giandomenico, F. (2023). Security-informed safety analysis of autonomous transport systems considering AI-powered cyberattacks and protection. *Entropy (Basel, Switzerland)*, *25*(8), 1123. DOI: 10.3390/e25081123 PMID: 37628153

Iyengar, A., & Portilla, I. (2020). Analytics at the Edge. In *IBM Blog*.

Jabeen, T., Jabeen, I., Ashraf, H., Ullah, A., Jhanjhi, N., Ghoniem, R. M., & Ray, S. K. (2023). Smart Wireless Sensor Technology for Healthcare Monitoring System Using Cognitive Radio Networks. *Sensors (Basel)*, *23*(13), 6104. DOI: 10.3390/s23136104 PMID: 37447952

Jagatheesaperumal, S. K., Rahouti, M., Chehri, A., Xiong, K., & Bieniek, J. (2025). Blockchain-Based Security Architecture for unmanned aerial vehicles in B5G/6G Services and Beyond: A Comprehensive approach. *IEEE Open Journal of the Communications Society, 1*, 1042–1069. Advance online publication. DOI: 10.1109/OJCOMS.2025.3528220

Jain, N., Gambhir, A., & Pandey, M. (2024). Unmanned Aerial Networks—UAVs and AI. In *Recent Trends in Artificial Intelligence Towards a Smart World* (pp. 321–351). DOI: 10.1007/978-981-97-6790-8_12

Jain, N., Gambhir, A., & Pandey, M. (2024). Unmanned aerial networks—UAVs and AI. In *Recent Trends in Artificial Intelligence Towards a Smart World: Applications in Industries and Sectors* (pp. 321-351). Springer Nature Singapore.

Janga, B., Asamani, G. P., Sun, Z., & Cristea, N. (2023, August 21). A Review of Practical AI for Remote Sensing in Earth Sciences. *Remote Sensing (Basel), 15*(16), 4112–4112. DOI: 10.3390/rs15164112

Jang, S., & Kim, H.-I. (2024). Efficient deep reinforcement learning under task variations via knowledge transfer for drone control. *ICT Express, 10*(3), 576–582. DOI: 10.1016/j.icte.2024.04.002

Javaid, M., Haleem, A., Khan, I H., & Suman, R. (2023, March 1). Understanding the potential applications of Artificial Intelligence in Agriculture Sector. Elsevier BV, 2(1), 15-30. DOI: 10.1016/j.aac.2022.10.001

Javaid, M., Haleem, A., Singh, R. P., & Ahmed, M. (2024). Computer Vision to Enhance Healthcare Domain: An Overview of features, implementation, and opportunities. *Intelligent Pharmacy, 2*(6), 792–803. DOI: 10.1016/j.ipha.2024.05.007

Javan, F. D., Samadzadegan, F., Toosi, A., & Van Der Meijde, M. (2024). Unmanned Aerial Geophysical Remote Sensing: A Survey. *Preprints.org*. DOI: 10.20944/preprints202411.0058.v1

Javed, S., Hassan, A., Ahmad, R., Ahmed, W., Ahmed, R., Saadat, A., & Guizani, M. (2024). State-of-the-Art and future research challenges in UAV swarms. *IEEE Internet of Things Journal, 11*(11), 19023–19045. DOI: 10.1109/JIOT.2024.3364230

Jhanjhi, N., Humayun, M., & Almuayqil, S. N. (2021). Cyber security and privacy issues in industrial internet of things. *Computer Systems Science and Engineering, 37*(3), 361–380. DOI: 10.32604/csse.2021.015206

Jhanjhi, N., Shah, I. A., & Brohi, S. N. (2025). *Hardware Vulnerabilities: Taxonomy and Business Security Models. Generative AI for Web Engineering Models*. IGI Global.

Jhanjhi, N., Shah, I. A., & Nawaz, S. (2025). Generative AI for Visualization. In *Generative AI for Web Engineering Models* (pp. 63–82). IGI Global.

Jiang, S., Jiang, W., & Wang, L. (2021). Unmanned Aerial Vehicle-Based Photogrammetric 3D Mapping: A survey of techniques, applications, and challenges. *IEEE Geoscience and Remote Sensing Magazine*, *10*(2), 135–171. DOI: 10.1109/MGRS.2021.3122248

Jiang, S., Xu, Q., Wang, W., Peng, P., & Li, J. (2025). Vehicle positioning systems in tunnel environments: A review. *Complex & Intelligent Systems*, *11*(2), 164. Advance online publication. DOI: 10.1007/s40747-024-01744-1

Jiao, L., Friedman, R., Fu, X., Secci, S., Smoreda, Z., & Tschofenig, H. (2013). Cloud-based computation offloading for mobile devices: State of the art, challenges and opportunities. 2013 Future Network & Mobile Summit, 1-11.

Ji, Y., Li, S., Peng, C., Xu, H., Cao, R., & Zhang, M. (2021). Obstacle detection and recognition in farmland based on fusion point cloud data. *Computers and Electronics in Agriculture*, *189*, 106409. DOI: 10.1016/j.compag.2021.106409

Joeaneke, P., Val, O. O., Olaniyi, O. O., Ogungbemi, O. S., Olisa, A. O., & Akinola, O. I. (2024). Protecting Autonomous UAVs from GPS Spoofing and Jamming: A Comparative Analysis of Detection and Mitigation Techniques. SSRN *Electronic Journal*. DOI: 10.2139/ssrn.4995938

Joshi, A., Spilbergs, A., & Miķelsone, E. (2024, June). AI-enabled drone autonomous navigation and decision-making for defence security. In *ENVIRONMENT. TECHNOLOGIES. RESOURCES.Proceedings of the International Scientific and Practical Conference* (Vol. 4, pp. 138-143). DOI: 10.17770/etr2024vol4.8237

Jouhari, M., Al-Ali, A., Baccour, E., Mohamed, A., Erbad, A., Guizani, M., & Hamdi, M. (2022, January 15). Distributed CNN Inference on Resource-Constrained UAVs for Surveillance Systems: Design and Optimization. *IEEE Internet of Things Journal*, *9*(2), 1227–1242. DOI: 10.1109/JIOT.2021.3079164

Jun, A. Y. M., Jinu, B. A., Seng, L. K., Maharaiq, M. H. F. B. Z., Khongsuwan, W., Junn, B. T. K., Hao, A. W., & Sindiramutty, S. R. (2024). Exploring the Impact of Crypto-Ransomware on Critical Industries: Case Studies and Solutions. *Preprints.org*. DOI: 10.20944/preprints202409.1325.v1

Kadam, P., Fang, G., & Zou, J. J. (2024). Object Tracking Using Computer Vision: A review. *Computers*, *13*(6), 136. DOI: 10.3390/computers13060136

Kamarulzaman, A. M. M., Jaafar, W. S. W. M., Said, M. N. M., Saad, S. N. M., & Mohan, M. (2023, May 30). UAV Implementations in Urban Planning and Related Sectors of Rapidly Developing Nations: A Review and Future Perspectives for Malaysia. *Remote Sensing (Basel)*, *15*(11), 2845–2845. DOI: 10.3390/rs15112845

Kangunde, V., Jamisola, R. S.Jr, & Theophilus, E. K. (2021). A review on drones controlled in real-time. *International Journal of Dynamics and Control*, *9*(4), 1832–1846. DOI: 10.1007/s40435-020-00737-5 PMID: 33425650

Kanungo, S. (2019). Edge-to-cloud intelligence: Enhancing iot devices with machine learning and cloud computing. *International Peer-Reviewed Journal*, *2*(12), 238–245.

Kashi, R. N., Prashanth, A., Kashi, S. R., & Prabhakara, G. (2023). A survey and analysis of drone detection systems using a systems approach superposed on scenarios. *Systems Engineering*, *27*(3), 598–636. DOI: 10.1002/sys.21735

Kasioulis, M., Symeonides, M., Ioannou, G., Pallis, G., & Dikaiakos, M. D. (2024). Energy modeling of inference workloads with ai accelerators at the edge: A benchmarking study. In *2024 IEEE International Conference on Cloud Engineering (IC2E)*, pages 189–196. IEEE. DOI: 10.1109/IC2E61754.2024.00028

Katna, S., Singh, S. K., Kumar, S., Manro, D., Chhabra, A., & Sharma, S. K. (2024). Communication systems for drone swarms and remote operations. In *Advances in computational intelligence and robotics book series* (pp. 115–136). DOI: 10.4018/979-8-3693-2707-4.ch006

Kaufmann, E., Bauersfeld, L., Loquercio, A., Müller, M., Koltun, V., & Scaramuzza, D. (2023, August 30). Champion-level drone racing using deep reinforcement learning. *Nature*, *620*(7976), 982–987. DOI: 10.1038/s41586-023-06419-4 PMID: 37648758

Kestner, P. (2024). Techniques and tactics of cyber warfare. In *Springer eBooks* (pp. 215–242). DOI: 10.1007/978-3-658-43879-1_7

Khang, A., Abdullayev, V., Litvinova, E., Chumachenko, S., Alyar, A. V., & Anh, P. T. N. (2024). Application of computer Vision (CV) in the healthcare ecosystem. In *CRC Press eBooks* (pp. 1–16). DOI: 10.1201/9781003429609-1

Khan, L. U., Saad, W., Han, Z., Hossain, E., & Hong, C. S. (2021). Federated learning for internet of things: Recent advances, taxonomy, and open challenges. *IEEE Communications Surveys and Tutorials*, *23*(3), 1759–1799. DOI: 10.1109/COMST.2021.3090430

Khan, M. A., Menouar, H., Eldeeb, A., Abu-Dayya, A., & Salim, F. D. (2022). On the Detection of Unauthorized Drones—Techniques and Future Perspectives: A Review. *IEEE Sensors Journal*, *22*(12), 11439–11455. DOI: 10.1109/JSEN.2022.3171293

Khan, N. A., Jhanjhi, N. Z., Brohi, S. N., Usmani, R. S. A., & Nayyar, A. (2020). Smart traffic monitoring system using unmanned aerial vehicles (UAVs). *Computer Communications*, *157*, 434–443. DOI: 10.1016/j.comcom.2020.04.049

Khan, S. K., Farasat, M., Naseem, U., & Ali, F. (2020). Performance evaluation of next-generation wireless (5G) UAV relay. *Wireless Personal Communications*, *113*(2), 945–960. DOI: 10.1007/s11277-020-07261-x

Khan, W. Z., Ahmed, E., Hakak, S., Yaqoob, I., & Ahmed, A. (2019). Edge computing: A survey. *Future Generation Computer Systems*, *97*, 219–235. DOI: 10.1016/j.future.2019.02.050

Khuwaja, A. A., Chen, Y., Zhao, N., Alouini, M.-S., & Dobbins, P. (2018). A survey of channel modeling for UAV communications. *IEEE Communications Surveys and Tutorials*, *20*(4), 2804–2821. DOI: 10.1109/COMST.2018.2856587

Kim, J., Guivant, J., Sollie, M. L., Bryne, T. H., & Johansen, T. A. (2021). Compressed pseudo-SLAM: Pseudorange-integrated compressed simultaneous localisation and mapping for unmanned aerial vehicle navigation. *Journal of Navigation*, *74*(5), 1091–1103. DOI: 10.1017/S037346332100031X

Kiran, S., Rajper, S., Shaikh, R. A., Shah, I. A., & Danwar, S. H. (2021). Categorization of CVE Based on Vulnerability Software By Using Machine Learning Techniques. *International Journal (Toronto, Ont.)*, *10*(3).

Kiruthiga, R., Nithya, B., et al. (2024). Advancing uav multi-object tracking: Integrating yolov8, nano instance segmentation, and dueling double deep q-network.

Kiyani, F. F., Hamid, B., Humayun, M., Sindiramutty, S. R. L., & Chowdhury, S. (2024). Discovery of Influential Publications Using Research Article's Usage Context. *2024 International Conference on Emerging Trends in Networks and Computer Communications (ETNCC)*, 1–7. DOI: 10.1109/ETNCC63262.2024.10767576

Kommineni, M., & Baseer, K. K. (2024). An Architecture and Review of Intelligence Based Traffic Control System for Smart Cities. *EAI Endorsed Transactions on Energy Web*, *11*. Advance online publication. DOI: 10.4108/ew.4964

Kong, X., Wang, K., Hou, M., Hao, X., Shen, G., Chen, X., & Xia, F. (2021a). A federated learning-based license plate recognition scheme for 5G-enabled internet of vehicles. *IEEE Transactions on Industrial Informatics*, *17*(12), 8523–8530. DOI: 10.1109/TII.2021.3067324

Kong, X., Zhu, B., Shen, G., Workneh, T. C., Ji, Z., Chen, Y., & Liu, Z. (2021b). Spatial-temporal-cost combination based taxi driving fraud detection for collaborative internet of vehicles. *IEEE Transactions on Industrial Informatics, 18*(5), 3426–3436. DOI: 10.1109/TII.2021.3111536

Kouris, A., Venieris, S. I., & Bouganis, C.-S. (2019). Towards efficient on-board deployment of dnns on intelligent autonomous systems. In *2019 IEEE Computer Society Annual Symposium on VLSI (ISVLSI)*, pages 568–573. IEEE. DOI: 10.1109/ISVLSI.2019.00107

Krishna, K. R. (2017). *Push button agriculture: robotics, drones, Satellite-Guided soil and crop management.*

Krishnamurthy, A. (2024). Beyond Rationality. *Wiley*, 409–427. DOI: 10.1002/9781394214334.ch17

Krishnan, S., & Murugappan, M. (2023). *Internet of drones: Applications, Opportunities, and Challenges.* CRC Press. DOI: 10.1201/9781003252085

Kumar, A. (2024). Geo-Tagged 3D geometric modeling of urban structures by mitigating reflected GPS signals using a laser range sensor. In *Lecture notes in networks and systems* (pp. 396–414). DOI: 10.1007/978-3-031-62281-6_29

Kumar, A., Tallapaneni, K., Yadav, A., & Gautam, H. (2024). Enhancing home security with AI-driven drone surveillance. *J Sen Net Data Comm, 4*(2), 01-10.

Kumar, A., & Singh, R. (2024). Intelligent traffic management systems: Integration of AI and traditional infrastructure. *Transportation Research Part C, Emerging Technologies, 148*, 103931.

Kumar, G., & Altalbe, A. (2024). Artificial intelligence (AI) advancements for transportation security: In-depth insights into electric and aerial vehicle systems. *Environment, Development and Sustainability*, •••, 1–51. DOI: 10.1007/s10668-023-04127-7

Kumar, M. S., Vimal, S., Jhanjhi, N., Dhanabalan, S. S., & Alhumyani, H. A. (2021). Blockchain based peer to peer communication in autonomous drone operation. *Energy Reports, 7*, 7925–7939. DOI: 10.1016/j.egyr.2021.08.073

Kumar, S., Poyyamozhi, M., Murugesan, B., Rajamanickam, N., Alroobaea, R., & Nureldeen, W. (2024). Investigation of unsafe construction site conditions using deep learning algorithms using unmanned aerial vehicles. *Sensors (Basel), 24*(20), 6737. DOI: 10.3390/s24206737 PMID: 39460217

Kuru, K. (2021). Planning the future of smart cities with swarms of fully autonomous unmanned aerial vehicles using a novel framework. *IEEE Access : Practical Innovations, Open Solutions*, *9*, 6571–6595. DOI: 10.1109/ACCESS.2020.3049094

Kurunathan, H., Huang, H., Li, K., Ni, W., & Hossain, E. (2023). Machine learning-aided operations and communications of unmanned aerial vehicles: A contemporary survey. *IEEE Communications Surveys and Tutorials*.

Kyrkou, C., Timotheou, S., Kolios, P., Theocharides, T., & Panayiotou, C. G. (2019, January 1). Drones: Augmenting Our Quality of Life. *IEEE Potentials*, *38*(1), 30–36. DOI: 10.1109/MPOT.2018.2850386

Laghari, A. A., Jumani, A. K., Laghari, R. A., Li, H., Karim, S., & Khan, A. A. (2024). *Unmanned aerial vehicles advances in object detection and communication security review*. Cognitive Robotics. DOI: 10.1016/j.cogr.2024.07.002

Lawrence, I. D., Pavitra, A. R. R., Karu, R., & Saravanan, M. (2024). An Advanced Application of UAV – Drone Technologies in Precision Agriculture for Seed Dropping, Fertilizers and Pesticides Spraying and Field Monitoring. *Wiley*, 139–157. DOI: 10.1002/9781394186686.ch7

Lee, J., Nazki, H., Baek, J., Hong, Y., & Lee, M. (2020). Artificial intelligence approach for tomato detection and mass estimation in precision agriculture. *Sustainability (Basel)*, *12*(21), 9138. DOI: 10.3390/su12219138

Le, N., Thai, H., Yoon, C., & Huh, J. (2024). Recent development of Drone Technology Software Engineering: A Systematic survey. *IEEE Access : Practical Innovations, Open Solutions*, *12*, 128729–128751. DOI: 10.1109/ACCESS.2024.3454546

Lenka, A. K., & Tripathy, H. K. (2024). 5 Computer vision for medical diagnosis and surgery. In *De Gruyter eBooks* (pp. 101–124). DOI: 10.1515/9783110750942-005

Lenssen, E., Nagtegaal, I., Van Oostveen, C., Sieben, A., Van Rijssen, L., & Weggelaar, A. M. (2025). Exploring nurses' leadership and resilience in a complex daily work environment: A qualitative study. *BMC Nursing*, *24*(1), 173. Advance online publication. DOI: 10.1186/s12912-025-02761-2 PMID: 39953459

Leong, P. Y., & Ahmad, N. S. (2024). LiDar-based Obstacle Avoidance with Autonomous Vehicles: A Comprehensive Review. *IEEE Access : Practical Innovations, Open Solutions*, *1*, 164248–164261. Advance online publication. DOI: 10.1109/ACCESS.2024.3493238

Lerro, A., Battipede, M., Gili, P., & Brandl, A. (2020). Aerodynamic angle estimation: Comparison between numerical results and operative environment data. *CEAS Aeronautical Journal*, *11*(1), 249–262. DOI: 10.1007/s13272-019-00417-x

Leslie, P., Chau, E., Watson, T., Furxhi, O., Jacobs, E., Driggers, R., & Conroy, J. (2024). Visible, swir, and lwir drone wide area coverage sensor systems. *Optical Engineering (Redondo Beach, Calif.)*, *63*(2), 025101–025101. DOI: 10.1117/1.OE.63.2.025101

Li, A., Sun, J., Wang, B., Duan, L., Li, S., Chen, Y., & Li, H. (2021, December). Lotteryfl: Empower edge intelligence with personalized and communication-efficient federated learning. In *2021 IEEE/ACM Symposium on Edge Computing (SEC)* (pp. 68-79). IEEE.

Lim, M., Abdullah, A., Jhanjhi, N., Khan, M. K., & Supramaniam, M. (2019). Link Prediction in Time-Evolving Criminal Network with deep Reinforcement learning technique. *IEEE Access : Practical Innovations, Open Solutions*, *7*, 184797–184807. DOI: 10.1109/ACCESS.2019.2958873

Lim, W. Y. B., Luong, N. C., Hoang, D. T., Jiao, Y., Liang, Y.-C., Yang, Q., Niyato, D., & Miao, C. (2020). Federated learning in mobile edge networks: A comprehensive survey. *IEEE Communications Surveys and Tutorials*, *22*(3), 2031–2063. DOI: 10.1109/COMST.2020.2986024

Lin, S., Yang, G., & Zhang, J. (2020, November). A collaborative learning framework via federated meta-learning. In 2020 IEEE 40th International Conference on Distributed Computing Systems (ICDCS) (pp. 289-299). IEEE.

Lin, J., Yu, W., Zhang, N., Yang, X., Zhang, H., & Zhao, W. (2017). A survey on internet of things: Architecture, enabling technologies, security and privacy, and applications. *IEEE Internet of Things Journal*, *4*(5), 1125–1142. DOI: 10.1109/JIOT.2017.2683200

Linqiang, Y., Sindiramutty, S. R. L., Ashraf, H., Muzammal, S. M., Balakrishnan, S. P., Gupta, S., & Kavita, N. (2024). Intelligent Household Waste Classification System Based on Machine Learning. *2024 International Conference on Emerging Trends in Networks and Computer Communications (ETNCC)*, 760–768. DOI: 10.1109/ETNCC63262.2024.10767563

Liu, S., Zha, J., Sun, J., Li, Z., & Wang, G. (2023). EdgeYOLO: An edge-real-time object detector. 2023 42nd Chinese Control Conference (CCC).

Liu, X., Li, X., Shi, Q., Xu, C., & Tang, Y. (2021). UAV attitude estimation based on MARG and optical flow sensors using gated recurrent unit. *International Journal of Distributed Sensor Networks*, *17*(4), 15501477211009814. DOI: 10.1177/15501477211009814

Liu, X., Wang, Y., & Thompson, R. (2023). Advances in computer vision for urban traffic management. *IEEE Access : Practical Innovations, Open Solutions*, *11*, 45672–45686.

Liu, Y., Huo, L., Wu, J., & Bashir, A. K. (2023). SWarm Learning-Based Dynamic optimal management for traffic congestion in 6G-Driven intelligent Transportation System. *IEEE Transactions on Intelligent Transportation Systems*, *24*(7), 7831–7846. DOI: 10.1109/TITS.2023.3234444

Liu, Z., An, P., Yang, Y., Qiu, S., Liu, Q., & Xu, X. (2024). Vision-Based Drone Detection in Complex Environments: A survey. *Drones (Basel)*, *8*(11), 643. DOI: 10.3390/drones8110643

Li, Y., Miao, N., Ma, L., Shuang, F., & Huang, X. (2023). Transformer for object detection: Review and benchmark. *Engineering Applications of Artificial Intelligence*, *126*, 107021. DOI: 10.1016/j.engappai.2023.107021

Li, Y., Shen, H., Fu, Y., & Wang, K. (2024). A method of dense point cloud SLAM based on improved YOLOV8 and fused with ORB-SLAM3 to cope with dynamic environments. *Expert Systems with Applications*, *255*, 124918. DOI: 10.1016/j.eswa.2024.124918

Lochan, K., Khan, A., Elsayed, I., Suthar, B., Seneviratne, L., & Hussain, I. (2024). Advancements in precision Spraying of Agricultural Robots: A Comprehensive review. *IEEE Access : Practical Innovations, Open Solutions*, *12*, 129447–129483. DOI: 10.1109/ACCESS.2024.3450904

Lopatin, E., & Poikonen, P. (2023). Cost-Effective Aerial Inventory of Spruce Seedlings Using Consumer Drones and Deep Learning Techniques with Two-Stage UAV Flight Patterns. *Forests*, *14*(5), 973. Advance online publication. DOI: 10.3390/f14050973

Lopes, D., & Silva, J. (2023, March 22). Urban air mobility (UAM) in the metropolitan region of São Paulo: Potential and threats. *OmniaScience*, *13*(1), 1–1. DOI: 10.3926/jairm.345

Luber, M., Arras, K. O., Plagemann, C., & Burgard, W. (2009). Classifying dynamic objects: An unsupervised learning approach. *Autonomous Robots*, *26*(2-3), 141–151. DOI: 10.1007/s10514-009-9112-4

Luong, N. C., Hoang, D. T., Gong, S., Niyato, D., Wang, P., Liang, Y.-C., & Kim, D. I. (2019). Applications of deep reinforcement learning in communications and networking: A survey. *IEEE Communications Surveys and Tutorials*, *21*(4), 3133–3174. DOI: 10.1109/COMST.2019.2916583

Luo, Q., Hu, S., Li, C., Li, G., & Shi, W. (2021). Resource scheduling in edge computing: A survey. *IEEE Communications Surveys and Tutorials*, *23*(4), 2131–2165. DOI: 10.1109/COMST.2021.3106401

Lyu, M., Zhao, Y., Huang, C., & Huang, H. (2023). Unmanned aerial vehicles for search and rescue: A survey. *Remote Sensing (Basel)*, *15*(13), 3266. DOI: 10.3390/rs15133266

Maatouk, N., Letaifa, A. B., & Rachedi, A. (2024). AI empowered data offloading in MEC enabled UAV networks. *Multimedia Tools and Applications*. Advance online publication. DOI: 10.1007/s11042-024-19859-5

Maguluri, L. P., Arularasan, A. N., & Boopathi, S. (2023). Assessing security concerns for AI-based drones in smart cities. In *Effective AI, Blockchain, and E-Governance Applications for Knowledge Discovery and Management* (pp. 27-47). IGI Global. DOI: 10.4018/978-1-6684-9151-5.ch002

MahmoudZadeh, S., Yazdani, A., Kalantari, Y., Ciftler, B., Aidarus, F., & Al Kadri, M. O. (2024). Holistic Review of UAV-Centric Situational Awareness: Applications, limitations, and Algorithmic challenges. *Robotics (Basel, Switzerland)*, *13*(8), 117. DOI: 10.3390/robotics13080117

Majumdar, P., Mitra, S., Bhattacharya, D., & Bhushan, B. (2024). Enhancing sustainable 5G powered agriculture 4.0 : Summary of low power connectivity, internet of UAV things, AI solutions and research trends. *Multimedia Tools and Applications*. Advance online publication. DOI: 10.1007/s11042-024-19728-1

Mallikarjunaradhya, V., Pothukuchi, A. S., & Kota, L. V. (2023). An overview of the strategic advantages of AI-powered threat intelligence in the cloud. *Journal of Science and Technology*, *4*(4), 1–12. DOI: 10.55662/JST.2023.4401

Manchuri, A., Kakera, A., Saleh, A., & Raja, S. (2024). pplication of Supervised Machine Learning Models in Biodiesel Production Research - A Short Review. *Borneo Journal of Sciences and Technology*. DOI: 10.35370/bjost.2024.6.1-10

Mandloi, D., Arya, R., & Verma, A. K. (2024). Internet of drones. In *Recent Trends in Artificial Intelligence Towards a Smart World: Applications in Industries and Sectors* (pp. 353-373). Springer Nature Singapore. DOI: 10.1007/978-981-97-6790-8_13

Mantau, A. B. J. Human observing for wide area surveillance using autonomous unmanned aerial vehicles (uavs).

Markovic, D. S., Zivkovic, D., Branovic, I., Popovic, R., & Cvetkovic, D. (2013). Smart power grid and cloud computing. *Renewable & Sustainable Energy Reviews*, *24*, 566–577. DOI: 10.1016/j.rser.2013.03.068

McEnroe, P., Wang, S., & Liyanage, M. (2022). A survey on the convergence of edge computing and AI for UAVs: Opportunities and challenges. *IEEE Internet of Things Journal*, *9*(17), 15435–15459. DOI: 10.1109/JIOT.2022.3176400

McEwen, A., Byrne, S., Hansen, C., Daubar, I., Sutton, S., Dundas, C., Bardabelias, N., Baugh, N., Bergstrom, J., Beyer, R., Block, K., Bray, V., Bridges, J., Chojnacki, M., Conway, S., Delamere, W., Ebben, T., Espinosa, A., Fennema, A., & Wilson, S. (2023). The high-resolution imaging science experiment (HiRISE) in the MRO extended science phases (2009–2023). *Icarus*, *419*, 115795. DOI: 10.1016/j.icarus.2023.115795

McMillen, P., & Levin, M. (2024). Collective intelligence: A unifying concept for integrating biology across scales and substrates. *Communications Biology*, *7*(1), 378. Advance online publication. DOI: 10.1038/s42003-024-06037-4 PMID: 38548821

McNabb, M. (2024, January 16). Drones in Disaster Response: Harnessing the potential of unmanned aerial systems. *DRONELIFE*. https://dronelife.com/2024/01/16/drones-in-disaster-response-harnessing-the-potential-of-unmanned-aerial-systems/

Mengozzi, S., Zanatta, L., Barchi, F., Bartolini, A., & Acquaviva, A. (2024). Towards nano-drones agile flight using deep reinforcement learning. In *2024 IEEE International Conference on Omni-layer Intelligent Systems (COINS)*, pages 1–6. IEEE. DOI: 10.1109/COINS61597.2024.10622558

Menichino, A., Di Vito, V., Dziugiel, B., Liberacki, A., Hesselink, H., & Giannuzzi, M. (2022, April). Urban air mobility perspectives over mid-term time horizon: main enabling technologies readiness review. In 2022 Integrated Communication, Navigation and Surveillance Conference (ICNS) (pp. 1-13). IEEE.

Merenda, M., Porcaro, C., & Iero, D. (2020). Edge machine learning for ai-enabled iot devices: A review. *Sensors (Basel)*, *20*(9), 2533. DOI: 10.3390/s20092533 PMID: 32365645

Merkert, R., & Bushell, J. (2020, October 1). Managing the drone revolution: A systematic literature review into the current use of airborne drones and future strategic directions for their effective control. Elsevier BV, 89, 101929-101929. DOI: 10.1016/j.jairtraman.2020.101929

Merkle, N., Bahmanyar, R., Henry, C., Azimi, S. M., Yuan, X., Schopferer, S., Gstaiger, V., Auer, S., Schneibel, A., Wieland, M., & Kraft, T. (2023, January 1). Drones4Good: Supporting Disaster Relief Through Remote Sensing and AI. Cornell University. https://doi.org/DOI: 10.48550/arXiv.2308

Min, W., Muthanna, M. S. A., Ibrahim, M., Alkanhel, R., Muthanna, A., & Laouid, A. (2025). Privacy-preserving federated UAV data collection framework for autonomous path optimization in maritime operations. *Applied Soft Computing*, *112906*, 112906. Advance online publication. DOI: 10.1016/j.asoc.2025.112906

Mittal, P., Singh, R., & Sharma, A. (2020). Deep learning-based object detection in low-altitude uav datasets: A survey. *Image and Vision Computing*, *104*, 104046. DOI 10.1016/j.imavis.2020.104046

Mocnej, J., Miškuf, M., Papcun, P., & Zolotová, I. (2018). Impact of edge computing paradigm on energy consumption in IoT. *IFAC-PapersOnLine*, *51*(6), 162–167. DOI: 10.1016/j.ifacol.2018.07.147

Mohamed, N., Ahmed, A. A., Alsharif, A., & ElKhozondar, H. J. (2023, November). Employing AI-driven drones and advanced cyber penetration tools for breakthrough criminal network surveillance. In *2023 IEEE 9th International Women in Engineering (WIE)Conference on Electrical and Computer Engineering (WIECON-ECE)* (pp. 1-6). IEEE. DOI: 10.1109/WIECON-ECE60392.2023.10456514

Mohamed, N., Rao, L. S., Sharma, M., & Shukla, S. K. (2023, May). In-depth review of integration of AI in cloud computing. In 2023 3rd international conference on advance computing and innovative technologies in engineering (ICACITE) (pp. 1431-1434). IEEE.

Mohamed, B. (2024). AI-empowered flying ad-hoc networks for dynamic connectivity. *International Journal of Computing and Digital Systems*, *15*(1), 167–177. DOI: 10.12785/ijcds/150114

Mohammed, S., Fang, W. C., & Ramos, C. (2023). Special issue on "'artificial intelligence in cloud computing.'". *Computing*, *105*(3), 507–511. DOI: 10.1007/s00607-021-00985-z

Mohsan, S. H., Othman, N. Q. H., Li, Y., Alsharif, M. H., & Khan, M. A. (2023). Unmanned aerial vehicles (UAVs): Practical aspects, applications, open challenges, security issues, and future trends. *Intelligent Service Robotics*. Advance online publication. DOI: 10.1007/s11370-022-00452-4 PMID: 36687780

Mohyuddin, G., Khan, M. A., Haseeb, A., Mahpara, S., Waseem, M., & Saleh, A. M. (2024). Evaluation of Machine Learning Approaches for Precision Farming in Smart Agriculture System: A Comprehensive Review. *IEEE Access : Practical Innovations, Open Solutions*, *12*, 60155–60184. DOI: 10.1109/ACCESS.2024.3390581

Molaei, A., Kolu, A., Lahtinen, K., & Geimer, M. (2024). Automatic recognition of excavator working cycles using supervised learning and motion data obtained from inertial measurement units (IMUs). *Construction Robotics*, *8*(2), 14. Advance online publication. DOI: 10.1007/s41693-024-00130-0

Munasinghe, I., Perera, A., & Deo, R. C. (2024). A Comprehensive review of UAV-UGV Collaboration: Advancements and challenges. *Journal of Sensor and Actuator Networks*, *13*(6), 81. DOI: 10.3390/jsan13060081

Muzammul, M., Algarni, A., Ghadi, Y. Y., & Assam, M. (2024). Enhancing UAV aerial image analysis: Integrating advanced SAHI techniques with real-time detection models on the VisDrone dataset. *IEEE Access : Practical Innovations, Open Solutions*, *12*, 21621–21633.

Nagendra, H., Lucas, R., Honrado, J. P., Jongman, R. H., Tarantino, C., Adamo, M., & Mairota, P. (2013). Remote sensing for conservation monitoring: Assessing pro- tected areas, habitat extent, habitat condition, species diversity, and threats. *Ecological Indicators*, *33*, 45–59. DOI: 10.1016/j.ecolind.2012.09.014

Nalini, M. (2024). AI-powered drones for healthcare applications. In *Artificial Intelligence for Autonomous Vehicles* (pp. 131-149). DOI: 10.1002/9781119847656.ch6

Naranjo, M., Fuentes, D., Muelas, E., Díez, E., Ciruelo, L., Alonso, C., Abenza, E., Gómez-Espinosa, R., & Luengo, I. (2023). Object Detection-Based System for Traffic Signs on Drone-Captured Images. *Drones (Basel)*, *7*(2), 112. Advance online publication. DOI: 10.3390/drones7020112

Natarajan, G., Elango, E., Sundaravadivazhagan, B., & Rethinam, S. (2024). Artificial Intelligence Algorithms and Models for Embodied Agents: Enhancing Autonomy in Drones and Robots. In *Building Embodied AI Systems: The Agents, the Architecture Principles, Challenges, and Application Domains* (pp. 103–132). DOI: 10.1007/978-3-031-68256-8_5

Nayyar, A., Gadhavi, L., & Zaman, N. (2021). Machine learning in healthcare: review, opportunities and challenges. In *Elsevier eBooks* (pp. 23–45). DOI: 10.1016/B978-0-12-821229-5.00011-2

Nguyen Hoang, A. M. (2019). Computer vision deployment on edge devices.

Nguyen, D. C., Ding, M., Pham, Q.-V., Pathirana, P. N., Le, L. B., Seneviratne, A., Li, J., Niyato, D., & Poor, H. V. (2021). Federated learning meets blockchain in edge computing: Opportunities and challenges. *IEEE Internet of Things Journal*, *8*(16), 12806–12825. DOI: 10.1109/JIOT.2021.3072611

Nguyen, D. D., Rohacs, J., & Rohacs, D. (2021). Autonomous flight trajectory control system for drones in smart city traffic management. *ISPRS International Journal of Geo-Information*, *10*(5), 338. DOI: 10.3390/ijgi10050338

Nguyen, V. T., Zhou, J., Dong, C., Cui, G., An, Q., & Vinnakota, S. (2024, June). Enhancing house inspections: UAVs integrated with LLMs for efficient AI-powered surveillance. In *2024 International Joint Conference on Neural Networks (IJCNN)* (pp. 1-8). IEEE. DOI: 10.1109/IJCNN60899.2024.10650902

Nithya, S., Sangeetha, M., Prethi, K. A., Sahoo, K. S., Panda, S. K., & Gandomi, A. H. (2020). SDCF: A software-defined cyber foraging framework for cloudlet environment. *IEEE Transactions on Network and Service Management*, *17*(4), 2423–2435. DOI: 10.1109/TNSM.2020.3015657

Nyaaba, A. A., & Ayamga, M. (2021). Intricacies of medical drones in healthcare delivery: Implications for Africa. *Technology in Society*, *66*, 101624. DOI: 10.1016/j.techsoc.2021.101624

Obiuto, N. C., Festus-Ikhuoria, I. C., Olajiga, O. K., & Adebayo, R. A. (2024). Reviewing the role of AI in drone technology and applications. *Computer Science & IT Research Journal*, *5*(4), 741–756. DOI: 10.51594/csitrj.v5i4.1019

Ocaña, R. T., Peñaranda, S. M., & Falcon, V. M. P. (2024). Smart Shopping Carts: Reducing Checkout Queues with RFID and AWS Facial Recognition. *2024 IEEE XXXI International Conference on Electronics, Electrical Engineering and Computing (INTERCON)*, 1–7. DOI: 10.1109/INTERCON63140.2024.10833475

Okulski, M., & Ławryńczuk, M. (2022). How Much Energy Do We Need to Fly with Greater Agility? Energy Consumption and Performance of an Attitude Stabilization Controller in a Quadcopter Drone: A Modified MPC vs. PID. *Energies*, *15*(4), 1380. DOI: 10.3390/en15041380

Osco, L P., Marcato, J., Ramos, A P M., Jorge, L A D C., Fatholahi, S N., Silva, J D A., Matsubara, E T., Pistori, H., Gonçalves, W N., & Li, J. (2021, October 1). A review on deep learning in UAV remote sensing. Elsevier BV, 102, 102456-102456. DOI: 10.1016/j.jag.2021.102456

Othman, M., Xia, F., & Khan, A. N. (2015). Context-aware mobile cloud computing and its challenges. *IEEE Cloud Computing*, *2*(3), 42–49. DOI: 10.1109/MCC.2015.62

Pahk, J., Park, S., Shim, J., Son, S., Lee, J., An, J., Lim, Y., & Choi, G. (2024). Lane segmentation data augmentation for heavy rain sensor blockage using realistically translated raindrop images and CARLA Simulator. *IEEE Robotics and Automation Letters*, *9*(6), 5488–5495. DOI: 10.1109/LRA.2024.3390587

Palmer, J., Samuelson, A. E., Gill, R. J., Leadbeater, E., & Jansen, V. A. (2024). Foraging distance distributions reveal how honeybee waggle dance recruitment varies with landscape. *Communications Biology*, *7*(1), 1306. Advance online publication. DOI: 10.1038/s42003-024-06987-9 PMID: 39394388

Papyan, N., Kulhandjian, M., Kulhandjian, H., & Aslanyan, L. (2024, March 1). AI-Based Drone Assisted Human Rescue in Disaster Environments: Challenges and Opportunities. *Pattern Recognition and Image Analysis*, *34*(1), 169–186. DOI: 10.1134/S1054661824010152

Peksa, J., & Mamchur, D. (2024). A review on the state of the art in copter drones and flight control systems. *Sensors (Basel)*, *24*(11), 3349. DOI: 10.3390/s24113349 PMID: 38894139

Peng, H., Ding, C., Geng, T., Choudhury, S., Barker, K., & Li, A. (2024). Evaluating emerging ai/ml accelerators: Ipu, rdu, and nvidia/amd gpus. In *Companion of the 15th ACM/SPEC International Conference on Performance Engineering*, pages 14–20.

Peng, J., Zhang, P., Zheng, L., & Tan, J. (2020). UAV positioning based on multi-sensor fusion. *IEEE Access : Practical Innovations, Open Solutions*, *8*, 34455–34467. DOI: 10.1109/ACCESS.2020.2974285

Pham, Q., Ruby, R., Fang, F., Nguyen, D. C., Yang, Z., Le, M., Ding, Z., & Hwang, W. (2022, June 1). Aerial Computing: A New Computing Paradigm, Applications, and Challenges. *IEEE Internet of Things Journal*, *9*(11), 8339–8363. DOI: 10.1109/JIOT.2022.3160691

Piñeros, V. J., Reveles-Espinoza, A. M., & Monroy, J. A. (2024). From remote sensing to artificial intelligence in coral reef monitoring. *Machines (Basel)*, *12*(10), 693. DOI: 10.3390/machines12100693

Ponte, S., Ariante, G., Papa, U., & Del Core, G. (2020). An embedded platform for positioning and obstacle detection for small unmanned aerial vehicles. *Electronics (Basel)*, *9*(7), 1175. DOI: 10.3390/electronics9071175

Prabhakar, P., Yokesh, V., Aruchamy, P., & Nanthakumar, S. (2025). Artificial Intelligence-Enabled fully echoed Q-Routing and adaptive directional medium access control protocol for flying Ad-Hoc networks. *International Journal of Communication Systems*, *38*(4), e6138. Advance online publication. DOI: 10.1002/dac.6138

Pratihar, D. K. (2010). *Intelligent Autonomous Systems: Foundations and Applications*. Springer Science & Business Media. DOI: 10.1007/978-3-642-11676-6

Pueyo, P., Dendarieta, J., Montijano, E., Murillo, A. C., & Schwager, M. (2024, January 1). CineMPC: A Fully Autonomous Drone Cinematography System Incorporating Zoom, Focus, Pose, and Scene Composition. *IEEE Transactions on Robotics*, *40*, 1740–1757. DOI: 10.1109/TRO.2024.3353550

Purcell, C. R., Walsh, A. J., Colefax, A. P., & Butcher, P. (2022). Assessing the ability of deep learning techniques to perform real-time identification of shark species in live streaming video from drones. *Frontiers in Marine Science*, *9*, 981897. Advance online publication. DOI: 10.3389/fmars.2022.981897

Qazi, S., Khawaja, B. A., & Farooq, Q. U. (2022). IoT-Equipped and AI-Enabled Next Generation Smart Agriculture: A Critical review, current challenges and future trends. *IEEE Access : Practical Innovations, Open Solutions*, *10*, 21219–21235. DOI: 10.1109/ACCESS.2022.3152544

Qian, B., Guo, X., Boateng, G. O., An, R., Liu, H., Lai, Z., & Chen, C. (2025). Magnetoquasistatic Positioning: Recent advances, applications, potential solutions, and future prospects. *IEEE Internet of Things Journal*, *1*(6), 6250–6274. Advance online publication. DOI: 10.1109/JIOT.2024.3521084

Qu, H., & Su, W. (2024). Deep Learning-Based Weed–Crop Recognition for Smart Agricultural Equipment: A review. *Agronomy (Basel)*, *14*(2), 363. DOI: 10.3390/agronomy14020363

Raivi, A. M., Huda, S. M. A., Alam, M. M., & Moh, S. (2023, January 28). Drone Routing for Drone-Based Delivery Systems: A Review of Trajectory Planning, Charging, and Security. *Sensors (Basel)*, *23*(3), 1463–1463. DOI: 10.3390/s23031463 PMID: 36772502

Rajagopal, M. K., & MS, B. M. (2023). Artificial Intelligence based drone for early disease detection and precision pesticide management in cashew farming. arXiv preprint arXiv:2303.08556.

Raj, M., & Seamans, R. (2019). Primer on artificial intelligence and robotics. *Journal of Organization Design*, *8*(1), 11. DOI: 10.1186/s41469-019-0050-0

Ramadan, M. N., Basmaji, T., Gad, A., Hamdan, H., Akgün, B. T., Ali, M. A., Alkhedher, M., & Ghazal, M. (2024). Towards early forest fire detection and prevention using AI-powered drones and the IoT. *Internet of Things : Engineering Cyber Physical Human Systems*, *27*, 101248. DOI: 10.1016/j.iot.2024.101248

Ran, W., Lan, Y., Dai, X., Gu, J., Liu, B., Geng, L., Han, X. & Yi, L. (2022). Obstacle detection system for autonomous vineyard robots based on passthrough filter. *International Journal of Precision Agricultural Aviation*, 5.

Ranaweera, P., Jurcut, A. D., & Liyanage, M. (2021). Survey on multi-access edge computing security and privacy. *IEEE Communications Surveys and Tutorials*, *23*(2), 1078–1124. DOI: 10.1109/COMST.2021.3062546

Raniga, D., Amarasingam, N., Sandino, J., Doshi, A., Barthelemy, J., Randall, K., Robinson, S. A., Gonzalez, F., & Bollard, B. (2024). Monitoring of Antarctica's fragile vegetation using Drone-Based remote sensing, multispectral imagery and AI. *Sensors (Basel)*, *24*(4), 1063. DOI: 10.3390/s24041063 PMID: 38400222

Rasti, S., Bleakley, C. J., Holden, N., Whetton, R., Langton, D., & O'Hare, G. (2022). A survey of high resolution image processing techniques for cereal crop growth monitoring. *Information Processing in Agriculture*, *9*(2), 300–315.

Ravichandran, N., Tewaraja, T., Rajasegaran, V., Kumar, S. S., Gunasekar, S. K. L., & Sindiramutty, S. R. (2024). Comprehensive Review Analysis and Countermeasures for Cybersecurity Threats: DDoS, Ransomware, and Trojan Horse Attacks. *Preprints.org*. DOI: 10.20944/preprints202409.1369.v1

Regona, M., Yiğitcanlar, T., Xia, B., & Li, R Y M. (2022, March 1). Opportunities and Adoption Challenges of AI in the Construction Industry: A PRISMA Review. Springer Science+Business Media, 8(1), 45-45. DOI: 10.3390/joitmc8010045

Rejeb, A., Rejeb, K., Simske, S., & Treiblmaier, H. (2021). Humanitarian Drones: A review and research agenda. *Internet of Things : Engineering Cyber Physical Human Systems*, *16*, 100434. DOI: 10.1016/j.iot.2021.100434

Ren, J., Yu, G., He, Y., & Li, G. Y. (2019). Collaborative cloud and edge computing for latency minimization. *IEEE Transactions on Vehicular Technology*, *68*(5), 5031–5044. DOI: 10.1109/TVT.2019.2904244

Reyes-Hung, L., Soto, I., & Majumdar, A. K. (2024). Neural network- based stress detection in crop multispectral imagery for precision agriculture. In *2024 14th International Symposium on Communication Systems, Networks and Digital Signal Processing (CSNDSP)*, pages 551–556. IEEE.

Rezaee, M. R., Hamid, N. W. A., Hussin, M., & Zukarnain, Z. A. (2024). Comprehensive Review of Drones Collision Avoidance Schemes: Challenges and open issues. *IEEE Transactions on Intelligent Transportation Systems*, *25*(7), 6397–6426. DOI: 10.1109/TITS.2024.3375893

Riordan, J., Manduhu, M., Black, J., Dow, A., Dooly, G., & Matalonga, S. (2021, June). LiDAR simulation for performance evaluation of UAS detect and avoid. In *2021 International Conference on Unmanned Aircraft Systems (ICUAS)* (pp. 1355-1363). IEEE.

Riza, A. Z. B. M., Jennsen, L., Anggani, P., Rafeen, A. I., Ruth, P. N. J., Sookun, D., Sookun, V., Yusri, N. Z. B. M., Sern, L. J., Luximon, L., Omer, M. L., & Sindiramutty, S. R. (2025). Leveraging Machine Learning and AI to Combat Modern Cyber Threats. *preprints.org*. DOI: 10.20944/preprints202501.0360.v1

Rocha, A., Monteiro, M., Mattos, C., Dias, M., Soares, J., Magalhães, R., & Macedo, J. (2024). Edge AI for Internet of Medical Things: A literature review. *Computers & Electrical Engineering*, *116*, 109202. DOI: 10.1016/j.compeleceng.2024.109202

Roghair, J., Niaraki, A., Ko, K., & Jannesari, A. (2021, August). A vision based deep reinforcement learning algorithm for uav obstacle avoidance. In *Proceedings of SAI Intelligent Systems Conference* (pp. 115-128). Cham: Springer International Publishing.

Rovira-Sugranes, A., Razi, A., Afghah, F., & Chakareski, J. (2022). A review of AI-enabled routing protocols for UAV networks: Trends, challenges, and future outlook. *Ad Hoc Networks*, *130*, 102790. DOI: 10.1016/j.adhoc.2022.102790

Sabatini, R., Bartel, C., Shaid, T., & Kaharkar, A. (2012). Low-cost Vision Sensors and Integrated Systems for Unmanned Aerial Vehicle Navigation. *Journal of Systems and Software*, *2*(11), 323–349. https://scientific-journals.org/journalofsystemsandsoftware/archive/vol2no11/vol2no11_5.pdf

Saeed, R. A., Omri, M., Abdel-Khalek, S., Ali, E. S., & Alotaibi, M. F. (2022). Optimal path planning for drones based on swarm intelligence algorithm. *Neural Computing & Applications*, *34*(12), 10133–10155. DOI: 10.1007/s00521-022-06998-9

Saha, B., Kunze, S., & Poeschl, R. (2024). Comparative study of deep learning model architectures for drone detection and classification. In *2024 IEEE International Mediterranean Conference on Communications and Networking (MeditCom)*, pages 167–172. IEEE. DOI: 10.1109/MeditCom61057.2024.10621214

Şahin, M. C., & Tarhan, A. K. (2025). Evaluation and selection of hardware and AI models for edge applications: A method and a case study on UAVs. *Applied Sciences (Basel, Switzerland)*, *15*(3), 1026. DOI: 10.3390/app15031026

Saif, A. A. (2016). *The Drone Eats with Me: A Gaza Diary*. Beacon Press.

Salama, R., Al-Turjman, F., & Culmone, R. (2023, March). AI-powered drone to address smart city security issues. In *International Conference on Advanced Information Networking and Applications* (pp. 292-300). Springer International Publishing. DOI: 10.1007/978-3-031-28694-0_27

Sandino, J., Maire, F., Caccetta, P., Sanderson, C., & Gonzalez, F. (2021). Drone-based autonomous motion planning system for outdoor environments under object detection uncertainty. *Remote Sensing (Basel)*, *13*(21), 4481. DOI: 10.3390/rs13214481

Sarkar, M., & Sahoo, P. K. (2024). Leveraging edge computing for video data streaming in uav-based emergency response systems. *Sensors (Basel)*, *24*(15), 5076. DOI: 10.3390/s24155076 PMID: 39124122

Sartika, D. (2025). Critical phishing defense strategies and digital asset protection. In *IGI Global eBooks* (pp. 213–220). DOI: 10.4018/979-8-3693-8784-9.ch010

Sarvajcz, K., Ari, L., & Menyhart, J. (2024). AI on the Road: NVIDIA Jetson Nano-Powered Computer Vision-Based System for Real-Time Pedestrian and Priority Sign Detection. *Applied Sciences (Basel, Switzerland)*, *14*(4), 1440. DOI: 10.3390/app14041440

Sathya, D., Thangamani, R., & Balaji, B. S. (2024). The revolution of edge computing in smart farming. In *Signals and communication technology* (pp. 351–389). DOI: 10.1007/978-3-031-51195-0_17

Satyanarayanan, M. (2017). The emergence of edge computing. *Computer*, *50*(1), 30–39. DOI: 10.1109/MC.2017.9

Savkin, A. V., & Huang, H. (2021). Navigation of a UAV network for optimal surveillance of a group of ground targets moving along a road. *IEEE Transactions on Intelligent Transportation Systems*, *23*(7), 9281–9285. DOI: 10.1109/TITS.2021.3077880

Schaefer, D., Hately, A., Ferrara, G., Lambers, D., Schweiger, K. & Rushton, A. P. (2022). The european concept of operations for U-space and urban air mobility. *SESAR Innovation Days 2022*.

Schedl, D. C., Kurmi, I., & Bimber, O. (2021). An autonomous drone for search and rescue in forests using airborne optical sectioning. *Science Robotics*, *6*(55), eabg1188. Advance online publication. DOI: 10.1126/scirobotics.abg1188 PMID: 34162744

Scriban, A., Nabeneza, S., Cornelis, D., Delay, E., Vayssières, J., Cesaro, J., & Salgado, P. (2024). GPS-Based hidden Markov models to document pastoral mobility in the Sahel. *Sensors (Basel)*, *24*(21), 6964. DOI: 10.3390/s24216964 PMID: 39517860

Sehrawat, N., Vashisht, S., Singh, A., Dhiman, G., Viriyasitavat, W., & Alghamdi, N. S. (2024). A power prediction approach for a solar-powered aerial vehicle enhanced by stacked machine learning technique. *Computers & Electrical Engineering*, *115*, 109128. DOI: 10.1016/j.compeleceng.2024.109128

Seng, Y. J., Cen, T. Y., Raslan, M. H. B. M., Subramaniam, M. R., Xin, L. Y., Kin, S. J., Long, M. S., & Sindiramutty, S. R. (2024). In-Depth Analysis and Countermeasures for Ransomware Attacks: Case Studies and Recommendations. *Preprint.org*. DOI: 10.20944/preprints202408.2261.v1

Shah, I. A. (2024a). Autonomous Shipping: Security Issues and Challenges. *Cybersecurity in the Transportation Industry*, 187-210.

Shah, I. A. (2024b). IoT-Based Smart Transportation Industry: Security Challenges. *Cybersecurity in the Transportation Industry*, 211-239.

Shah, I. A. (2024b). Privacy and security challenges in unmanned aerial vehicles (UAVs). *Cybersecurity in the Transportation Industry*, 93-115.

Shah, I. A. (2024c). Privacy and security challenges in unmanned aerial vehicles (UAVs). *Cybersecurity in the Transportation Industry*, 93-115.

Shah, I. A., Jhanjhi, N. Z., & Laraib, A. (2022). Cybersecurity and blockchain usage in contemporary business. In *Advances in information security, privacy, and ethics book series* (pp. 49–64). DOI: 10.4018/978-1-6684-5284-4.ch003

Shah, I. A., Jhanjhi, N. Z., & Ray, S. K. (2024). IoT Devices in Drones: Security Issues and Future Challenges. In *Cybersecurity Issues and Challenges in the Drone Industry* (pp. 217-235). IGI Global.

Shah, I. A., Jhanjhi, N., & Ashraf, H. (2024). Logistics With the Internet of Things: Challenges, Perspectives, and Applications. In *Navigating Cyber Threats and Cybersecurity in the Logistics Industry* (pp. 172-195). IGI Global.

Shah, I. A., & Jhanjhi, N. Z. (2024). Cybersecurity Issues and Challenges in the Drone Industry.

Shah, I. A., Jhanjhi, N. Z., & Brohi, S. N. (2024). Use of AI-based drones in smart cities. In *Cybersecurity Issues and Challenges in the Drone Industry* (pp. 362–380). IGI Global. DOI: 10.4018/979-8-3693-0774-8.ch015

Shah, I. A., Jhanjhi, N. Z., & Ujjan, R. M. A. (2024b). *Use of AI applications for the drone industry. Cybersecurity Issues and Challenges in the Drone Industry*. IGI Global.

Shah, I. A., Jhanjhi, N., & Ashraf, H. (2024a). *Logistics With the Internet of Things: Challenges, Perspectives, and Applications. Navigating Cyber Threats and Cybersecurity in the Logistics Industry*. IGI Global. DOI: 10.4018/979-8-3693-3816-2.ch006

Shah, I. A., Jhanjhi, N., & Brohi, S. N. (2024b). *IoT Smart Healthcare Security Challenges and Solutions. Advances in Computational Intelligence for the Healthcare Industry 4.0*. IGI Global Scientific Publishing.

Shah, I. A., Jhanjhi, N., & Brohi, S. N. (2025). *Machine Learning Models for Detecting Software Vulnerabilities. Generative AI for Web Engineering Models*. IGI Global.

Shah, I. A., Laraib, A., & Hussain, F. (2024b). *Logistics Industry in the Context of the Blockchain Technology. Navigating Cyber Threats and Cybersecurity in the Logistics Industry*. IGI Global.

Shah, I. A., Murugesan, R. K., & Ashraf, H. (2024c). *Advances in Health With the Help of Explainable AI. Advances in Computational Intelligence for the Healthcare Industry 4.0*. IGI Global.

Shah, I., Jhanjhi, N., & Rajper, S. (2024a). Use of Deep Learning Applications for Drone Technology. Cybersecurity Issues and Challenges in the Drone Industry. *IGI Global.DOI*, *10*, 979–8.

Shah, I., Jhanjhi, N., & Ray, S. (2024a). IoT Devices in Drones: Security Issues and Future Challenges. Cybersecurity Issues and Challenges in the Drone Industry. *IGI Global.DOI*, *10*, 979–8.

Shahmoradi, J., Talebi, E., Roghanchi, P., & Hassanalian, M. (2020, July 15). A Comprehensive Review of Applications of Drone Technology in the Mining Industry. *Drones (Basel)*, *4*(3), 34–34. DOI: 10.3390/drones4030034

Shahzadi, R., Ali, M., Khan, H. Z., & Naeem, M. (2021). UAV assisted 5G and beyond wireless networks: A survey. *Journal of Network and Computer Applications*, *189*, 103114. DOI: 10.1016/j.jnca.2021.103114

Shaiba, O. A. (2020). *Update urban basemap by using the LiDAR mobile mapping system : the case of Abu Dhabi municipal system*. DOI: 10.5821/dissertation-2117-330389

Shaikh, T. A., Mir, W. A., Rasool, T., & Sofi, S. (2022). Machine Learning for smart agriculture and Precision Farming: Towards making the fields talk. *Archives of Computational Methods in Engineering*, *29*(7), 4557–4597. DOI: 10.1007/s11831-022-09761-4

Shakeri, R., Al-Garadi, M. A., Badawy, A., Mohamed, A., Khattab, T., Al-Ali, A. K., Harras, K. A., & Guizani, M. (2019). Design Challenges of Multi-UAV Systems in Cyber-Physical Applications: A Comprehensive Survey and Future Directions. *IEEE Communications Surveys and Tutorials*, *21*(4), 3340–3385. DOI: 10.1109/COMST.2019.2924143

Shakhatreh, H., Sawalmeh, A. H., Al-Fuqaha, A., Dou, Z., Almaita, E., Khalil, I., Othman, N. S., Khreishah, A., & Guizani, M. (2019). Unmanned aerial vehicles (UAVs): A survey on civil applications and key research challenges. *IEEE Access : Practical Innovations, Open Solutions*, *7*, 48572–48634. DOI: 10.1109/ACCESS.2019.2909530

Shang, Y., Wang, H., Qin, W., Wang, Q., Liu, H., Yin, Y., Song, Z., & Meng, Z. (2023). Design and test of obstacle detection and harvester pre-collision system based on 2D lidar. *Agronomy (Basel)*, *13*(2), 388. DOI: 10.3390/agronomy13020388

Shankar, K., Wang, P., Xu, R., Mahgoub, A., & Chaterji, S. (2020). Janus: Benchmarking commercial and open-source cloud and edge platforms for object and anomaly detection workloads. In *2020 IEEE 13th International Conference on Cloud Computing (CLOUD)*, pages 590–599. IEEE.

Sharma, R., Singh, A., Kavita, N., Jhanjhi, N. Z., Masud, M., Jaha, E. S., & Verma, S. (2021). Plant disease diagnosis and image classification using deep learning. *Computers, Materials & Continua/Computers. Materials & Continua (Print)*, *71*(2), 2125–2140. DOI: 10.32604/cmc.2022.020017

Shen, J., & Brower, J. (2021). Access and Edge Network Architecture and Management. In *Future Networks, Services and Management: Underlay and Overlay, Edge, Applications, Slicing, Cloud, Space, AI/ML, and Quantum Computing*. DOI: 10.1007/978-3-030-81961-3_5

Sheng, L. T., Azman, A. B., Khan, N. A., Sindiramutty, S. R., Singh, S., & Verma, S. (2024). Reinforcement learning-based traffic signal control. *2024 International Conference on Emerging Trends in Networks and Computer Communications (ETNCC)*, 792–798. DOI: 10.1109/ETNCC63262.2024.10767501

Shi, T., Ding, Y., & Zhu, W. (2023). YOLOv5s_2E: Improved YOLOv5s for aerial small target detection. *IEEE Access : Practical Innovations, Open Solutions*, *11*, 80479–80490. DOI: 10.1109/ACCESS.2023.3300372

Shi, W., Cao, J., Zhang, Q., Li, Y., & Xu, L. (2016). Edge computing: Vision and challenges. *IEEE Internet of Things Journal*, *3*(5), 637–646. DOI: 10.1109/JIOT.2016.2579198

Shi, Y., Yang, K., Jiang, T., Zhang, J., & Letaief, K. B. (2020). Communication-efficient edge AI: Algorithms and systems. *IEEE Communications Surveys and Tutorials*, *22*(4), 2167–2191. DOI: 10.1109/COMST.2020.3007787

Shokri, D., Larouche, C., & Homayouni, S. (2024). Proposing an Efficient Deep Learning Algorithm Based on Segment Anything Model for Detection and Tracking of Vehicles through Uncalibrated Urban Traffic Surveillance Cameras. *Electronics (Basel), 13*(14), 2883. DOI: 10.3390/electronics13142883

Shraim, H., Awada, A., & Youness, R. (2018). A survey on quadrotors: ConFigureurations, modeling and identification, control, collision avoidance, fault diagnosis and tolerant control. *IEEE Aerospace and Electronic Systems Magazine, 33*(7), 14–33. DOI: 10.1109/MAES.2018.160246

Sier, H., Yu, X., Catalano, I., Queralta, J. P., Zou, Z., & Westerlund, T. (2023, June). Uav tracking with lidar as a camera sensor in gnss-denied environments. In *2023 International Conference on Localization and GNSS (ICL-GNSS)* (pp. 1-7). IEEE.

Siewert, S. B., Andalibi, M., Bruder, S., Buchholz, J. M., Chamberlain, D., Lindsey, A. L., & Stockhouse, D. (2020). Comparison of RADAR, passive optical with acoustic, and fused multi-modal active and passive sensing for UAS traffic management compliance and urban air mobility safety. In *AIAA Scitech 2020 Forum* (p. 1456).

Simsek, M., Aijaz, A., Dohler, M., Sachs, J., & Fettweis, G. (2016). 5G-enabled tactile internet. *IEEE Journal on Selected Areas in Communications, 34*(3), 460–473. DOI: 10.1109/JSAC.2016.2525398

Sindiramutty, S. R., Jhanjhi, N. Z., Tan, C. E., Yun, K. J., Manchuri, A. R., Ashraf, H., Murugesan, R. K., Tee, W. J., & Hussain, M. (2024). Data security and privacy concerns in drone operations. In *Advances in information security, privacy, and ethics book series* (pp. 236–290). DOI: 10.4018/979-8-3693-0774-8.ch010

Sindiramutty, S. R., Prabagaran, K. R. V., Akbar, R., Hussain, M., & Malik, N. A. (2024). Overview of Generative AI techniques for cybersecurity. In *Advances in information security, privacy, and ethics book series* (pp. 1–52). DOI: 10.4018/979-8-3693-5415-5.ch001

Sindiramutty, S. R., Prabagaran, K. R. V., Jhanjhi, N. Z., Ghazanfar, M. A., Malik, N. A., & Soomro, T. R. (2024). Security Considerations in Generative AI for web Applications. In *Advances in information security, privacy, and ethics book series* (pp. 281–332). DOI: 10.4018/979-8-3693-5415-5.ch009

Sindiramutty, S. R., Prabagaran, K. R. V., Jhanjhi, N. Z., Murugesan, R. K., Brohi, S. N., & Masud, M. (2024). Generative AI in network security and intrusion detection. In *Advances in information security, privacy, and ethics book series* (pp. 77–124). DOI: 10.4018/979-8-3693-5415-5.ch003

Sindiramutty, S. R., Prabagaran, K. R. V., Jhanjhi, N. Z., Murugesan, R. K., Brohi, S. N., & Wei, G. W. (2024). Generative AI for threat intelligence and information sharing. In *Advances in digital crime, forensics, and cyber terrorism book series* (pp. 191–234). DOI: 10.4018/979-8-3693-8944-7.ch006

Sindiramutty, S. R., Prabagaran, K. R. V., Jhanjhi, N. Z., Murugesan, R. K., Malik, N. A., & Hussain, M. (2024). Ethics and transparency in secure web model generation. In *Advances in information security, privacy, and ethics book series* (pp. 411–464). DOI: 10.4018/979-8-3693-5415-5.ch012

Singh, T., Kumar, S., Singh, S. K., Priyanshu, N., Gupta, B. B., Wu, J., & Castiglione, A. (2024). Enhancing autonomous system security with AI and secure computation technologies. In *Advances in computational intelligence and robotics book series* (pp. 159–186). DOI: 10.4018/979-8-3693-2707-4.ch008

Singhal, V., Jain, S. S., Anand, D., Singh, A., Verma, S., Kavita, N., Rodrigues, J. J. P. C., Jhanjhi, N. Z., Ghosh, U., Jo, O., & Iwendi, C. (2020). Artificial Intelligence Enabled Road Vehicle-Train Collision Risk Assessment Framework for Unmanned railway level crossings. *IEEE Access : Practical Innovations, Open Solutions*, *8*, 113790–113806. DOI: 10.1109/ACCESS.2020.3002416

Singh, M. K., Pippal, S. K., & Sharma, V. (2024). Lightweight blockchain mechanism for secure data transmission in healthcare system. *Biomedical Signal Processing and Control*, *102*, 107411. DOI: 10.1016/j.bspc.2024.107411

Singh, M., Patel, K., & Johnson, R. (2024). Deep learning for traffic signal optimization: A reinforcement learning approach. *Transportation Research Part B: Methodological*, *169*, 234–248.

Singh, R., & Gill, S. S. (2023). Edge AI: A survey. *Internet of Things and Cyber-Physical Systems*, *3*, 71–92. DOI: 10.1016/j.iotcps.2023.02.004

Singh, S. (2017). Optimize cloud computations using edge computing. *2017 International Conference on Big Data, IoT and Data Science, BID 2017, 2018-January*, 49–53. DOI: 10.1109/BID.2017.8336572

Singh, S., Zuber, M., Hamidon, M. N., Mazlan, N., Basri, A. A., & Ahmad, K. A. (2022). Classification of actuation mechanism designs with structural block diagrams for flapping-wing drones: A comprehensive review. *Progress in Aerospace Sciences*, *132*, 100833. DOI: 10.1016/j.paerosci.2022.100833

Singh, T., Sinjana, Y., & Kapoor, A. P. (2024). Integrating Mobile Edge Computing in Drone Delivery: A Holistic Service Provision Strategy for Drone-Based Healthcare Logistics. *SSRN*. DOI: 10.2139/ssrn.4994845

Siriwardhana, Y., Porambage, P., Liyanage, M., & Ylianttila, M. (2021). *A Survey on Mobile Augmented Reality with 5G Mobile Edge Computing: Architectures*. Applications, and Technical Aspects.

Skorobogatov, G., Barrado, C., & Salamí, E. (2020). Multiple UAV systems: A survey. *Unmanned Systems*, *8*(2), 149–169. DOI: 10.1142/S2301385020500090

Smeesters, L., Verbaenen, J., Schifano, L., Vervaeke, M., Thienpont, H., Teti, G., Forconi, A., & Lulli, F. (2023, February 23). Wide-Field-of-View Multispectral Camera Design for Continuous Turfgrass Monitoring. *Sensors (Basel)*, *23*(5), 2470–2470. DOI: 10.3390/s23052470 PMID: 36904674

Soori, M., Arezoo, B., & Dastres, R. (2023). Artificial intelligence, machine learning and deep learning in advanced robotics, a review. *Cognitive Robotics*, *3*, 54–70. DOI: 10.1016/j.cogr.2023.04.001

Sten, G., Feng, L., & Möller, B. (2025). Enhancing Off-Road Topography Estimation by Fusing LIDAR and Stereo Camera Data with Interpolated Ground Plane. *Sensors (Basel)*, *25*(2), 509. DOI: 10.3390/s25020509 PMID: 39860879

Suanpang, P., & Jamjuntr, P. (2024). Optimizing Autonomous UAV Navigation with D* Algorithm for Sustainable Development. *Sustainability (Basel)*, *16*(17), 7867. DOI: 10.3390/su16177867

Suganya, B., Gopi, R., Kumar, A. R., & Singh, G. (2024). Dynamic task offloading edge-aware optimization framework for enhanced uav operations on edge computing plat- form. *Scientific Reports*, *14*(1), 16383. DOI: 10.1038/s41598-024-67285-2 PMID: 39013972

Sulaiman, M. M., Matloub, F. K., Ajjam, S. K. A., & Alwan, H. H. (2024). Innovative: International Multi-disciplinary Journal of Applied Technology. *Journal of Applied Technology*, *2*, 2024.

Sun, C., Fontanesi, G., Canberk, B., Mohajerzadeh, A., Chatzinotas, S., Grace, D., & Ahmadi, H. (2024). Advancing UAV Communications: A Comprehensive survey of Cutting-Edge Machine Learning Techniques. *IEEE Open Journal of Vehicular Technology*, *5*, 825–854. DOI: 10.1109/OJVT.2024.3401024

Sun, H., Xue, J., Zhang, Y., Li, H., Liu, R., Song, Y., & Liu, S. (2024). Novel method of rapid and accurate tree trunk location in pear orchard combining stereo vision and semantic segmentation. *Measurement*, *242*, 116127. DOI: 10.1016/j.measurement.2024.116127

Sun, J., & Hubbard, S. (2025). An examination of UAS incidents: Characteristics and safety considerations. *Drones (Basel)*, *9*(2), 112. DOI: 10.3390/drones9020112

Sun, J., Li, B., Jiang, Y., & Wen, C. (2016). A Camera-Based Target Detection and Positioning UAV system for search and rescue (SAR) purposes. *Sensors (Basel)*, *16*(11), 1778. DOI: 10.3390/s16111778 PMID: 27792156

Sun, L., Wang, J., Wan, L., Li, K., Wang, X., & Lin, Y. (2024). Human-UAV interaction assisted heterogeneous UAV swarm scheduling for target searching in communication denial environment. *IEEE Transactions on Automation Science and Engineering*, •••, 1–16. DOI: 10.1109/TASE.2024.3412005

Suryanarayana, G. (2023). AI applications of drones. In *Drone Technology* (p. 153). Future Trends and Practical Applications.

Su, X., Xue, S., Liu, F., Wu, J., Yang, J., Zhou, C., Hu, W., Paris, C., Nepal, S., & Jin, D. (2022). A comprehensive survey on community detection with deep learning. *IEEE Transactions on Neural Networks and Learning Systems*. PMID: 35263257

Svanstrom, F., Englund, C., & Alonso-Fernandez, F. (2021). Real-Time drone detection and tracking with visible, thermal and acoustic sensors. *2022 26th International Conference on Pattern Recognition (ICPR)*, 7265–7272. DOI: 10.1109/ICPR48806.2021.9413241

Tahir, A., Böling, J., Haghbayan, M.-H., Toivonen, H. T., & Plosila, J. (2019). Swarms of unmanned aerial vehicles—A survey. *Journal of Industrial Information Integration*, *16*, 100106. DOI: 10.1016/j.jii.2019.100106

Tallapaneni, K., Kumar, A., Gautam, H., Yadav, A., Kumar, A., & Yadav, H. (2024). Enhancing home security with AI-driven drone surveillance. *Authorea Preprints*. DOI: 10.36227/techrxiv.171665618.82092467/v1

Talwandi, N. S., Thakur, P., & Khare, S. (2024). An Automatic Navigation System for New Technical Advanced Drones for Different Alpplications. *2024 IEEE International Conference on Computing, Power and Communication Technologies (IC2PCT)*, 736–741. DOI: 10.1109/IC2PCT60090.2024.10486524

Tan, Y., & Zheng, Z. (2013). Research advance in swarm robotics. *Defence Technology*, *9*(1), 18–39. DOI: 10.1016/j.dt.2013.03.001

Tarekegn, G. B., Juang, R., Lin, H., Munaye, Y. Y., Wang, L., & Bitew, M. A. (2022). Deep-Reinforcement-Learning-Based drone base station deployment for wireless communication services. *IEEE Internet of Things Journal*, *9*(21), 21899–21915. DOI: 10.1109/JIOT.2022.3182633

Tariq, M. U. (2024). Harnessing Persuasive Technologies for Enhanced Learner Engagement and Motivation. In M. Sanmugam, D. Lim, N. Mohd Barkhaya, W. Wan Yahaya, & Z. Khlaif (Eds.), *Power of Persuasive Educational Technologies in Enhancing Learning* (pp. 30-62). IGI Global. https://doi.org/DOI: 10.4018/979-8-3693-6397-3.ch002

Tariq, M. U. (2024). Smart Transportation Systems: Paving the Way for Sustainable Urban Mobility. In Munuhwa, S. (Ed.), *Contemporary Solutions for Sustainable Transportation Practices* (pp. 254–283). IGI Global., DOI: 10.4018/979-8-3693-3755-4.ch010

Tariq, M. U. (2025). Enhancing Pandemic Preparedness With AI-Driven Risk Prediction Models. In Shah, I., Sial, Q., & Fateh, S. (Eds.), *Generative AI Techniques for Sustainability in Healthcare Security* (pp. 77–94). IGI Global Scientific Publishing., DOI: 10.4018/979-8-3693-6577-9.ch005

Tariq, M. U. (2025). Navigating the Ethical Frontier-Human Oversight in AI-Driven Decision-Making System. In Hai-Jew, S. (Ed.), *Enhancing Automated Decision-Making Through AI* (pp. 425–448). IGI Global Scientific Publishing., DOI: 10.4018/979-8-3693-6230-3.ch013

Tariq, M. U., & Sergio, R. P. (2025). Ethical Considerations and Social Impact of AI and Computational Intelligence in Society 5.0. In Pandey, R., Srivastava, N., Prasad, R., Prasad, J., & Garcia, M. (Eds.), *Open AI and Computational Intelligence for Society 5.0* (pp. 225–248). IGI Global Scientific Publishing., DOI: 10.4018/979-8-3693-4326-5.ch010

Telli, K., Kraa, O., Himeur, Y., Ouamane, A., Boumehraz, M., Atalla, S., & Mansoor, W. (2023, January 1). A Comprehensive Review of Recent Research Trends on UAVs. Cornell University. https://doi.org/DOI: 10.48550/arXiv.2307

Thakur, A., & Mishra, S. K. (2024). An in-depth evaluation of deep learning-enabled adaptive approaches for detecting obstacles using sensor-fused data in autonomous vehicles. *Engineering Applications of Artificial Intelligence*, *133*, 108550. DOI: 10.1016/j.engappai.2024.108550

Thangamani, R., Suguna, R. K., & Kamalam, G. K. (2024). Drones and Autonomous Robotics Incorporating Computational Intelligence. *Wiley*, 243–296. DOI: 10.1002/9781394175437.ch8

Thomas, G. S., Muthukaruppasamy, S., Kumar, S. S., Karthikeyan, B. J., & Saravanan, K. (2024). Navigating the Nexus: Unravelling Challenges, Ethics, and Applications of Embodied AI in Drone Technology Through the Lens of Computer Vision. In *Building Embodied AI Systems: The Agents, the Architecture Principles, Challenges, and Application Domains* (pp. 61–78). DOI: 10.1007/978-3-031-68256-8_3

Tlili, F., Ayed, S., & Fourati, L. C. (2024). Advancing UAV security with artificial intelligence: A comprehensive survey of techniques and future directions. *Internet of Things : Engineering Cyber Physical Human Systems*, *27*, 101281. DOI: 10.1016/j.ict.2024.101281

Tomasicchio, G., Andolfi, L., Brancati, M., Di Donna, A. M., Giannattasio, S., Del Frete, R., Ostrogovich, L., Renga, A., Grassi, M., Ceresoli, M., Silvestrini, S., & Lavagna, M. (2024). Multi-Sensor fusion and resilient PVT techniques for safe lunar landing missions. *Proceedings of the Satellite Division's International Technical Meeting (Online)/Proceedings of the Satellite Division's International Technical Meeting (CD-ROM)*, 3125–3143. DOI: 10.33012/2024.19849

Topham, G., Weaver, M., & Siddique, H. (2018, December 21). Runway reopens after days of drone disruption at Gatwick. *The Guardian*. https://www.theguardian.com/uk-news/2018/dec/20/tens-of-thousands-of-passengers-stranded-by-gatwick-airport-drones

Trawny, N., Johnson, A. E., Bailey, E. S., Massone, G., Reid, M., Setterfield, T. P., Cheng, Y., Sellar, G., & Soto, J. (2024). The Enhanced Lander Vision System for Mars sample return lander entry descent and landing. *AIAA SCITECH 2022 Forum*. DOI: 10.2514/6.2024-0314

Tsouros, D. C., Bibi, S., & Sarigiannidis, P. G. (2019). A review on UAV-based applications for precision agriculture. *Information (Basel)*, *10*(11), 349. DOI: 10.3390/info10110349

Tuśnio, N., & Wróblewski, W. (2021). The Efficiency of Drones Usage for Safety and Rescue Operations in an Open Area: A Case from Poland. *Sustainability (Basel)*, *14*(1), 327. DOI: 10.3390/su14010327

Ullah, F., Saqib, S., & Xiong, Y. (2024). Integrating artificial intelligence in biodiversity conservation: Bridging classical and modern approaches. *Biodiversity and Conservation*. Advance online publication. DOI: 10.1007/s10531-024-02977-9

Ullah, Z., Al-Turjman, F., & Mostarda, L. (2020). Cognition in UAV-aided 5G and beyond communications: A survey. *IEEE Transactions on Cognitive Communications and Networking*, *6*(3), 872–891. DOI: 10.1109/TCCN.2020.2968311

Vaezi, M., Lin, X., Zhang, H., Saad, W., & Poor, H. V. (2023). Deep Reinforcement Learning for Interference Management in UAV-Based 3D Networks: Potentials and challenges. *IEEE Communications Magazine*, *62*(2), 134–140. DOI: 10.1109/MCOM.001.2200973

Vailshery, L. S. (2021). Internet of Things (IoT) and non-IoT active device connections worldwide from 2010 to 2025. *Online,https://bit. ly/2SwKVuB*

Van Groesen, W., & Pauwels, P. (2022). Tracking prefabricated assets and compliance using quick response (QR) codes, blockchain and smart contract technology. *Automation in Construction*, *141*, 104420. DOI: 10.1016/j.autcon.2022.104420

Varghese, B., Wang, N., Barbhuiya, S., Kilpatrick, P., & Nikolopoulos, D. S. (2016). *Challenges and opportunities in edge computing. 2016 IEEE international conference on smart cloud (SmartCloud)*. IEEE.

Veprytska, O., & Kharchenko, V. (2024). Analysis of AI-powered attacks and protection of UAV assets: Quality model-based assessing cybersecurity of mobile systems for demining. In *IntelITSIS* (pp. 356-374).

Verma, A., & Verma, V. (n.d.). *Comparative Study of Cloud Computing and Edge Computing: Three Level Architecture Models and Security Challenges*. https://ssrn.com/abstract=3908915

Vijayalakshmi, B., Ramar, K., Jhanjhi, N. Z., Verma, S., Kaliappan, M., Vijayalakshmi, K., Vimal, S., Kavita, , & Ghosh, U. (2021). An attention-based deep learning model for traffic flow prediction using spatiotemporal features towards sustainable smart city. *International Journal of Communication Systems*, *34*(3), e4609. DOI: 10.1002/dac.4609

Vishweshwaran, M., & Sujatha, E. R. (2024). *A review on Applications of drones in geotechnical engineering*. Indian Geotechnical Journal., DOI: 10.1007/s40098-024-01071-9

Vrochidou, E., Oustadakis, D., Kefalas, A., & Papakostas, G. A. (2022). Computer vision in Self-Steering tractors. *Machines (Basel)*, *10*(2), 129. DOI: 10.3390/machines10020129

Waheed, A., Seegolam, B., Jowaheer, M. F., Sze, C. L. X., Hua, E. T. F., & Sindiramutty, S. R. (2024). Zero-Day Exploits in Cybersecurity: Case Studies and Countermeasure. *Preprints.org*. DOI: 10.20944/preprints202407.2338.v1

Wang, J., Liu, W., Zhang, W., & Liu, B. (2022). LV-YOLOv5: A light-weight object detector of Vit on Drone-captured Scenarios. 2022 16th IEEE International Conference on Signal Processing (ICSP)

Wang, B., Chen, L., & Li, W. (2022). Smart traffic management: Current challenges and future directions. *Journal of Intelligent Transport Systems*, *26*(4), 382–396.

Wang, C., Zhang, S., Ma, T., Xiao, Y., Chen, M. Z., & Wang, L. (2024). Swarm intelligence: A survey of model classification and applications. *Chinese Journal of Aeronautics*, 102982. Advance online publication. DOI: 10.1016/j.cja.2024.03.019

Wang, H., Liu, J., & Zhang, W. (2023). Multi-object tracking for intelligent traffic systems: A hybrid approach. *IEEE Transactions on Vehicular Technology*, *72*(5), 5678–5691.

Wang, J., Liu, Y., & Song, H. (2021). Counter-unmanned aircraft system (s)(C-UAS): State of the art, challenges, and future trends. *IEEE Aerospace and Electronic Systems Magazine*, *36*(3), 4–29. DOI: 10.1109/MAES.2020.3015537

Wang, K., Kooistra, L., Pan, R., Wang, W., & Valente, J. (2024). UAV-based simultaneous localization and mapping in outdoor environments: A systematic scoping review. *Journal of Field Robotics*, *41*(5), 1617–1642. DOI: 10.1002/rob.22325

Wang, S., & Ahmad, N. S. (2024). A comprehensive review on sensor fusion techniques for localization of a dynamic target in GPS-Denied environments. *IEEE Access : Practical Innovations, Open Solutions*, *1*. Advance online publication. DOI: 10.1109/ACCESS.2024.3519874

Wang, T., Huang, P., & Dong, G. (2021). Cooperative persistent surveillance on a road network by multi-UGVs with detection ability. *IEEE Transactions on Industrial Electronics*, *69*(11), 11468–11478. DOI: 10.1109/TIE.2021.3121729

Wang, X., Han, Y., Leung, V. C., Niyato, D., Yan, X., & Chen, X. (2020). Convergence of edge computing and deep learning: A comprehensive survey. *IEEE Communications Surveys and Tutorials*, *22*(2), 869–904. DOI: 10.1109/COMST.2020.2970550

Wang, X., Luo, B., & Zhang, Z. (2021, April). Application of UAV target tracking based on computer vision. [). IOP Publishing.]. *Journal of Physics: Conference Series*, *1881*(4), 042053.

Wang, X., Zheng, S., Lin, X., Zhang, Q., & Liu, X. (2024). Robust loop closure detection and relocalization with semantic-line graph matching constraints in indoor environments. *International Journal of Applied Earth Observation and Geoinformation*, *129*, 103844. DOI: 10.1016/j.jag.2024.103844

Wang, Y., He, Y., Yu, F. R., Song, B., & Leung, V. C. (2023). Efficient resource allocation for building the metaverse with uavs: A quantum collective reinforcement learning approach. *IEEE Wireless Communications*, *30*(5), 152–159. DOI: 10.1109/MWC.009.2300029

Weiqi, X., Hooi, S. T. C., Sindiramutty, S. R. L., Asirvatham, D. L., Kumar, D., & Verma, S. (2024). Surface Anomaly Detection Using Machine Learning Technique. *2024 International Conference on Emerging Trends in Networks and Computer Communications (ETNCC)*, 1–7. DOI: 10.1109/ETNCC63262.2024.10767562

Wei, Y., Zhou, S., Leng, S., Maharjan, S., & Zhang, Y. (2021). Federated learning empowered end-edge-cloud cooperation for 5G HetNet security. *IEEE Network*, *35*(2), 88–94. DOI: 10.1109/MNET.011.2000340

Westheider, J., Ruckin, J., & Popovic, M. (2023). Multi-UAV Adaptive Path Planning Using Deep Reinforcement Learning. *IEEE International Conference on Intelligent Robots and Systems*, 649–656. DOI: 10.1109/IROS55552.2023.10342516

Wisskirchen, G., Biacabe, B. T., Bormann, U., Muntz, A., Niehaus, G., Soler, G. J., & Von Brauchitsch, B. (2017). Artificial intelligence and robotics and their impact on the workplace. *IBA Global Employment Institute*, *11*, 49–67.

Witt, R. R., Beranek, C. T., Howell, L. G., Ryan, S. A., Clulow, J., Jordan, N. R., Denholm, B., & Roff, A. (2020). Real-time drone derived thermal imagery outperforms traditional survey methods for an arboreal forest mammal. *PLoS One*, *15*(11), e0242204. DOI: 10.1371/journal.pone.0242204 PMID: 33196649

Wolniak, R., Stecuła, K., & Aydın, B. (2024). Digital Transformation of Grocery In-Store Shopping-Scanners, Artificial intelligence, Augmented Reality and beyond: A review. *Foods*, *13*(18), 2948. DOI: 10.3390/foods13182948 PMID: 39335877

Wu, J., Ye, Y., & Du, J. (2024). Autonomous drones in urban navigation: Autoencoder learning fusion for aerodynamics. *Journal of Construction Engineering and Management*, *150*(7), 04024067. DOI: 10.1061/JCEMD4.COENG-14787

Wu, Q., Xu, J., Zeng, Y., Ng, D. W. K., Al-Dhahir, N., Schober, R., & Swindlehurst, A. L. (2021). A comprehensive overview on 5G-and-beyond networks with UAVs: From communications to sensing and intelligence. *IEEE Journal on Selected Areas in Communications*, *39*(10), 2912–2945. DOI: 10.1109/JSAC.2021.3088681

Wu, Y., Dai, H.-N., & Wang, H. (2020). Convergence of blockchain and edge computing for secure and scalable IIoT critical infrastructures in industry 4.0. *IEEE Internet of Things Journal*, *8*(4), 2300–2317. DOI: 10.1109/JIOT.2020.3025916

Wu, Y., Zhang, K., & Zhang, Y. (2021). Digital twin networks: A survey. *IEEE Internet of Things Journal*, *8*(18), 13789–13804. DOI: 10.1109/JIOT.2021.3079510

Xia, F., Jedari, B., Yang, L. T., Ma, J., & Huang, R. (2016a). A signaling game for uncertain data delivery in selfish mobile social networks. *IEEE Transactions on Computational Social Systems*, *3*(2), 100–112. DOI: 10.1109/TCSS.2016.2584103

Xia, F., Liu, L., Jedari, B., & Das, S. K. (2016b). PIS: A multi-dimensional routing protocol for socially-aware networking. *IEEE Transactions on Mobile Computing, 15*(11), 2825–2836. DOI: 10.1109/TMC.2016.2517649

Xia, F., Wang, J., Kong, X., Zhang, D., & Wang, Z. (2019). Ranking station importance with human mobility patterns using subway network datasets. *IEEE Transactions on Intelligent Transportation Systems, 21*(7), 2840–2852. DOI: 10.1109/TITS.2019.2920962

Xiao, J., Zhang, R., Zhang, Y., & Feroskhan, M. (2023, January 1). Vision-based Learning for Drones: A Survey. Cornell University. https://doi.org/DOI: 10.48550/arXiv.2312

Xiao, Y., Xu, T., Yu, X., Fang, Y., & Li, J. (2024). A Lightweight Fusion Strategy with Enhanced Inter-layer Feature Correlation for Small Object Detection. *IEEE Transactions on Geoscience and Remote Sensing, 62*, 1–11. DOI: 10.1109/TGRS.2024.3457155

Xie, W., Wan, Y., Wu, G., Li, Y., Zhou, F., & Wu, Q. (2024). A RF-Visual Directional fusion framework for precise UAV positioning. *IEEE Internet of Things Journal, 11*(22), 36736–36747. DOI: 10.1109/JIOT.2024.3424271

Xi, H., Ru, L., Tian, J., Lu, B., Hu, S., Wang, W., & Luan, X. (2025). URADV: A novel framework for generating Ultra-Robust adversarial patches against UAV object detection. *Mathematics, 13*(4), 591. DOI: 10.3390/math13040591

Xu, K., Li, Y., Liu, C., Liu, X., Hao, X., Gao, J., & Maropoulos, P. (2020, May 25). Advanced Data Collection and Analysis in Data-Driven Manufacturing Process. China Machine Press, 33(1). DOI: 10.1186/s10033-020-00459-x

Xu, Y., Pan, L., Du, C., Li, J., Jing, N., & Wu, J. (2018). Vision-based UAVs Aerial Image Localization. *ACM*, 9–18. DOI: 10.1145/3281548.3281556

Xun, A. T., En, L. Z., Shen, L. T., Xin, A. N., Soon, W. H., Jun, W. Z., Ramachandra, H., Xinghao, G., Khant, N. M., Weitao, F., & Sindiramutty, S. R. (2025). Building Trust in Cloud Computing:Strategies for Resilient Security. *Preprints.org*. DOI: 10.20944/preprints202501.0716.v1

Yadav, I. C., Devi, N. L., Syed, J. H., Cheng, Z., Li, J., Zhang, G., & Jones, K. C. (2014). Current status of persistent organic pesticides residues in air, water, and soil, and their possible effect on neighboring countries: A comprehensive review of India. *The Science of the Total Environment, 511*, 123–137. DOI: 10.1016/j.scitotenv.2014.12.041 PMID: 25540847

Yadav, R., Dahal, K. R., Adhikari, B. B., Ghimire, P., & Belbase, N. P. (2024). Effect of Nutrient Sources on Productivity of Spring Maize (Zea Mays L.) at Khairahani, Chitwan, Nepal. *Multi-Disciplinary Explorations: The Kasthamandap College Journal*, *2*(2), 149–160. DOI: 10.3126/mdekcj.v2i2.74190

Yan, C., Fu, L., Zhang, J., & Wang, J. (2019). A comprehensive survey on UAV communication channel modeling. *IEEE Access : Practical Innovations, Open Solutions*, *7*, 107769–107792. DOI: 10.1109/ACCESS.2019.2933173

Yang, C., Li, X., Zhang, H., Meng, Y., Zhang, R., & Yuan, R. (2024). Uav small target detection in complex scenes based on improved yolov8s. 2024 39th Youth Academic Annual Conference of Chinese Association of Automation (YAC)

Yang, J., Zhang, J., Ye, F., & Cheng, X. (2019). A uav based multi-object detection scheme to enhance road condition monitoring and control for future smart transportation. Artificial Intelligence for Communications and Networks: First EAI International Conference, AICON 2019, Harbin, China, May 25–26, 2019, Proceedings, Part II DOI: 10.1007/978-3-030-22971-9_23

Yang, R., Yu, F. R., Si, P., Yang, Z., & Zhang, Y. (2019). Integrated blockchain and edge computing systems: A survey, some research issues and challenges. *IEEE Communications Surveys and Tutorials*, *21*(2), 1508–1532. DOI: 10.1109/COMST.2019.2894727

Yang, S., Bailey, E., Yang, Z., Ostrometzky, J., Zussman, G., Seskar, I., & Kostic, Z. (2020). COSMOS Smart Intersection: Edge Compute and Communications for Bird's Eye Object Tracking. *2020 IEEE International Conference on Pervasive Computing and Communications Workshops, PerCom Workshops 2020*. DOI: 10.1109/PerComWorkshops48775.2020.9156225

Yang, Y., Gao, X., Wang, Y., & Song, S. (2022). VAMYOLOX: An accurate and efficient object detection algorithm based on visual attention mechanism for UAV optical sensors. *IEEE Sensors Journal*, *23*(11), 11139–11155. DOI: 10.1109/JSEN.2022.3219199

Yang, Y., Khalife, J., Morales, J. J., & Kassas, Z. M. (2021). UAV waypoint opportunistic navigation in GNSS-denied environments. *IEEE Transactions on Aerospace and Electronic Systems*, *58*(1), 663–678. DOI: 10.1109/TAES.2021.3103140

Yasin, J. N., Mohamed, S. A., Haghbayan, M.-H., Heikkonen, J., Tenhunen, H., & Plosila, J. (2020). Unmanned aerial vehicles (uavs): Collision avoidance systems and approaches. *IEEE Access : Practical Innovations, Open Solutions*, *8*, 105139–105155. DOI: 10.1109/ACCESS.2020.3000064

Yazid, Y., Ez-Zazi, I., Guerrero-González, A., El Oualkadi, A., & Arioua, M. (2021). Uav-enabled mobile edge-computing for iot based on ai: A comprehensive review. In *Drones* (Vol. 5, Issue 4). MDPI. DOI: 10.3390/drones5040148

Ye, D., Yu, R., Pan, M., & Han, Z. (2020). Federated learning in vehicular edge computing: A selective model aggregation approach. *IEEE Access : Practical Innovations, Open Solutions*, *8*, 23920–23935. DOI: 10.1109/ACCESS.2020.2968399

Yedilkhan, D., Kyzyrkanov, A. E., Kutpanova, Z. A., Aljawarneh, S., & Atanov, S. K. (2024). Intelligent obstacle avoidance algorithm for safe urban monitoring with autonomous mobile drones. *Journal of Electronic Science and Technology*, *22*(4), 100277. DOI: 10.1016/j.jnlest.2024.100277

Ying, X., Murugesan, R. K., Sindiramutty, S. R., Wei, G. W., Balakrishnan, S., Kumar, D., & Verma, S. (2024). Scene Text Recognition using Deep Learning Techniques. *2024 International Conference on Emerging Trends in Networks and Computer Communications (ETNCC)*, 1–9. DOI: 10.1109/ETNCC63262.2024.10767484

Yi, S., Li, C., & Li, Q. (2015, June). A survey of fog computing: concepts, applications and issues. In *Proceedings of the 2015 workshop on mobile big data* (pp. 37-42).

You, H., Li, S., Xu, Y., He, Z., & Wang, D. (2021). Tree Extraction from Airborne Laser Scanning Data in Urban Areas. *Remote Sensing (Basel)*, *13*(17), 3428. DOI: 10.3390/rs13173428

Youvan, D. C. (n.d.). *AI at the Edge vs. AI in the Cloud: A Comparative Analysis of High-End and Edge AI Systems*. https://doi.org/DOI: 10.13140/RG.2.2.18865.80488

Yu, J., Yang, L., Xu, N., Yang, J., & Huang, T. (2018). Slimmable neural networks. *arXiv preprint arXiv:1812.08928.*

Yu, P. (2024). Self-driving car implementation guided by computer vision for detection. *Applied and Computational Engineering*, *74*(1), 1–6. DOI: 10.54254/2755-2721/74/20240420

Yu, X., & Zhang, Y. (2015). Sense and avoid technologies with applications to unmanned aircraft systems: Review and prospects. *Progress in Aerospace Sciences*, *74*, 152–166. DOI: 10.1016/j.paerosci.2015.01.001

Zaman, N., Ghazanfar, M. A., Anwar, M., Lee, S. W., Qazi, N., Karimi, A., & Javed, A. (2023). Stock market prediction based on machine learning and social sentiment analysis. *Authorea Preprints.*

Zhang, S., & Li, L. (2023, March 23). The Multi-visits Drone-Vehicle Routing Problem with Simultaneous Pickup and Delivery Service. Springer Science+Business Media. DOI: 10.1007/s40305-023-00471-7

Zhang, X., Wang, Y., Lu, S., Liu, L., & Shi, W. (2019). OpenEI: An open framework for edge intelligence. 2019 IEEE 39th International Conference on Distributed Computing Systems (ICDCS),. IEEE, 1840-1851.

Zhang, L., Yu, X., Adu-Gyamfi, Y., & Sun, C. (2023). Spatio-Temporal fusion of LiDAR and camera data for omnidirectional depth perception. *Transportation Research Record: Journal of the Transportation Research Board*, *2678*(4), 308–322. DOI: 10.1177/03611981231184187

Zhang, M., & Li, S. (2023). Economic impact of urban traffic congestion: A global perspective. *Transportation Research Part A, Policy and Practice*, *167*, 170–185.

Zhang, T., Qin, L., Zou, Q., Zhang, L., Wang, R., & Zhang, H. (2024). Crackscopenet: A lightweight neural network for rapid crack detection on resource-constrained drone platforms. *Drones (Basel)*, *8*(9), 417. DOI: 10.3390/drones8090417

Zhang, Z., Liu, X., Zhou, H., Xu, S., & Lee, C. (2023). Advances in Machine-Learning Enhanced Nanosensors: From cloud artificial intelligence toward future edge computing at chip level. *Small Structures*, *5*(4), 2300325. Advance online publication. DOI: 10.1002/sstr.202300325

Zhao, J. (2024). Drone technology for precision agriculture: Advancements and optimization strategies. *Highlights in Science. Engineering and Technology*, *111*, 185–191.

Zhao, Z., Hwang, K., & Villeta, J. (2012, June). Game cloud design with virtualized CPU/GPU servers and initial performance results. In *Proceedings of the 3rd workshop on Scientific Cloud Computing* (pp. 23-30).

Zhong, Q. (2024). The analysis of drone track planning based on machine-visual. *High- lights in Science. Engineering and Technology*, *103*, 178–183.

Zhou, P., Yang, L., & Chen, X. (2023). Robust traffic object detection under variable environmental conditions. *IEEE Transactions on Intelligent Transportation Systems*, *24*(8), 7890–7904.

Zhou, S., Ng, S. T., Yang, Y., & Xu, J. F. (2021). Integrating computer vision and traffic modeling for near-real-time signal timing optimization of multiple intersections. *Sustainable Cities and Society*, *68*, 102775. DOI: 10.1016/j.scs.2021.102775

Zhou, X., Liang, W., Ma, J., Yan, Z., Kevin, I., & Wang, K. (2022). 2D federated learning for personalized human activity recognition in cyber-physical-social systems. *IEEE Transactions on Network Science and Engineering*, *9*(6), 3934–3944. DOI: 10.1109/TNSE.2022.3144699

Zhou, Y., Rao, B., & Wang, W. (2020). UAV swarm intelligence: Recent advances and future trends. *IEEE Access : Practical Innovations, Open Solutions*, *8*, 183856–183878. DOI: 10.1109/ACCESS.2020.3028865

Zhou, Z., Chen, X., Li, E., Zeng, L., Luo, K., & Zhang, J. (2019). Edge intelligence: Paving the last mile of artificial intelligence with edge computing. *Proceedings of the IEEE*, *107*(8), 1738–1762. DOI: 10.1109/JPROC.2019.2918951

Zhu, P., Wen, L., Du, D., Bian, X., Fan, H., Hu, Q., & Ling, H. (2021). Detection and Tracking Meet Drones challenge. *IEEE Transactions on Pattern Analysis and Machine Intelligence*, *44*(11), 7380–7399. DOI: 10.1109/TPAMI.2021.3119563 PMID: 34648430

Zinno, R., Haghshenas, S. S., Guido, G., & Vitale, A. (2022, January 1). Artificial Intelligence and Structural Health Monitoring of Bridges: A Review of the State-of-the-Art. *IEEE Access : Practical Innovations, Open Solutions*, *10*, 88058–88078. DOI: 10.1109/ACCESS.2022.3199443

About the Contributors

Qurban Ali is pursuing a PhD in Computer Science at the University of Milano-Bicocca, Italy. Previously, he worked as a lecturer at Sindh Madressatul Islam University, Karachi. He earned an MS in Computer Science with a specialization in Computer Networks from Sukkur IBA University in 2017. His research interests include software testing in Cyber-Physical Systems (CPS), with a focus on Automated Driving Systems (ADS). Current interests involve regression testing, with a particular focus on road format interoperability for different autonomous driving systems. He has more than four years of experience in research and teaching and has been involved in several research projects, both international and national, in close collaboration with leading partners from industry and academia.

Goh Wei Wei is the Programme Director of the Dual Awards program in the School of Computing and IT, Faculty of Innovation and Technology in Taylor's University. Her expertise is in e-Learning Technologies. Her research interests includes E-Learning technologies, emerging web technologies, critical thinking skills, learning theories, MOOC, ontological based webpage segmentation, tools and applications, social media marketing and communications as well as persuasive messages and communications in social media. Some of her professional affiliations includes the International Association of Computer Science and Information Technology (IACSIT) and IEEE. She can be reached at weiwei.goh@taylors.edu.my

Rituraj Jain is currently working in the Department of Information Technology, Marwadi University, Rajkot, GUjarat, India. He has published many papers in International Journals and Conferences. He has published 4 books. He is an Oracle Database SQL Certified Expert. He is also associated with professional bodies like the Indian Society of Technical Education (ISTE), Computer Science Teachers Association (CSTA), Academy & Industry Research Collaboration Center (AIRCC), International Association of Computer Science and Information Technology (IACSIT), International Association of Engineers (IAENG). His research interests include Cloud Computing, Machine Learning, Deep Learning, Fuzzy Logic, IoT, Genetic Algorithms, Big Data and Software Engineering.

Vishal Jain is presently working as Professor at the Department of Computer Science and Engineering, School of Engineering and Technology, Sharda University, Greater Noida, India. Before that, he worked for several years as an Associate Professor at Bharati Vidyapeeth's Institute of Computer Applications and Management (BVICAM), New Delhi. He has more than 18 years of experience in the academics. He has earned degrees: Ph.D (CSE), M.Tech (CSE), MBA (HR), MCA, MCP, and CCNA. He has more than 2200 research citations with Google Scholar (h-index score 24 and i-10 index 55) and has authored more than 150 research papers in professional journals and conferences. He has authored and edited more than 60 books (most of them are indexed at the Scopus) with various reputed publishers, including Elsevier, Springer, IET, Apple Academic Press, CRC, Taylor and Francis Group, Scrivener, Wiley, Emerald, NOVA Science, River Publishers, IGI-Global and Bentham Science. He has more than 200 Scopus publications. His research areas include machine learning, information retrieval, semantic web, ontology engineering, data mining, ad hoc networks, sensor networks and network security.

Damodharan Palaniappan works as an Associate Professor & Head in the Department of Information Technology, Marwadi University, Rajkot, Gujarat. He completed his Ph.D. at Anna University, Chennai, M.E. in Computer Science and Engineering at C.I.E.T, Coimbatore and B.E. in Computer Science and Engineering at CSI College of Engineering, Ketty, Nilgiri. He has 19 years' experience of teaching and research. Currently, he guides 6 Ph.D. research scholars. He has authored 8 book chapters, and he has published research papers in reputed journals. His interested areas of research are Data Mining, Image Processing, Software, Machine Learning and Deep Learning.

Muhammad Usman Tariq has more than 16+ year's experience in industry and academia. He has authored more than 200+ research articles, 110+ case studies, 120+ book chapters and several books other than 4 patents. He is founder and CEO of The Case HQ, a unique repository for courses, narrative and video case studies. He has been working as a consultant and trainer for industries representing six sigma, quality, health and safety, environmental systems, project management, and information security standards. His work has encompassed sectors in aviation, manufacturing, food, hospitality, education, finance, research, software and transportation. He has diverse and significant experience working with accreditation agencies of ABET, ACBSP, AACSB, WASC, CAA, EFQM and NCEAC. Additionally, Dr. Tariq has operational experience in incubators, research labs, government research projects, private sector startups, program creation and management at various industrial and academic levels. He is Certified Higher Education Teacher from Harvard University, USA, Certified Online Educator from HMBSU, Certified Six Sigma Master Black Belt and has been awarded PFHEA, SFSEDA, SMIEEE, and CMBE.

Index

A

AI-powered drones 29, 72, 74, 79, 80, 109, 113, 115, 116, 117, 119, 123, 127, 231, 237, 279, 280, 283, 284, 288
Architecture 6, 7, 8, 10, 25, 36, 87, 91, 95, 100, 101, 152, 163, 164, 185, 188, 196, 201, 203, 209, 212, 216, 219, 220, 257, 258, 260, 263, 278, 296, 298
Artificial Intelligence 1, 2, 54, 56, 61, 62, 65, 67, 69, 70, 71, 72, 73, 74, 75, 77, 78, 79, 81, 83, 85, 86, 87, 89, 90, 91, 96, 98, 105, 106, 107, 108, 109, 110, 113, 115, 117, 118, 119, 121, 122, 123, 124, 125, 126, 128, 130, 131, 136, 151, 153, 155, 156, 163, 164, 166, 183, 188, 191, 193, 194, 195, 196, 198, 206, 219, 220, 224, 225, 226, 229, 231, 232, 233, 237, 239, 240, 241, 245, 246, 247, 248, 251, 252, 254, 261, 267, 268, 270, 277, 286, 288, 291, 293, 295, 298, 299, 301, 302
Autonomous Navigation 29, 34, 39, 42, 46, 52, 54, 109, 110, 111, 112, 115, 117, 123, 126, 128, 153, 191, 223, 228, 244, 245, 253
Autonomous UAVs 224, 296

C

Cloud AI 195, 196, 198, 199, 202, 203, 204, 206, 207, 208, 209, 210, 211, 212, 214, 216, 217, 218
Computer Vision 1, 2, 27, 29, 32, 33, 34, 35, 38, 39, 40, 42, 51, 53, 54, 56, 59, 60, 61, 62, 63, 64, 65, 66, 67, 69, 70, 73, 74, 75, 78, 81, 82, 83, 109, 110, 111, 112, 113, 115, 121, 123, 128, 129, 130, 132, 133, 135, 136, 137, 139, 140, 147, 148, 149, 150, 151, 153, 156, 157, 158, 161, 164, 165, 166, 183, 184, 187, 188, 192, 197, 198, 223, 225, 226, 227, 228, 229, 240, 243, 247, 267, 270
Cybersecurity 79, 101, 102, 125, 127, 128, 151, 160, 162, 165, 181, 193, 207, 242, 269, 299, 300

D

data security 187, 215, 260, 300
Decentralized Decision-Making 287
Deep Learning 1, 25, 27, 32, 33, 34, 38, 39, 42, 46, 47, 49, 55, 56, 58, 59, 86, 101, 102, 132, 139, 140, 145, 147, 148, 149, 151, 152, 153, 155, 157, 160, 162, 163, 164, 166, 203, 206, 207, 211, 220, 224, 226, 240, 243, 244, 254, 262, 268, 295, 300, 303
disaster management 29, 33, 73, 114, 119, 167, 168, 176, 179, 203, 209, 213, 215, 217, 219, 228, 236, 245, 291
Disaster Response 30, 34, 37, 40, 42, 50, 53, 54, 115, 121, 124, 139, 147, 173, 207, 213, 246, 266, 275, 283, 287, 289, 290, 298
Drone Industry 64, 65, 69, 72, 91, 101, 102, 127, 152, 167, 168, 171, 172, 174, 177, 181, 183, 185, 187, 188, 193, 196, 225, 257, 258, 259, 260, 264, 265
Drone Navigation 82, 129, 130, 131, 132, 133, 134, 135, 136, 139, 140, 144, 145, 146, 147, 149, 150, 152, 275
Drones 29, 30, 31, 32, 33, 34, 35, 36, 37, 38, 39, 40, 41, 42, 43, 44, 45, 46, 47, 48, 49, 50, 51, 52, 53, 54, 55, 56, 57, 58, 59, 60, 61, 62, 63, 64, 65, 66, 67, 68, 69, 70, 71, 72, 73, 74, 75, 76, 77, 78, 79, 80, 81, 82, 83, 84, 85, 86, 87, 90, 91, 92, 93, 97, 98, 105, 106, 107, 108, 109, 110, 111, 112, 113, 114, 115, 116, 117, 118, 119, 120, 121, 122, 123, 124, 125, 126, 127, 128, 129, 130, 131, 132, 133, 134, 135, 137, 139, 140, 141, 142, 143, 144, 145, 146, 147, 148, 149, 150, 151, 152, 153, 154, 155, 157, 158, 159, 160, 161, 163, 164, 165, 166, 168, 169, 170, 171, 172, 174, 175, 178, 179, 181, 184, 188, 190, 193, 196,

197, 198, 200, 201, 202, 203, 205, 210, 211, 213, 215, 216, 217, 218, 219, 220, 221, 223, 224, 225, 226, 227, 228, 229, 230, 231, 232, 233, 234, 235, 236, 237, 238, 239, 240, 241, 242, 243, 244, 245, 246, 247, 248, 249, 250, 251, 252, 253, 255, 257, 258, 260, 261, 262, 263, 264, 265, 266, 267, 268, 271, 272, 273, 274, 275, 277, 278, 279, 280, 281, 282, 283, 284, 285, 286, 287, 288, 289, 290, 291, 292, 293, 294, 298, 299, 301, 302, 303

Drone Technology 60, 62, 72, 77, 84, 87, 89, 90, 91, 92, 94, 96, 97, 98, 101, 105, 108, 112, 121, 127, 130, 158, 164, 167, 171, 184, 188, 189, 195, 196, 203, 219, 232, 239, 243, 274, 292

E

Edge AI 52, 94, 95, 145, 183, 188, 190, 191, 195, 196, 200, 201, 202, 203, 206, 207, 208, 209, 210, 211, 212, 213, 214, 216, 217, 218, 220, 221, 269, 271, 274, 275, 279, 280, 281, 282, 283, 284, 285, 289, 295, 299

Edge Computing 7, 25, 29, 33, 35, 36, 37, 38, 48, 50, 51, 54, 55, 57, 59, 61, 62, 63, 64, 65, 66, 67, 69, 70, 73, 75, 77, 78, 79, 80, 81, 82, 83, 89, 91, 92, 94, 95, 96, 97, 99, 100, 101, 102, 103, 129, 132, 141, 143, 146, 147, 149, 151, 154, 159, 166, 174, 183, 184, 185, 186, 187, 188, 189, 190, 191, 196, 201, 202, 203, 204, 208, 212, 219, 220, 223, 225, 228, 240, 244, 255, 257, 258, 259, 260, 261, 263, 264, 266, 267, 268, 269, 270, 274, 280, 299, 301

Emerging Trends 89, 91, 157, 158, 162, 165, 166, 167, 172, 295, 296, 297, 300, 302, 303

ethical considerations 72, 128, 249

F

future 25, 28, 52, 54, 57, 62, 70, 72, 75, 78, 81, 82, 83, 86, 95, 99, 100, 109, 110, 121, 122, 125, 127, 128, 130, 132, 133, 144, 147, 150, 155, 157, 158, 159, 160, 161, 162, 163, 166, 167, 168, 169, 172, 173, 176, 177, 181, 182, 183, 187, 191, 192, 193, 194, 217, 218, 219, 220, 224, 225, 226, 236, 237, 239, 240, 243, 247, 249, 250, 252, 253, 268, 275, 278, 282, 283, 287, 289, 290, 291, 292, 293, 295, 296, 299, 300, 301, 302, 303

I

Internet of Things 56, 84, 99, 100, 101, 102, 122, 127, 128, 153, 155, 156, 159, 160, 164, 186, 192, 193, 220, 251, 252, 254, 261, 268, 269, 270, 294, 295, 296, 301, 302

M

Manufacturing 61, 62, 63, 64, 65, 66, 67, 69, 70, 71, 72, 73, 74, 75, 76, 77, 78, 80, 81, 82, 83, 86, 87, 107, 114, 118, 122, 136, 155, 185, 186, 187, 201, 228, 247, 258

Multi-Drone Coordination 271, 273, 274, 275, 278, 284, 285, 289

O

Object Detection 1, 2, 3, 5, 7, 8, 13, 24, 25, 28, 32, 33, 34, 38, 39, 42, 48, 49, 50, 51, 52, 54, 56, 58, 59, 70, 86, 126, 128, 129, 132, 139, 144, 153, 155, 156, 166, 192, 193, 194, 197, 203, 206, 213, 220, 239, 267, 297

Obstacle Avoidance 32, 33, 38, 39, 40, 41, 44, 47, 48, 52, 53, 117, 158, 211, 228, 244, 252, 269, 303

P

Precision Agriculture 49, 53, 54, 59, 60, 68, 71, 95, 141, 153, 171, 188, 189, 191, 193, 194, 197, 204, 213, 219, 227, 266, 274, 297

R

Real-Time Data Processing 36, 92, 96, 97, 116, 117, 119, 125, 167, 176, 189, 201, 208, 228, 266, 282

S

security and surveillance 71, 105, 107, 109, 110, 112, 113, 115, 116, 117, 118, 120, 121, 123, 124, 196, 273
Sensor Fusion 40, 81, 130, 133, 142, 144, 148, 150, 151, 164, 165, 180, 271
Smart Traffic Management 3, 28
Swarm Intelligence 122, 224, 244, 271, 272, 274, 275, 276, 277, 278, 281, 282, 283, 284, 286, 289, 291, 299, 302, 303

T

Terrain Mapping 32, 33, 34, 38, 39, 40, 144
Traffic Management 1, 2, 3, 4, 5, 7, 25, 26, 27, 28, 50, 59, 82, 83, 113, 178, 181, 184, 212, 245

U

UAV 30, 56, 58, 59, 62, 68, 69, 73, 77, 83, 84, 85, 87, 93, 95, 106, 125, 128, 132, 137, 144, 152, 153, 154, 156, 158, 159, 161, 163, 165, 166, 168, 169, 170, 173, 175, 176, 178, 179, 180, 181, 182, 183, 187, 188, 189, 190, 191, 192, 193, 194, 195, 196, 202, 203, 204, 205, 207, 208, 209, 210, 211, 212, 213, 214, 215, 216, 217, 218, 219, 220, 221, 246, 251, 253, 254, 267, 268, 269, 270, 292, 293, 294, 296, 297, 298, 300, 301, 302, 303
UAVs 27, 36, 55, 57, 58, 62, 67, 69, 73, 79, 84, 85, 86, 87, 90, 91, 92, 93, 94, 105, 108, 113, 114, 125, 127, 132, 137, 144, 151, 155, 156, 159, 166, 167, 168, 169, 170, 171, 172, 173, 175, 176, 177, 178, 181, 182, 183, 184, 186, 187, 188, 189, 190, 191, 193, 194, 196, 197, 203, 204, 206, 207, 208, 209, 210, 211, 212, 213, 214, 215, 216, 217, 218, 219, 223, 224, 227, 246, 248, 252, 253, 254, 257, 258, 260, 261, 262, 263, 267, 268, 269, 270, 291, 292, 294, 296, 299
Unmanned Aerial Vehicles 27, 36, 55, 56, 58, 62, 63, 67, 68, 70, 75, 85, 86, 91, 105, 114, 118, 121, 125, 126, 152, 153, 154, 157, 158, 159, 169, 178, 180, 181, 182, 184, 186, 188, 192, 193, 196, 209, 219, 224, 239, 252, 253, 257, 260, 267, 269, 270, 295, 296, 297

Y

Yolo 7, 9, 47, 49, 50, 206

www.ingramcontent.com/pod-product-compliance
Lightning Source LLC
LaVergne TN
LVHW080601300926
849162LV00005B/1018